T0345010

Handbook of Formal Analysis and Verification in Cryptography

This handbook addresses topics that are very important for secure communication and processing of information. It introduces readers to several formal verification methods and software used to analyze cryptographic protocols. The chapters give readers general knowledge of formal methods focusing on cryptographic protocols.

Handbook of Formal Analysis and Verification in Cryptography includes major formalisms and tools used for formal verification of cryptography, with a spotlight on new-generation cryptosystems such as post-quantum, and presents a connection between formal analysis and cryptographic schemes. The book offers formal methods to show whether security assumptions are valid and compares the most prominent formalism and tools as they outline common challenges and future research directions.

Graduate students, researchers, and engineers worldwide will find this an exciting read.

Prospects in Pure and Applied Mathematics

Series Editors: Mohammad M. Banat and Özen Özer

This new series is intended for the publication of top quality books in various domains and applications of engineering mathematics, like, but not limited to Mathematical Techniques in Engineering, Optimization Techniques in Engineering, Modeling and Simulation Techniques, Computer-Aided Design, Numerical Methods in Engineering, and so on. Books are required to present a clear focus on future prospects in this exciting subject. An application-oriented approach, sufficient attention to application domains is required.

Handbook of Formal Analysis and Verification in Cryptography
Sedat Akleylek and Besik Dundua

Handbook of Formal Analysis and Verification in Cryptography

Edited by
Sedat Akleylek
Besik Dundua

CRC Press is an imprint of the
Taylor & Francis Group, an **informa** business

Designed cover image: Shutterstock

First edition published 2024
by CRC Press
6000 Broken Sound Parkway NW, Suite 300, Boca Raton, FL 33487-2742

and by CRC Press
4 Park Square, Milton Park, Abingdon, Oxon, OX14 4RN

CRC Press is an imprint of Taylor & Francis Group, LLC

ISBN: 978-0-367-54665-6 (hbk)
ISBN: 978-0-367-54666-3 (pbk)
ISBN: 978-1-003-09005-2 (ebk)

DOI: 10.1201/9781003090052

Typeset in Nimbus font
by KnowledgeWorks Global Ltd.

Publisher's note: This book has been prepared from camera-ready copy provided by the authors.

Contents

Foreword

Over the years, there has been a substantial amount of research on the application of formal methods to applications that require that some fundamental assumptions about the environment be taken into account. The most prominent example has been the analysis of cryptographic protocols, in which it is assumed that principals communicate over a network controlled by a hostile intruder who has compromised some of the principals. The development of specialized techniques for reasoning about protocols that obey this assumption has led to important contributions of formal methods to cryptographic protocol standards. Moreover, this is not the only area in which this approach has proved fruitful. For example, a substantial amount of research exists on the development of formal languages for quantum programs, in which the special features of quantum computation must be taken into account. In another example, formal analysis of cryptographic Web applications requires that the highly distributed nature of Web applications be taken into account, not only for communication, but for storage and computation. However, it is not always easy to find places where information about various approaches is gathered together in one place. This handbook, which contains a collection of surveys of some of these areas with an emphasis of on cryptographic protocol analysis, is intended to address this deficiency.

The handbook can be divided into three parts, the first two on formal analysis of cryptographic protocols, and the last one giving surveys of other areas.

The first part, consisting of Chapters 1 and 2, gives background and a broad overview of formal analysis of cryptographic protocols. Chapter 1 is a survey of cryptography since Shannon's groundbreaking work in the 1940s. It goes on to cover relevant topics including secret-key cryptography, public-key cryptography, post-quantum cryptography, and homomorphic encryption, giving details about specific algorithms along the way. This gives a context from which to view the other chapters in the book. Chapter 2 then gives an overview of the various theories and techniques used in formal analysis of cryptographic protocols, showing how techniques such as model checking, belief logics, and model extraction have been used in this area.

Once the reader has been introduced to the basics, the next part, consisting of Chapters 3 and 4, introduces specific tools and techniques. Chapter 3 covers tools that have been developed specifically for the analysis of cryptographic protocols, covering six different such tools. Chapter 4 demonstrates how an existing formal methods tool, the Isabelle/HOL theorem prover, can be used for the formal analysis of cryptographic protocols, taking advantage of the prover's support for inductive reasoning. Finally, Chapter 5 covers special topics in the formal analysis of cryptographic protocols, in particular, how to specify protocols, and the use of multiple analysis tools.

The last parts, consisting of Chapters 6 and 7, give in-depth explorations of the applications of formal methods to two other areas. Chapter 6 presents an overview of the application of formal methods to the development of secure cryptographic Web applications, taking into account the highly distributed nature of Web services and their effects on security. This includes not only formal verification of the protocols the Web applications use, but the software in which the applications are implemented. Chapter 7 gives a survey of formal quantum programming algorithms, showing how the mechanisms of quantum computation are captured formally.

I'm pleased to see this presentation of the different ways in which formal methods can contribute to different areas of computation, and how the need to take into account basic assumptions about the environment can lead to whole new areas of study.

Enjoy!

November 26, 2022
Catherine Ann Meadows

Preface

Cryptographic protocols have a significant role to provide security in our daily life such as, for instance, online banking or instant messaging. Information security concepts are satisfied by cryptographic protocols such as, e.g., SSL/TLS and SSH. These protocols include cryptographic primitives, which become essential for the security of cryptographic protocols. Some cryptographic primitives are key exchange/encapsulation, authentication, digital signature, and encryption. When a cryptosystem is designed, at first a computationally hard problem should be defined. However, satisfying the requirements of computationally hard problems does not yet mean that the cryptographic protocol is secure. Therefore, implementation attacks should be checked. Eavesdropping is an example of this kind of attack. By using the man-in-the-middle attack approach, one can obtain the secret keys without solving those computationally hard problems. Therefore, any cryptographic protocol should provide some countermeasures to resist those kinds of attacks.

Applications of formal methods in cryptography aim at guaranteeing security at a high level. In recent years, formal methods for the analysis of cryptographic protocols are experiencing a renewed period of growth with the emergence of new concepts and systems that allow a better understanding and usability. Formal analysis of cryptographic protocols is used to model a cryptographic protocol and its security properties to verify that a protocol is secure. Such a formal analysis revealed security flaws in some protocols. For example, in 1995 Lowe found an attack on Needham-Schroeder public-key protocol, which was "proven" to be safe back in 1978. Along with the evolution of communication systems, new protocols need to be designed for secure information transfer. Obviously, before using such protocols, it is necessary to verify that they are secure. The main focus of this book is the analysis and verification of security protocols to provide readers with various aspects of formal methods used in cyber security.

This book introduces several formal verification methods and software used nowadays for modeling and analyzing cryptographic protocols from various perspectives. Connections between formal analysis/verification/ modeling and cryptographic schemes are presented. Since designing a secure cryptographic scheme is a very challenging issue, the system should be thoroughly checked for vulnerability to adversarial attacks. Formal methods are crucial in checking whether the security assumptions are valid or not for a cryptographic scheme.

We hope that this book will be useful for anyone who wants to do study in formal analysis and verification of cryptographic protocols. The handbook contains seven chapters and is structured as follows:

- Chapter 1 surveys the history and evolution of cryptography since Shannon. It provides readers with a general background and main concepts of cryptography, which are useful to understand the following chapters. It also discusses the recent developments in cryptography such as post-quantum cryptography and homomorphic encryption.
- Chapter 2 discusses formal methods to ensure the security of cryptographic protocols. It has both theoretical and implementation aspects. This is an introductory chapter where readers can learn basic techniques used for the design, verification, and analysis of cryptographic protocols. It also gives the frameworks for cryptographic protocol engineering from the formal verification point of view.
- Chapter 3 describes various automated tools for formal verification of cryptographic protocols. In particular, CasperFDR, ProVerif, Scyther, Tamarin, Avispa, and Maude-NPA protocol verification tools are recalled and compared to each other. Needham-Schroeder Public Key (NSPK) protocol is used as a means of comparison.
- Chapter 4 introduces techniques to verify cryptographic protocols using interactive theorem provers, namely with Isabelle/HOL. It provides more details on how to use and benefit from them.
- Chapter 5 presents different approaches for modeling and verification of security protocols. How protocols are modeled in a formal framework is detailed. It also discusses the theoretical and practical challenges in automated verification. In particular, it includes case studies, demonstrating the practical applicability of these tools in two different application fields: epayments and blockchain (bitcoin payment protocol, e-commerce protocols, iKP protocol, SET purchase protocol).
- Chapter 6 overviews Web security by considering cryptographic methods. It covers cryptographic Web applications with real-world case scenarios. It discusses a high-level description of the TLS protocol and an encrypted Web storage protocol from the formal analysis perspective.

- Chapter 7 provides a state-of-the-art introduction to formal methods for quantum algorithms. It also discusses solutions for the formal verification of quantum compilation and the equivalence of quantum programs. It analyzes the solutions for formally verified quantum programming languages.

Acknowledgments We thank our colleagues for pointing out errors and suggesting improvements. In particular, we express our thanks to Catherine Meadows, for her Foreword. We would like to thank Taylor & Francis Group/CRC Press. We acknowledge the support from the Scientific and Technological Research Council of Turkey (TUBITAK) under grant no. 118E312 and the Shota Rustaveli National Science Foundation of Georgia under grant no. 04/03.

May 19, 2023

Sedat Akleylek
Cyber Security and Information Technologies
Research and Development Center and Department of
Computer Engineering Ondokuz Mayis University
Samsun, Turkey
Chair of Security and Theoretical Computer Science
University of Tartu
Tartu, Estonia

Besik Dundua
Department of Programming
Ilia Vekua Institute of Applied Mathematics of
Ivane Javakhishvili Tbilisi State University
Tbilisi, Georgia
School of Mathematics and Computer Science
Kutaisi International University
Kutaisi, Georgia

Contributors

Sébastien Bardin
Université Paris-Saclay
Palaiseau, France

Daniele Bringhenti
Politecnico di Torino
Torino, Italy

Michele Bugliesi
Universitá Ca' Foscari Venezia
Italy

Stefano Calzavara
Universitá Ca' Foscari Venezia
Italy

Christophe Chareton
Université Paris-Saclay
Palaiseau, France

Rémi Garcia
Teesside University
Middlesbrough, United Kingdom

Çetin Kaya Koç
Iğdır University, NUAA and University of California, Santa Barbara
Iğdır, Turkey, China and USA

Pascal Lafourcade
University Clermont Auvergne
Clermont-Ferrand, France

Dongho Lee
Université Paris-Saclay and Inria
Palaiseau and Gif-sur-Yvette, France

Paolo Modesti
Teesside University
Middlesbrough, United Kingdom

Murat Moran
Giresun University
Giresun, Turkey

Pasquale Noce
HID Global
Italy

Funda Özdemir
Istinye University
Istanbul, Turkey

Maxime Puys
University Clermont Auvergne
Clermont-Ferrand, France

Alvise Rabitti
Universitá Ca' Foscari Venezia
Italy

Riccardo Sisto
Politecnico di Torino
Torino, Italy

Fulvio Valenza
Politecnico di Torino
Torino, Italy

Benoit Valiron
Université Paris-Saclay and Inria
Palaiseau and Gif-sur-Yvette, France

Renault Vilmart
Université Paris-Saclay and Inria
Palaiseau and Gif-sur-Yvette, France

David Williams
University of Portsmouth
Portsmouth, United Kingdom

Zhaowei Xu
Université Paris-Saclay and University of Tartu
Gif-sur-Yvette, France and Estonia

Jalolliddin Yusupov
Turin Polytechnic University
Tashkent, Uzbekistan

Readers

Rohid Chadha
University of Missouri
Columbia, USA

Miguel Correia
Universidade de Lisboa
Lisboa, Portugal

Barbara Fila
INSA and IRISA
Rennes, France

Debasis Giri
Maulana Abul Kalam Azad
University of Technology
West Bengal, India

Daniele Gorla
Sapienza Universitá di Roma
Roma, Italy

Wai-Kong Lee
Gachon University
Seongnam, South Korea

Catherine Meadows
Naval Research Laboratory
Washington, DC, USA

Livinus Obiora Nweke
Norwegian University of Science
and Technology
Trondheim, Norway

Lawrence C. Paulson
University of Cambridge
Cambridge, United Kingdom

Robert Rand
University of Chicago
Chicago, USA

Zulfukar Saygi
TOBB University of Economics
and Technology
Ankara, Turkey

Anders Schlichtkrull
Aalborg University
Copenhagen, Denmark

Sonja Smets
University of Amsterdam
Amsterdam, The Netherlands

Roberto Zunino
Universitá degli Studi di Trento
Trento, Italy

1 Development of Cryptography since Shannon

Çetin Kaya Koç
Iğdır University, NUAA and University of California, Santa Barbara, Iğdır, Turkey, China and USA

Funda Özdemir
Istinye University, Istanbul, Turkey

CONTENTS

DOI: 10.1201/9781003090052-1

Cryptographic algorithms of the past millennia were formulated under a single model by Claude Shannon in 1948, marking the beginning of modern cryptography. His clarification of cryptographic security allowed the banking and finance industry to establish secure communication between geographically distributed branches. The second phase of modern cryptography called public key cryptography started with Diffie-Helman and RSA algorithms in 1977, which made secure communication between server and client computers possible. Moreover, the theoretical introduction of quantum computers in 1994 by Peter Shor and subsequent technical developments required researchers to upgrade public-key cryptography for the post-quantum world, giving birth to research efforts for post-quantum cryptography. The third and current phase of cryptographic revolution will be made possible by homomorphic encryption and its applications in privacy. Various types of homomorphic encryption allow us to encrypt everything and work with encrypted data so that neither the computers nor the network need to be trusted.

1.1 INTRODUCTION

Shannon's work [83] was a turning point and marked the closure of classical cryptography and the beginning of modern cryptography. Indeed, starting from 1949, cryptography theory and applications have gone through significant progress, certainly much faster than the previous several centuries.

Humans' interest in cryptography is as old as the invention of writing. While we have good information and insights about cryptographic methods in the past 2 millennia, we surmise that older algorithms like their more recent successors were all letter- or word-based "codes" in which one substitutes each letter or word with the corresponding code-letter or code-word found in the codebook, according to a selection algorithm. Sender and receiver must share the codebook to "encode" or "decode" the messages.

On the other hand, based on the frequencies of the code-letters, cryptanalysts attempted to make sense of the encoded (encrypted) message without having access to the codebook. The competition between the cryptographers and cryptanalysts has been real and fierce, especially when applications involved state or military data. We do get into history of classical cryptography due to the lack of space in this paper and recommend a new and well-written book for interested readers, *History of Cryptography and Cryptanalysis* [34].

The interplay between cryptographic theory and applications opened up new areas of applications and also motivated the practitioners of the theory to develop new methods and algorithms. There is much to write about cryptography, but given the space, we will limit our focus to secret-key cryptography, public-key cryptography, post-quantum cryptography and homomorphic encryption, which are also section headers in this paper.

The development of **secret-key cryptography** started soon after Shannon's insights how one builds complex, usable and efficient secret-key cryptographic algorithms. Horst Feistel's [39, 38, 56] at IBM, followed up by US NIST Data Encryption Standard [66], and a plethora of academic, commercial, cyberpunk algorithms and standards, to finally [67] which is another US standard. These algorithms were all built upon Shannon's ideas. The driving factor comes from banking application, which is for our need to relay confidential financial information. This is an ongoing work, and the academic, industrial and government bodies will continue to develop newer secret-key cryptographic algorithms.

A second revolution in cryptography happened somewhere between 1976 and 1978, interestingly right around time when the secret-key cryptographic algorithm was standardized by the US. While trying to address the problem of how to share secret keys between two or more parties, researchers at Stanford and MIT invented **public-key cryptography**. The Diffie-Hellman key exchange algorithm [33] and the RSA public-key cryptographic algorithm [79] have indeed changed cryptography as significantly as Shannon's contribution. In the ensuing years, practical solutions to key exchange between parties, digital signatures, and methods allowed us to build trust architectures into Internet-connected servers, desktop and mobile computers. The public-key cryptography provided techniques, mechanisms and tools for private and authenticated communication, and for performing secure and authenticated transactions over the Internet as well as other open networks. This infrastructure was needed to carry over the legal and contractual certainty from our paper-based offices to our virtual offices existing in cyberspace. The timing of the invention of public-key cryptography was near perfect!

The first two decades of the 21st century presented two challenges for cryptographers. The first and formidable challenge was that quantum computers were becoming feasible. Experimental quantum computers developed or sponsored by two major companies (IBM and Microsoft) and several research institutes in major research universities are available for researchers to test their quantum algorithms [91]. It has already been established by Peter Shor [84, 86] that a quantum computer (if available) with several thousands of quantum bits (qubits) can be programmed to break public-key almost all of the public-key cryptographic algorithms. Therefore academic and governmental efforts started to design public-key cryptographic algorithms that would be resistant to quantum computing attacks, which gave birth to **post-quantum cryptography**. In April 2015, the US NIST held a "Workshop on Cybersecurity in a Post-Quantum World" to discuss cryptographic algorithms for public-key-based key agreement and digital signatures that are not susceptible to cryptanalysis by quantum algorithms. In this direction, NIST recently launched the so-called "Post-Quantum Cryptography Standardization" process, a multiyear effort aimed at selecting the next generation of quantum-resistant public-key cryptographic algorithms for standardization.

Another formidable challenge has been the desire to compute with the encrypted text without decrypting, which is termed as **homomorphic encryption**. The potential applications of homomorphic encryption were recognized and appreciated almost about the same time as the first public-key cryptographic algorithm RSA was invented, which is multiplicatively homomorphic. The ensuing 30 years have brought on several additively or multiplicatively homomorphic encryption functions with increasing algorithmic complexity. In 2009, Craig Gentry and several other authors later on proposed fully (both additively and multiplicatively) homomorphic encryption algorithms and addressed issues related to their formulation, arithmetic and security. We now have a variety of fully homomorphic encryption algorithms that can be applied to various private computation problems in healthcare, finance and national security.

We start with Shannon's ideas in Section 1.2 and show how Feistel used them to create his seminal cryptographic algorithm LUCIFER. In this paper, we focus only on the DES and AES, the US standardized algorithms since the first was the chosen algorithm for applications ranging from banking to Internet for 2 decades, while latter has been in use as its replacement for more than 2 decades, going into the 3rd.

Section 1.3 covers public-key cryptography which tackles key management, public-key encryption and digital signatures, providing authentication and nonrepudiation properties for the exchanged data and communicating parties. We will cover the basic ideas and algorithms of public-key cryptography briefly, and move into post-quantum cryptography in Section 1.4. The advent of quantum computing is bound to change public-key cryptography, and the changes have already started. We will give an overview of post-quantum cryptography in this section.

Finally, Section 1.5 covers homomorphic encryption which will bring a kind of luxury to data science such that we can keep *everything encrypted* and still accomplish the necessary computations for maintaining the data as well as inferring from it. This will indeed be a revolution and will bring nearly-absolute security for our precious data.

1.2 SECRET-KEY CRYPTOGRAPHY

Claude Shannon wrote his "secrecy" paper in 1945; however, it was declassified and published only in 1949 [83]. Shannon suggested that cryptanalysis using statistical methods might be defeated by the mixing or iteration of non-commutative operations. Shannon refers to these operations as confusion (or substitution) and diffusion (transposition). His ideas were used in the design of the top 3 encryption algorithms in the following 5 decades. The LUCIFER [39, 38] and DES S-boxes and P-boxes [65, 66] are Feistel's interpretation of Shannon's confusion and diffusion. Similarly, AES or Rinjdael also uses many rounds to mix confusion and diffusion [67, 30].

1.2.1 LUCIFER

Horst Feistel changed cryptology. Historically (pre-Shannon days), encryption algorithms have been dictated by the hardware available. The pinwheels of the 1934 Hagelin machine, the rotors of the 1918 German Enigma machine, the telephone dial switches used in the 1937 Japanese Purple machine and nonlinear versions of the linear feedback shift register were subsequently based on the 1947 transistor breakthrough [56]. Horst knew about some of these, but he realized that hardware was a limitation; a program could directly implement encryption, and so he started in the reverse direction. When asked by about the idea behind his algorithm LUCIFER, Horst said "The Shannon secrecy paper [10] reveals all" [56]. He understood the power of Shannon's idea, followed the master's advice, leading to LUCIFER and DES.

LUCIFER was the very first encryption algorithm designed for software. In fact, LUCIFER is the name for the software implementation of the block cipher described in the 1971 patent designated by Feistel. Coded in the APL language, LUCIFER originally was stored in the APL directory (folder) with the intended name DEMONSTRATION. Early versions of

APL limited the character length of a file name, and a colleague suggested the name DEMON, modified by Horst to LUCIFER. Shannon's secrecy paper alone may have provided the real inspiration for a person of Horst's creative genius. Still, Horst Feistel went in his own cryptographic direction, providing a fresh point of view. A modified LUCIFER became the Data Encryption Standard (DES), affirmed in 1976 as a Federal Information Processing Standard (FIPS 46-1). AES became the new FIPS, replacing DES in 2000 as a standard. The Triple DES-variant (3DES) continues to be used for authenticated transactions in banking [54, 55].

1.2.2 DES

The Data Encryption Standard is a US standard that provided confidentiality for financial transactions from the 1970s till to the end of the 1990s. It was developed by IBM, based on ideas of Horst Feistel, and submitted to the National Bureau of Standards (the precursor of the National Institute of Standards and Technology) following an invitation to propose a candidate for the protection of sensitive, unclassified electronic data. After consulting with the National Security Agency (NSA), the NBS eventually selected a slightly modified version, which was published as an official Federal Information Processing Standard (FIPS) for the United States in 1977, with the number FIPS 46. It quickly became an international standard and enjoyed widespread deployment.

However, there was also some controversy about the DES for several years. The design philosophy of certain elements (S-boxes) was never explained (classified), and its key length was unnaturally short (56 bits) while it could have been 64 bits. The NSA involvement was found suspicious by some researchers, especially on its key length [49]. There were also conspiracy theories about the DES having a "backdoor" for easy decryption (which was never proven to-day). Academic community approached the DES with caution; however, in the end, it significantly contributed to the development of modern cryptography for our communication and computing systems.

The fundamental building block in DES is a substitution followed by a permutation on the text based on the key. This is called a round function. DES has 16 rounds.

A cryptographic algorithm should be a good pseudorandom generator in order to foil key clustering attacks. DES was designed so that all distributions were as uniform as possible. For example, changing 1 bit of the plaintext or the key causes the ciphertext to change in approximately 32 of its 64 bits in a seemingly unpredictable and random manner.

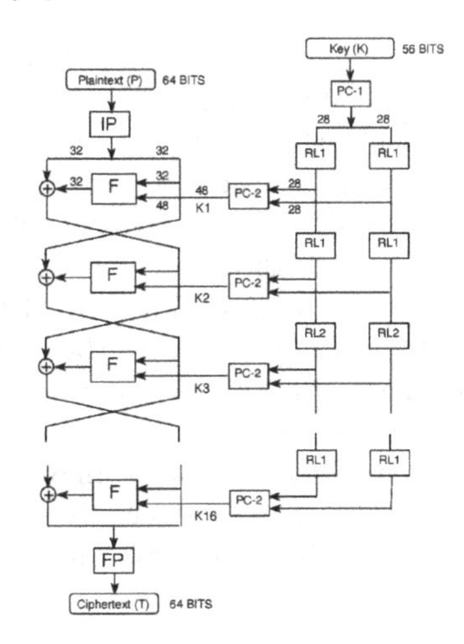

Figure 1.1 The 16 rounds of DES.

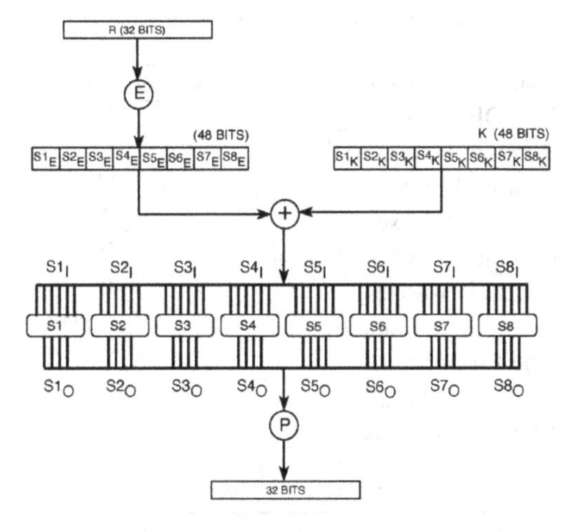

Figure 1.2 The round function of DES.

However, Biham and Shamir [8] observed that with a fixed key, the differential behavior of DES does not exhibit pseudorandomness. If we fix the XOR of two plaintexts P and P^* at P', then T' (which is equal to $T \oplus T^*$) is not uniformly distributed. In contrast, the XOR of two uniformly distributed random numbers would itself be uniformly distributed. The attack (called differential cryptanalysis) based on the nonrandom behavior of the DES still could not break DES, primarily due to the fact that 16 rounds made the tracing of the differences of the plaintexts and ciphertexts very difficult. Biham and Shamir showed that DES reduced to 6 rounds can be broken by a chosen plaintext attack in less 0.3 seconds on a PC using 240 ciphertexts; the known plaintext version requires 2^{36} ciphertexts. On the other hand, DES reduced to 8 rounds can be broken by a chosen plaintext attack in less than 2 minutes on a PC by analyzing about 2^{14} ciphertexts; the known plaintext attack needs about 2^{38} ciphertexts. Yet, the full DES (16 rounds) can only be broken by analyzing 2^{36} ciphertexts from a larger pool of 2^{47} chosen plaintexts using 2^{37} times. The differential cryptanalysis confirmed the importance of the number of rounds and the method by which the S-boxes are constructed.

On the other hand, the variations on DES turn out to be easier to cryptanalyze than the original DES. Most importantly, certain changes in the structure of DES have catastrophic results, as shown in Table 1.1.

Table 1.1
Effectiveness of differential cryptanalysis.

Modified Operation	Chosen Plaintexts
Full DES (no change)	2^{47}
Random P permutation	2^{47}
Identity P permutation	2^{19}
Order of S-boxes	2^{38}
Change XOR by addition	2^{31}
Random S-boxes	2^{21}
Random permutation	$2^{44} \sim 2^{48}$
One Entry S-box	2^{33}
Uniform S-boxes	2^{26}
Eliminate expansion E	2^{26}
Order of E and subkey XOR	2^{44}

Feistel ciphers take an important part in secret key cryptography from both theoretical and practical point of view. After DES, new schemes have been published, like GOST in Russia, IDEA, and RC-6 in the United States. From a practical point of view, Feistel ciphers had their days of glory with the DES algorithm and its variants (3DES with two or three keys, XDES, etc.) that were the most widely used secret key algorithms around the world between 1977 and 2000. After 2000, the AES algorithm, which is not a direct Feistel cipher, but still based on Shannon's ideas (confusion and diffusion) which were very effectively utilized by Feistel, has become the standard for secret key encryption.

1.2.3 AES

AES is a key-alternating block cipher, with plaintext encrypted in blocks of 128 bits. The key sizes in AES can be 128, 192 or 256 bits. It is an iterated block cipher because a fixed encryption process, usually called a round, is applied a number of times to the block of bits. Finally, we mean by key-alternating that the cipher key is XORed to the state (the running version of the block of input bits) alternately with the application of the round transformation. The original Rijndael design allows for any choice of block length and key size between 128 and 256 in multiples of 32 bits. In this sense, Rijndael is a superset of AES; the two are not identical, but the difference is only in these configurations initially put into Rijndael but not used in AES [30].

The state matrix of AES is formed from the input data as a 4×4, 4×6 and 4×8 matrices, for 128, 192 or 256 bits, respectively. Given the 128-bit data $(A_0 A_1 A_2 \cdots A_{14} A_{15})$ such that each of A_i is 8 bits (1 byte), the 4×4 state matrix is formed as

$$\begin{bmatrix} A_0 & A_4 & A_8 & A_{12} \\ A_1 & A_5 & A_9 & A_{13} \\ A_2 & A_6 & A_{10} & A_{14} \\ A_3 & A_7 & A_{11} & A_{15} \end{bmatrix}$$

The 8-bit (1-byte) binary data is usually represented in hexadecimal, such as $(\texttt{a3}) = (1010\ 0011)$. While the 8-bit input data block is a binary number in its most generic form, the Rijndael/AES treats each one of the bytes in the state matrix as elements of the Galois field GF(2^8). The irreducible polynomial of the field GF(2^8) is $p(x) = x^8 + x^4 + x^3 + x + 1$. A field element $a(x) \in \mathrm{GF}(2^8)$ is represented using a polynomial of degree at most 7 with coefficients $a_i \in \mathrm{GF}(2)$ such that $\sum_{i=0}^{7} a_i x^i = a_7x^7 + x_6x^6 + a_5x^5 + a_4x^4 + a_3x^3 + a_2x^2 + a_1x + a_0$. For example, $(\texttt{a3}) = (1010\ 0011) = x^7 + x^5 + x + 1$. AES has 4 sub-rounds, named as AddRoundKey, SubBytes, ShiftRows, MixColumn. Except the ShiftRows operation, all of them involve finite field addition, inversion and linear and nonlinear operations in the field GF(2^8).

Here we describe only the MixColumn operation which multiplies a fixed 4×4 matrix with every 4×1 column vector of the 4×4 state matrix. The MixColumn matrix M in hex and polynomial representation is

$$\begin{bmatrix} \texttt{02} & \texttt{03} & \texttt{01} & \texttt{01} \\ \texttt{01} & \texttt{02} & \texttt{03} & \texttt{01} \\ \texttt{01} & \texttt{01} & \texttt{02} & \texttt{03} \\ \texttt{03} & \texttt{01} & \texttt{01} & \texttt{02} \end{bmatrix} = \begin{bmatrix} x & x+1 & 1 & 1 \\ 1 & x & x+1 & 1 \\ 1 & 1 & x & x+1 \\ x+1 & 1 & 1 & x \end{bmatrix}$$

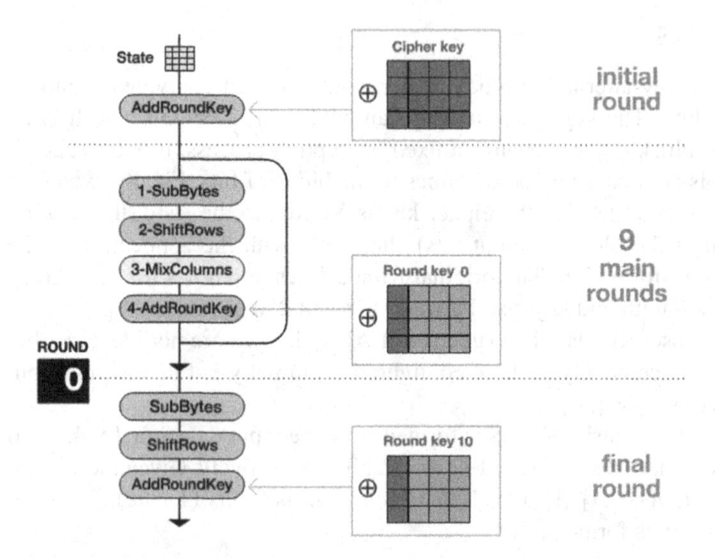

Figure 1.3 The 10 rounds of AES and the round function.

Given a 4×1 column vector u of the state matrix, such that each vector entry is an element of the finite field $GF(2^8)$, we perform a matrix-vector multiplication operation Mu using field multiplications and additions to compute the new column vector of the state matrix. We give an example of the MixColumn operation example below:

$$\begin{bmatrix} \texttt{02} & \texttt{03} & \texttt{01} & \texttt{01} \\ \texttt{01} & \texttt{02} & \texttt{03} & \texttt{01} \\ \texttt{01} & \texttt{01} & \texttt{02} & \texttt{03} \\ \texttt{03} & \texttt{01} & \texttt{01} & \texttt{02} \end{bmatrix} \begin{bmatrix} \texttt{d4} \\ \texttt{bf} \\ \texttt{5d} \\ \texttt{30} \end{bmatrix} = \begin{bmatrix} \texttt{04} \\ \texttt{66} \\ \texttt{81} \\ \texttt{e5} \end{bmatrix}$$

We show the computation of the first entry of the resulting vector, in other words, the computation of

$$[\texttt{02 03 01 01}] \begin{bmatrix} \texttt{d4} \\ \texttt{bf} \\ \texttt{5d} \\ \texttt{30} \end{bmatrix} = (\texttt{04})$$

By representing the column vector $[\texttt{d4 bf 5d 30}]^T$ as a polynomial vector, we can write this MixColumn operation in polynomial representation as

$$[x \quad x+1 \quad 1 \quad 1]\begin{bmatrix} x^7+x^6+x^4+x^2 \\ x^7+x^5+x^4+x^3+x^2+x+1 \\ x^6+x^4+x^3+x^2+1 \\ x^5+x^4 \end{bmatrix}$$

To compute the inner product, we perform polynomial multiplications, additions and reductions modulo $p(x)$ whenever necessary:

$$\begin{gathered} x\cdot(x^7+x^6+x^4+x^2)+ \\ (x+1)\cdot(x^7+x^5+x^4+x^3+x^2+x+1)+ \\ 1\cdot(x^6+x^4+x^3+x^2+1)+ \\ 1\cdot(x^5+x^4) \end{gathered}$$

The first product $x\cdot(x^7+x^6+x^4+x^2)$ needs to be computed and reduced if necessary. Here, we need reduction modulo the irreducible polynomial $p(x)$ since the resulting polynomial would be of degree 8

$$\begin{aligned} x\cdot(x^7+x^6+x^4+x^2) &= x^8+x^7+x^5+x^3 \\ &= (x^4+x^3+x+1)+x^7+x^5+x^3 \\ &= x^7+x^5+x^4+x+1 \\ &= (1011\ 0011) = (\texttt{b3}) \end{aligned}$$

After the polynomial multiplication, we reduced the highest degree term (which is x^8) by substituting it with x^4+x^3+x+1, which is the lower half of the irreducible polynomial $p(x)=x^8+x^4+x^3+x+1$.

The second product, after the multiplication gives

$$(x+1)\cdot(x^7+x^5+x^4+x^3+x^2+x+1) = x^8+x^7+x^6+1$$

We also need to reduce it modulo $p(x)$ since its degree is larger than 8. By substituting x^8 with x^4+x^3+x+1, we obtain

$$(x^4+x^3+x+1)+x^7+x^6+1 = x^7+x^6+x^4+x^3+x$$

which is equal to $(1101\ 1010)=(\texttt{da})$. However, we do not need reductions for the third and fourth products:

$$\begin{aligned} 1\cdot(x^6+x^4+x^3+x^2+1) &= x^6+x^4+x^3+x^2+1 &= (\texttt{5d}) \\ 1\cdot(x^5+x^4) &= x^5+x^4 &= (\texttt{30}) \end{aligned}$$

Finally, adding all 4 resulting polynomials, we obtain the top entry as

$$\begin{array}{rc}
(\texttt{02})\cdot(\texttt{d4})=(\texttt{b3}) & x^7+x^5+x^4+x+1 \\
(\texttt{03})\cdot(\texttt{bf})=(\texttt{da}) & x^7+x^6+x^4+x^3+x \\
(\texttt{01})\cdot(\texttt{5d})=(\texttt{5d}) & x^6+x^4+x^3+x^2+1 \\
(\texttt{01})\cdot(\texttt{30})=(\texttt{30}) & x^5+x^4 \\
\hline
(\texttt{04}) & x^2
\end{array}$$

The remaining 3 entries are obtained by repeating the above operations by multiplying the second, third, and fourth rows of the fixed MixColumn matrix with the same state column.

1.3 PUBLIC-KEY CRYPTOGRAPHY

In public-key cryptography, the encryption $E_{K_e}(M)$ and the decryption $D_{K_d}(C)$ functions are inverses of one another, and use different keys

$$C = E_{K_e}(M) \text{ and } M = D_{K_d}(C) \ .$$

These processes are asymmetric, and the keys are not equal, i.e., $K_e \neq K_d$. The naming conventions are

- K_e is the public key, which is expected to be known by anyone;
- K_d is the private key, known only to the user;
- K_e may be easily deduced from K_d;
- However, K_d is not easily deduced from K_e.

The User publishes his own public key K_e, so that anyone can obtain it and can encrypt a message M, and send the resulting ciphertext to the User $C = E_{K_e}(M)$. The private key K_d is known only to the User and only the User can decrypt the ciphertext to get the message $M = D_{K_d}(C)$. The adversary may be able to block the ciphertext, but it cannot decrypt. A public-key cryptographic algorithm is based on a function $y = f(x)$ such that given x, computing y is easy; while, given y, computing x is hard:

Such functions are called **one-way functions**. In order to decide which function is hard according to this criteria, we can resort to the theory of complexity. However, a one-way function is difficult for anyone to invert, including the receiver of the encrypted text. Instead, we need a function that is easy to invert for the legitimate receiver of the encrypted message, but hard for everyone else. Such functions are called **one-way trapdoor functions.**

In order to build a public-key encryption algorithm, we need a one-way trapdoor function. As this fact is understood around 1975-1976, researchers

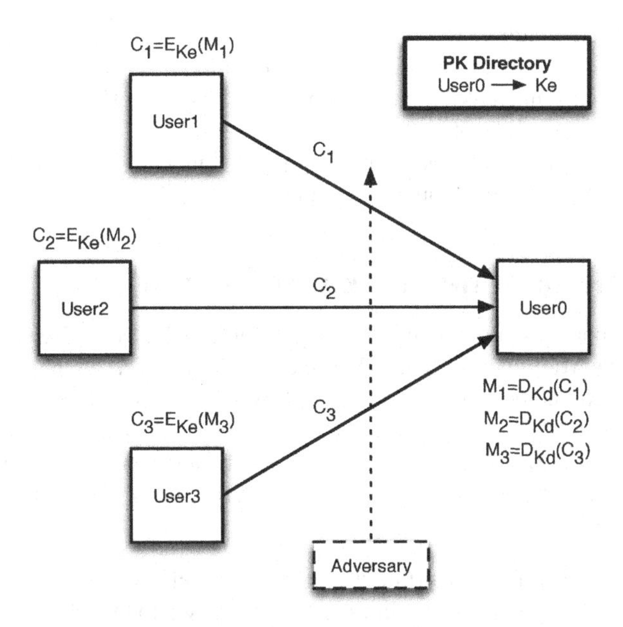

Figure 1.4 The general concept of public-key encryption.

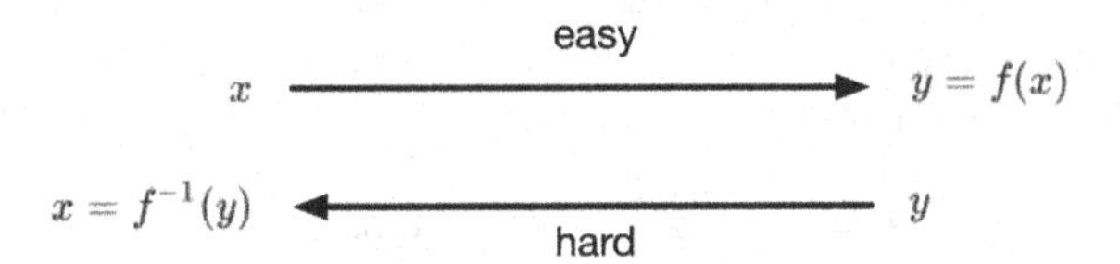

Figure 1.5 The general concept of a one-way function.

at Stanford and MIT ([79], [33]) were looking for such special functions which are either based on the known one-way functions or some other "unknown" constructions. Since then the following one-way functions have been identified, allowing us to build public-key encryption algorithms with the help of trapdoor mechanisms:

- Discrete Logarithm:
  Given p, g, and x, computing y in $y = g^x \pmod{p}$ is EASY
  Given p, g, y, computing x in $y = g^x \pmod{p}$ is HARD
- Factoring:
  Given p and q, computing n in $n = p \cdot q$ is EASY
  Given n, computing p or q in $n = p \cdot q$ is HARD

- Discrete Square Root:
  Given x and y, computing y in $y = x^2 \pmod{n}$ is EASY
  Given y and n, computing x in $y = x^2 \pmod{n}$ is HARD
- Discrete eth Root:
  Given x, n and e, computing y in $y = x^e \pmod{n}$ is EASY
  Given y, n and e, computing x in $y = x^e \pmod{n}$ is HARD

1.3.1 THE DIFFIE-HELLMAN KEY EXCHANGE METHOD

The first one-way function in this list gave birth to the **Diffie-Hellman Key Exchange Method**, whose trapdoor mechanism is based on the commutativity of exponentiation $(g^a)^b = (g^b)^a$. It was invented by Martin Hellman and Whitfield Diffie who published their paper "New Directions in Cryptography" in 1976 [33], introducing a radically new method of distributing cryptographic keys, and thus solving one of the fundamental problems of cryptography.

- A and B agree on a prime p and a primitive element g of $\mathscr{Z}_p^*$. This is accomplished in public: p and g are known to the adversary
- A selects $a \in \mathscr{Z}_p^*$, computes $r = g^a \pmod{p}$, and sends r to B
- B selects $b \in \mathscr{Z}_p^*$, computes $s = g^b \pmod{p}$, and sends s to A
- A (having received s) computes $K = s^a \pmod{p}$
- B (having received r) computes $K = r^b \pmod{p}$
- These two quantities are equal since

$$K = r^a = (g^a)^b = g^{ab} \pmod{p}\ ,$$
$$K = s^b = (g^b)^a = g^{ab} \pmod{p}\ .$$

At the end of these computation and communication steps, the parties A and B have the key value K, which is known to them but computing K is hard by anyone who sees and records all communicated values. The difficulty of computing the key K depends on the Discrete Logarithm Problem, whose general definition is given as: The computation of $x \in \mathscr{Z}_p^*$ in $y = g^x \pmod{p}$, given p, g, and y.

For example, given $p = 23$ and $g = 5$, can we find x such that $7 = 5^x \pmod{23}$? The answer in this case easy $x = 19$ since we can find it by trying all possible values in $\mathscr{Z}_{23}^* = \{1, 2, \ldots, 22\}$. However, the difficulty of computing the discrete logarithm for a larger p will significantly higher; consider $p = 158(2^{800} + 25) + 1 =$

105354628039501697530461658293395873194887181492591348934260873425871788357518586730038628773770557793738292587376

245199045043066135085968269741025626827114728303489756321
430023716636917406661590717647254947008311310713818992128
0884003892629359

and $g = 17$, and the computation of $x \in \mathscr{Z}_p^*$ such that $2 = 17^x \pmod p$. Such x exists since 17 is a primitive root of p; however, the number of trials to find it will require insurmountable time and energy.

If the Discrete Logarithm Problem is difficult in a group (such as $\mathscr{Z}_p^*$), we can use it to implement not only the Diffie-Hellman key exchange method, but also several other public-key cryptographic algorithms, such as the ElGamal public-key encryption method and the Digital Signature Algorithm. As we described, we can resort to the exhaustive search of the unknown value by trying all possible values of $x \in \mathscr{Z}_p^*$ iteratively.

```
z = g
for i = 2 to p − 1
        z = g · z (mod p)
        if y = z
                return x = i
```

This algorithm requires $p-2$ multiplications. However, it is an exponential algorithm in terms of the input size, which is the number of bits in the prime p. Since, the multiplications of two k-bit operands are of order $O(k^2)$, the search complexity is exponential in k, as $O(pk^2) = O(2^k k^2)$. There are better algorithms, such as Shanks Algorithm, Pollard Rho algorithm, Pohlig-Hellman algorithm, and the Index Calculus Algorithm; the first three algorithms are still of exponential complexity. The analysis is of the Index Calculus Algorithm is more complicated and is estimated to be

$$O\left(e^{c\cdot(\log p)^{1/3}(\log\log p)^{2/3}}\right).$$

This time complexity is sub-exponential since it is faster than exponential (in $\log p$) but slower than polynomial. Therefore, the Discrete Logarithm Problem remains to be a hard problem on a digital computer, making the Diffie-Hellman key exchange method a strong public-key cryptographic algorithm. Currently, much of wireless communication and internet security depends on it.

1.3.2 THE RSA ALGORITHM

The second important algorithm in the search for one-way trapdoor functions came from the 3 MIT professors, Ronald Rivest, Adi Shamir, and Leonard Adleman in the Summer and Fall of 1976. Their paper was published in 1978 [79], and MIT patented the method 1983 (which ended in

2000). The Rivest-Shamir-Adleman Algorithm or briefly as the **RSA Algorithm** constructs public and private keys for the User as follows:

- The User generates 2 large, about same size random primes: p and q
- The modulus n is the product of these two primes: $n = p \cdot q$
- Euler's totient function of n is given by $\phi(n) = (p-1) \cdot (q-1)$
- The User selects e as $1 < e < \phi(n)$ such that $\gcd(e, \phi(n)) = 1$ and computes $d = e^{-1} \pmod{\phi(n)}$ using the extended Euclidean algorithm.
- The **public key**: The modulus n and the public exponent e.
- The **private key**: The private exponent d, the primes p and q, and $\phi(n) = (p-1) \cdot (q-1)$

Once the keys are available, the encryption and decryption operations are performed by computing

$$
\begin{aligned}
C &= M^e \pmod{n}, \\
M &= C^d \pmod{n},
\end{aligned}
$$

where M, C are the plaintext and ciphertext such that $0 \leq M, C < n$.

The security of the RSA Algorithm depends on the discrete eth root problem, i.e., given y, n and e, computing x in $y = x^e \pmod{n}$ is known to be a hard problem. One can attempt to break the RSA algorithm in several ways:

- Compute eth Root of $M^e \pmod{n}$ and obtain M
- Factor $n = pq$, compute $d = e^{-1} \bmod (p-1)(q-1)$
- Obtain $\phi(n)$ by some method, and compute $d = e^{-1} \bmod \phi(n)$

There is no known algorithm for computing discrete eth root mod n directly, and it is obvious factoring n indeed breaks the RSA encryption algorithm. However, "Breaking RSA" does not mean that we can factor n. There is no general proof for such a claim.

1.3.3 DIGITAL SIGNATURES

A digital signature or digital signature algorithm is a mathematical method for demonstrating the authenticity of a digital message or document. A valid digital signature gives a recipient reason to believe that the message was created by a known sender (authentication) such that he cannot deny sending it (non-repudiation) and that the message was not altered in transit (integrity). Digital signatures are commonly used for software

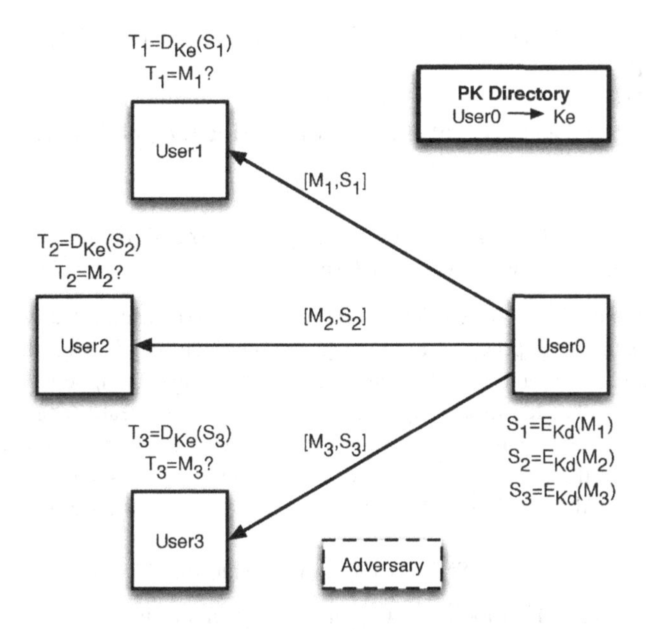

Figure 1.6 The general concept of digital signatures.

distribution, financial transactions, and in other cases where it is important to detect forgery or tampering. A public-key encryption algorithm is also a digital signature algorithm, the most notable example being the RSA algorithm.

Diffie and Hellman first described the concept of a digital signature scheme, and they conjectured that such methods exist. The RSA algorithm can be used as a public-key encryption method and as a digital signature algorithm

However, the plain RSA signatures have certain security problems. Other digital signature algorithms have been developed after the RSA: Lamport signatures, Merkle signatures, and Rabin signatures. Several more digital signature algorithms followed up and are in use today: ElGamal, the Digital Signature Algorithm (DSA), the elliptic curve DSA (ECDSA).

The steps of the (plain) RSA signatures follows as:

- User A has an RSA public key (n,e) and private key (n,d)
- User A creates a message $M < n$, and **encrypts the message using the private key** to obtain the signature S as

$$S = M^d \pmod{n}$$

and sends the message (plaintext) and the signature $[M, S]$ to User B

- User B receives $[M, S]$, obtains User A's public key from the directory, and **decrypts the signature using the public key**:

$$T = S^e \pmod{n}$$

If $T = M$, then the User B decides that the signature S on the message M was created by User A

The plain RSA signatures have several problems to be used directly as a signature scheme in practice. First of all, the message length is limited to the modulus length, and longer messages cannot be directly signed. A biggest concern is that legitimate signatures can be used to create **forged** signatures. Consider that $[M, S]$ is a legitimate pair of message and signature, created by the owner of the public and private key pair such that $S = M^d \pmod{n}$ and $M = S^e \pmod{n}$. The pair $[M^2 \bmod n,\ S^2 \bmod n]$ also verifies:

$$(S^2)^e = (S^e)^2 = M^2 \pmod{n}$$

It appears that $[M^2 \bmod n,\ S^2 \bmod n]$ is a legitimate signature.

The solution of these problems with plain RSA signatures are avoided by employing a hash function $H(\cdot)$. Instead of encrypting M with the private key, we encrypt $H(M)$: the hash of M

$$h = H(M) \ \rightarrow \ S = h^d \pmod{n} \ \rightarrow \ [M, S]$$

The receiving party verifies the message and signature pair $[M, S]$ using

$$h = H(M) \ \rightarrow \ T = S^e \pmod{n} \ \rightarrow \ T \stackrel{?}{=} h$$

The cryptographic hash function $H(\cdot)$ is a publicly available function, and does not involve a secret key.

The Diffie-Hellman and RSA algorithms opened up new avenues for cryptography, particularly in internet security and wireless communication. The next 4 decades from 1980s to now have seen their proliferation and implementations. New methods and standards have been developed by NIST, as well as banking, communication, and internet communities. Public-key cryptography has become an household term, including the software packages and communication utilities, such as SSL and https.

1.4 POST-QUANTUM CRYPTOGRAPHY

The quantum computer was developed based on the principles of quantum physics to perform computations. Classical computers use bits which is either 0 or 1 whereas quantum computers use quantum bits (called qubits) which can be either in a quantum state (0 or 1) or in a superposition of these states. A quantum computer is useful only if a quantum algorithm which solves a particular problem exists. It is important to distinguish between Quantum Cryptography and Post-Quantum Cryptography (PQC). Quantum cryptography refers to quantum mechanical solutions to achieve communication secrecy or quantum key distribution. On the other hand, post-quantum cryptography aims to design and deploy algorithms that are secure against classical and quantum attacks. The security proofs of current widely used public-key cryptosystems (namely, RSA, Diffie-Hellman and ECC) are based on the hardness of integer factorization, discrete logarithm and elliptic curve discrete logarithm problems. Solving these problems using classical computation technology, even with hardware accelerators, takes hundreds of years. However, in 1994 Peter Shor [85] proposed an algorithm which solves these problems in polynomial time with a large-scale (a few thousands of qubits) quantum computer. Although the key sizes for RSA and ECC used today are resistant against currently available small-scale quantum computers, the transition from classical public-key cryptography to post-quantum cryptography is needed in the near future, before any large-scale computers are built. Compared with public-key cryptography, symmetric cryptography is less affected by quantum attacks like Grover's algorithm which halves the security level.

In 2016, National Institute of Standards and Technology (NIST) released a report that announced a standardization plan for PQC and called for new quantum-resistant cryptographic algorithms for key encapsulation mechanisms (KEM), public-key encryption (PKE) and digital signatures. The evaluation criteria used throughout the NIST PQC standardization process [70] are: 1) security, 2) cost and performance, and 3) algorithm and implementation properties.

Among the 82 received submissions by the November 2017 deadline, 69 of them were accepted into the first round of the standardization process in December 2017 as they met the submission requirements and minimum acceptability criteria. In January 2019, NIST selected 26 algorithms to advance to the second round, after considering the public feedback and internal reviews of candidates. 17 of them were KEMs/PKEs and 9 were digital signatures. Four of the 26 candidates were mergers of the first round algorithms. In July 2020, NIST announced the 15 candidates moved on to the third round of the standardization process. Of the 15 advancing candidates, seven have been selected as finalists and eight as alternate candidates. The alternate candidates are considered as potential candidates for future standardization, most likely after another round of evaluation.

Table 1.2
NIST 3rd round finalists (Cb: Code-based; Lb: Lattice-based; Mv: Multivariate).

Scheme	Type	Security Problem
Classic McEliece	KEM, Cb	Decoding Goppa codes
CRYSTALS-Kyber	KEM, Lb	Module-LWE
NTRU	KEM, Lb	NTRU problem
SABER	KEM, Lb	Module-LWE
CRYSTALS-Dilithium	Sign, Lb	Module-LWE & Module-SIS
FALCON	Sign, Lb	R-SIS over NTRU lattices
Rainbow	Sign, Mv	Unbalanced Oil-Vinegar

Table 1.3
NIST third round alternate (Cb: Code-based; Lb: Lattice-based; Ib: Isogeny-based; Mv: Multivariate; Sym: Symmetric; Hb: Hash-based).

Scheme	Type	Security Problem
BIKE	KEM, Cb	Decoding QC-MDPC codes
HQC	KEM, Cb	Decisional QCSD with parity
FrodoKEM	KEM, Lb	LWE
NTRU Prime	KEM, Lb	NTRU
SIKE	KEM, Ib	Isogenies of elliptic curves
GeMSS	Sign, Mv	Hidden Field Equation (HFE)
Picnic	Sign, Sym	ZKP
SPHINCS+	Sign, Hb	Security of the hash functions

There are five competing families of PQC algorithms: Code-based encryption, Isogeny-based encryption, Lattice-based encryption and signatures, Multivariate signatures, and Hash-based signatures.

1.4.1 CODE-BASED CRYPTOGRAPHY

Code-based cryptography uses error-correcting codes to build public-key encryption algorithms. The first code-based cryptosystem was proposed by Robert J. McEliece in 1978 [63]. Athough it is as old as RSA and has much stronger security than RSA, due to large key sizes (large matrices as its public and private keys), it was not deployed in practical applications so far. However, today it is a strong candidate for PQC as it is resistant to attacks using Shor's algorithm.

The security of McEliece cryptosystem is based on the hardness of efficient decoding of a selected linear code. A decoding algorithm corrects errors which might have occurred during the transmission of a message over a communication channel. The classical decoding problem is to find the closest codeword $\mathbf{c} \in C$ to a given $\mathbf{y} \in \mathbb{F}_q^n$ assuming that there is a unique closest codeword. Berlekamp et al. [5] showed that the general decoding problem for binary linear codes (over $\mathbb{F}_2$) is NP-complete. The original McEliece cryptosystem uses secretly generated random binary Goppa codes [47] which can be efficiently decoded with the algebraic decoding algorithm of Patterson [75]. Before presenting the algorithms of McEliece, we give a brief description of linear codes.

Let $\mathbb{F}_q$ be the finite field with q elements, where q is a prime power. A q-ary *linear code* of length n and dimension k is a k-dimensional vector subspace of $\mathbb{F}_q^n$. The elements of the code are called *codewords*. The *minimum distance* of the code is the minimum *weight* of its nonzero codewords, where the weight of a codeword is the number of its nonzero coordinates. A linear code of length n, dimension k and minimum distance d is referred to as an $[n,k,d]$ code. A code of minimum distance $d \geq 2t+1$ can correct up to t errors, i.e., C is a $\lfloor (d-1)/2 \rfloor$-error correcting code. A vector with more errors will likely get decoded incorrectly.

Since a linear code is a vector space, it admits a basis. Any codeword can be expressed as the linear combination of these basis vectors. A *generator matrix* G of an $[n,k,d]$ code C is a $k \times n$ matrix whose rows form a basis for C. Namely, $C = \{\mathbf{x}G : \mathbf{x} \in \mathbb{F}_q^k\}$. A *parity-check matrix* of C is an $(n-k) \times n$ matrix H such that $\{\mathbf{c} \in \mathbb{F}_q^n : H\mathbf{c}^T = 0\}$ where $\mathbf{c}^T$ is the transpose of c. If G has the form $[I_k|A]$, where I_k is the $k \times k$ identity matrix, then G is said to be in *systematic form*. The matrix $H = [A^T|I_{n-k}]$ is then a parity-check matrix for C. There are many generator matrices for a linear code, but there is a unique one in systematic form if it exists.

The algorithms of the original McEliece cryptosystem is as follows:
KeyGen: A t-error correcting binary $[n,k,d]$ linear code C with a generator matrix G' is picked. Further, a $k \times k$ random binary invertible matrix S and an $n \times n$ random binary permutation matrix P are chosen. Multiplying

a vector by a permutation matrix, which has exactly one 1 in each row and each column and 0s elsewhere, permutes the entries of the vector. The public key is the pair $(G = SG'P, t)$ and the secret key is the triple (G', S, P) with an efficient decoding algorithm for C.

Enc: To encrypt a plaintext $\mathbf{m} \in \mathbb{F}_2^k$, a random vector $\mathbf{e} \in \mathbb{F}_2^n$ of weight t is chosen and the ciphertext is computed as

$$\mathbf{c} = \mathbf{m}G + \mathbf{e}.$$

Dec: To decrypt a ciphertext $\mathbf{c} \in \mathbb{F}_2^n$ the legitimate receiver, who knows the matrices S, G', P and an efficient decoding algorithm for C, computes first

$$\mathbf{c}P^{-1} = \mathbf{m}GP^{-1} + \mathbf{e}P^{-1} = \mathbf{m}SG'PP^{-1} + \mathbf{e}P^{-1} = \mathbf{m}SG' + \mathbf{e}P^{-1}.$$

Note that the weight of $\mathbf{e}P^{-1}$ is t and since C is a t-error correcting code, the codeword $\mathbf{m}SG'$ is obtained. Using the decoding algorithm for C, the legitimate receiver recovers $\mathbf{m}S$ and then covers $\mathbf{m}$ by multipliying the inverse of S.

McEliece's original parameter set with $n = 1024, k = 524, t = 50$ were designed for 2^{64} security, but it was broken in 2008 [6] with approximately 2^{60} CPU cycles. Further, the new parameters were designed to minimize public-key size while achieving 80-bit, 128-bit, or 256-bit security against known attacks [6]. For a detailed security analysis of these codes, we refer the reader to [58].

1.4.2 ISOGENY-BASED CRYPTOGRAPHY

Isogeny-based cryptography uses maps between elliptic curves, called isogenies, to build public-key cryptography. The first such cryptosystem was discovered by Couveignes in 1997, but became better known in 2006 [29]. This system further developed by Rostovtsev and Stolbunov in [80] and Stolbunov in [90]. All these proposed systems are based on the difficulty of computing isogenies between ordinary elliptic curves. This hardness assumption is totally different from the hardness of the elliptic curve discrete logarithm problem for security. Therefore, Shor's quantum algorithm [85] cannot break these systems. However, Couveignes–Rostovtsev–Stolbunov (CRS) cryptosystem based on ordinary elliptic curves can be broken with a subexponential quantum attack [24]. In 2011, Jao and De Feo [52] used isogenies between supersingular elliptic curves rather than ordinary ones to construct a novel key-exchange protocol, called *Supersingular Isogeny Diffie-Hellman (SIDH)*. The extended version of SIDH was later released by Jao, De Feo and Plût with [40]. SIDH addressed both the performance and security drawbacks of CRS system. Thenceforth, SIDH has attracted almost all research focus of isogeny-based cyrptography. SIDH resists against the attack proposed in [24] which exploits the commutativity of the endomorphism ring of an ordinary elliptic curve, since SIDH is constructed using the isogenies between supersingular elliptic curves whose endomorphism ring is non-commutative. Currently known fastest classical and quantum attacks against SIDH are both exponential. The SIDH algorithm also provides perfect forward secrecy which improves the long-term security of encrypted communications. Further, compromise of a key does not affect the security of the past communication.

SIDH was used to build the key encapsulation mechanism SIKE (Supersingular Isogeny Key Encapsulation) [51] based on pseudo-random walks in supersingular isogeny graphs, that was submitted to the NIST standardization process on post-quantum cryptography and selected as a third round alternate candidate. One of the main advantages to SIKE is that it has the smallest public key sizes of all the encryption and KEM schemes, as well as very small ciphertext sizes. Among all the post-quantum cryptosystems, isogeny-based systems are the most recent and their security against quantum attacks needs to be further studied.

In 2018, Castryck et al. [17] presented CSIDH (Commutative Supersingular Isogeny Diffie-Hellman) which directly adopts the CRS cryptosystem based on ordinary elliptic curves to supersingular case. CSIDH is vulnerable to the attack proposed in [24]. On the performance side, CSIDH is much faster than CRS while it is slower than SIDH. CSIDH has not been submitted to NIST's standardization process since it was designed after the submission deadline date.

Table 1.4
Instantiations of Diffie-Hellman.

	DH	**ECDH**	**SIDH**
elements	integers g modulo prime	points P in curve group	curves $\mathscr{E}$ in isogeny class
secrets	exponents x	scalars k	isogenies ϕ
computations	$g, x \mapsto g^x$	$k, P \mapsto [k]P$	$\phi, \mathscr{E} \mapsto \phi(\mathscr{E})$
hard problem	given g, g^x find x	given P, $[k]P$ find k	given $\mathscr{E}$, $\phi(\mathscr{E})$ find ϕ

In this section, original SIDH key-exchange protocol will be explained. Before that, we first briefly introduce supersingular elliptic curves over finite fields and isogenies. For more details on elliptic curves and their use in cryptography, we refer the interested readers to [87, 41].

1.4.2.1 Supersingular Elliptic Curves and Isogenies

Let $\mathbb{F}_q$ be a finite field of q elements where q is a prime power, namely $q = p^n$ for $n > 0$ and $p > 3$. An elliptic curve $\mathscr{E}$ defined over $\mathbb{F}_q$, denoted as $\mathscr{E}/\mathbb{F}_q$, is given by an equation in short Weierstrass form

$$\mathscr{E} : y^2 = x^3 + ax + b, \quad a, b \in \mathbb{F}_q \text{ and } 4a^3 + 27b^2 \neq 0.$$

The set of points on $\mathscr{E}$ over $\mathbb{F}_q$ are the set of pairs $(x, y) \in \mathbb{F}_q^2$ satisfying the curve equation

$$\mathscr{E}(\mathbb{F}_q) = \{(x, y) \in \mathbb{F}_q^2 : y^2 = x^3 + ax + b\} \cup \{\mathscr{O}_{\mathscr{E}}\},$$

where $\mathscr{O}_{\mathscr{E}} = (\infty, \infty)$ is the *point at infinity* , which is also considered to be a solution to the Weierstrass equation. The set of points on an elliptic curve $\mathscr{E}$ is an abelian group with the identity element $\mathscr{O}_{\mathscr{E}}$ under the "chord and tangent rule". The number of points on $\mathscr{E}/\mathbb{F}_q$ is $\#\mathscr{E}(\mathbb{F}_q) = q + 1 - t$ for an integer t lying in the interval $[-2\sqrt{q}, 2\sqrt{q}]$. An elliptic curve is called **supersingular** if $t \equiv 0 \bmod p$, or equivalently $\#\mathscr{E}(\mathbb{F}_q) = 1 \bmod q$ and is called **ordinary** otherwise.

For $k \in \mathbb{N}$ and $P \in \mathscr{E}(\mathbb{F}_q)$, we define $[k]P = P + P + \cdots + P$ (n times). The order of P is k if $[k]P = \mathscr{O}_{\mathscr{E}}$. Since $\mathscr{E}(\mathbb{F}_q)$ is a finite group, the order of any point $P \in \mathscr{E}(\mathbb{F}_q)$ is finite and divides the group order $\#\mathscr{E}(\mathbb{F}_q)$.

Let $\mathscr{E}_1$ and $\mathscr{E}_2$ be two elliptic curves over $\mathbb{F}_q$. An **isogeny** $\phi : \mathscr{E}_1 \to \mathscr{E}_2$ is a non-constant rational function which is a group homomorphism (i.e., compatible with the group operations) satisfying $\phi(\mathscr{O}_{\mathscr{E}_1}) = \mathscr{O}_{\mathscr{E}_2}$. Two elliptic curves are **isogenous** if there is an isogeny between them. **Endomorphisms** are a special class of isogenies where the domain and co-domain are the same curve. The endomorphism ring of $\mathscr{E}$ is the set of isogenies from $\mathscr{E}$ to itself, along with the zero map $0 : \mathscr{E} \to \mathscr{E}$ given by $0(P) = \mathscr{O}_{\mathscr{E}}$ for all points P on $\mathscr{E}$. In a set notation, $\mathrm{End}(E) = \{\phi : \mathscr{E} \to \mathscr{E}\} \cup \{0\}$. **Isomorphisms** also forms special class of isogenies where the kernel is trivial. If there is a pair of isogenies $\phi : \mathscr{E}_1 \to \mathscr{E}_2$ and $\psi : \mathscr{E}_2 \to \mathscr{E}_1$ such that both $\phi \circ \psi$ and $\psi \circ \phi$ are the identity, then ϕ and ψ are isomorphisms and so $\mathscr{E}_1$ and $\mathscr{E}_2$ are isomorphic curves. Elliptic curves up to isomorphism forms the isomorphism classes. The typical representation for isomorphism classes is the j-invariant which is

$$j(\mathscr{E}) = 1728 \frac{4a^3}{4a^3 + 27b^2}$$

for elliptic curves in short Weierstrass form. The SIDH algorithm establishes the secret key by computing the j-invariant of two isomorphic supersingular elliptic curves generated by the two communicating parties that happens to be isogenous to an initial supersingular curve.

A theorem of Tate states that $\mathscr{E}_1$ and $\mathscr{E}_2$ are isogenous if and only if they have the same number of points over $\mathbb{F}_q$ (indeed over any finite extension of $\mathbb{F}_q$), i.e., $\#\mathscr{E}_1(\mathbb{F}_q) = \#\mathscr{E}_2(\mathbb{F}_q)$ [93]. The set of curves that are isogenous to an elliptic curve $\mathscr{E}$ is called the **isogeny class** of $\mathscr{E}$. Note that if $\mathscr{E}$ is supersingular then all curves in its isogeny class are supersingular; similarly, isogeny class of an ordinary curve consists of ordinary curves. It is well known that every supersingular curve is isomorphic to one defined over $\mathbb{F}_{p^2}$. From now on, we consider the supersingular curves only.

The cryptography community is interested in separable isogenies, which does not factor through Frobenius map $(x,y) \mapsto (x^q, y^q)$ over $\mathbb{F}_q$. The **degree** of a separable isogeny is the number of points in its kernel. An isogeny is defined by its kernel in the sense that for every finite subgroup G of $\mathscr{E}_1$, there is a unique $\mathscr{E}_2$ (up to isomorphism) and a separable isogeny $\phi : \mathscr{E}_1 \to \mathscr{E}_2$ such that $\text{Ker } \phi = G$. Instead of $\mathscr{E}_2$, we sometimes write $\mathscr{E}_1/G$. Given a finite subgroup G of $\mathscr{E}_1$, an isogeny $\phi : \mathscr{E}_1 \to \mathscr{E}_2$ with kernel G can be constructed by Vélu's formulas [95]. Notice that the number of distinct isogenies of degree ℓ, called as ℓ-isogenies, is equal to the number of distinct subgroups of $\mathscr{E}_1$ of order ℓ.

As an example, for each $m \in \mathbb{Z}$ such that $p \nmid m$ and an elliptic curve $\mathscr{E}$ over $\mathbb{F}_q$, consider the following separable isogeny

$$\begin{aligned} [m] : \quad & \mathscr{E} \to \mathscr{E} \\ & P \mapsto [m]P. \end{aligned}$$

The kernel of this isogeny is the m-torsion subgroup of $\mathscr{E}$, denoted by $\mathscr{E}[m]$, which is the set of points on $\mathscr{E}$ of order m,

$$\mathscr{E}[m] = \{P \in \mathscr{E} : [m]P = \mathscr{O}_{\mathscr{E}}\} = \mathbb{Z}/m\mathbb{Z} \times \mathbb{Z}/m\mathbb{Z}.$$

The degree of the above isogeny is equal to $\#\mathscr{E}[m] = m^2$.

1.4.2.2 Supersingular Isogeny Diffie-Hellman (SIDH)

First, the domain (public) parameters are fixed. Let p be a prime of the form ${\ell_A}^{e_A} {\ell_B}^{e_B} f \pm 1$, where ℓ_A and ℓ_B are small primes, e_A and e_B are positive integers and f is a small cofactor such that p is a prime. A supersingular elliptic curve $\mathscr{E}$ defined over $\mathbb{F}_q = \mathbb{F}_{p^2}$ is constructed (this can be done via an efficient algorithm due to Broker [16]) such that it has cardinality $({\ell_A}^{e_A} {\ell_B}^{e_B} f)^2$. Elliptic curve points $P_A, Q_A \in \mathscr{E}[\ell_A^{e_A}]$ are chosen such that the group $\langle P_A, Q_A \rangle$ generated by P_A and Q_A is the entire group $\mathscr{E}[\ell_A^{e_A}]$, i.e., $\{P_A, Q_A\}$ form a basis for $\mathscr{E}[\ell_A^{e_A}]$. In a similar way, elliptic curve points $P_B, Q_B \in \mathscr{E}[\ell_B^{e_B}]$ are chosen such that the group $\langle P_B, Q_B \rangle$ generated by P_B and Q_B is the entire group $\mathscr{E}[\ell_B^{e_B}]$.

The SIDH key exchange protocol between two parties A and B works as follows:

- A picks two random integers $0 \leq m_A, n_A < \ell_A^{e_A}$ such that $\ell_A \nmid m_A, n_A$ and computes $[m_A]P_A + [n_A]Q_A$. Similarly, B picks two random integers $0 \leq m_B, n_B < \ell_B^{e_B}$ such that $\ell_B \nmid m_B, n_B$ and computes $[m_B]P_B + [n_B]Q_B$.
- A creates a secret isogeny $\phi_A : \mathscr{E} \to \mathscr{E}_A$ with kernel generated by the point $[m_A]P_A + [n_A]Q_A$ by using Vélu's formulas. Then $\mathscr{E}_A = \phi_A(\mathscr{E}) = \mathscr{E}/K_A$ where $K_A = \langle [m_A]P_A + [n_A]Q_A \rangle$ is the kernel. Similarly, B creates a secret isogeny $\phi_B : \mathscr{E} \to \mathscr{E}_B$ for which the kernel is $K_B = \langle [m_B]P_B + [n_B]Q_B \rangle$ and $\mathscr{E}_B = \phi_B(\mathscr{E}) = \mathscr{E}/K_B$.
- In the exchange step of the protocol, A and B publishes the messages $(\mathscr{E}_A, \phi_A(P_B), \phi_A(Q_B))$ and $(\mathscr{E}_B, \phi_B(P_A), \phi_B(Q_A))$, respectively.
- Upon receiving B's message, A computes an isogeny $\phi'_A : \mathscr{E}_B \to \mathscr{E}_{AB}$ with kernel $\langle [m_A]\phi_B(P_A) + [n_A]\phi_B(Q_A) \rangle = \langle \phi_B([m_A]P_A + [n_A]Q_A \rangle = \phi_B(K_A)$. Here, $\mathscr{E}_{AB} = \mathscr{E}_B/\phi_B(K_A)$. Similarly, having received A's message, B computes an isogeny $\phi'_B : \mathscr{E}_A \to \mathscr{E}_{BA}$ with kernel $\langle [m_B]\phi_A(P_B) + [n_B]\phi_A(Q_B) \rangle = \phi_A(K_B)$. Here, $\mathscr{E}_{BA} = \mathscr{E}_A/\phi_A(K_B)$.
- The elliptic curves $\mathscr{E}_{AB}$ and $\mathscr{E}_{BA}$ computed by A and B are isomorphic as they are both isomorphic to $\mathscr{E}/\langle K_A, K_B \rangle$, so they have the same j-invariant. This common j-invariant is the shared secret key.

Given the curves $\mathscr{E}_A$, $\mathscr{E}_B$ and the points $\phi_A(P_B), \phi_A(Q_B), \phi_B(P_A), \phi_B(Q_A)$ as described in the above protocol, finding the j-invariant of $\mathscr{E}/\langle K_A, K_B \rangle$ is called as *supersingular computational Diffie-Hellman (SSCDH) problem.* The security of SIDH is based on this problem. SSCDH is more special (due to auxiliary information) than the main problem in this area known as *supersingular isogeny problem* which is described as follows: Given a finite field K and two supersingular elliptic curves $\mathscr{E}_1$, $\mathscr{E}_2$ defined over K such that $|\mathscr{E}_1| = |\mathscr{E}_2|$, compute an isogeny $\phi : \mathscr{E}_1 \to \mathscr{E}_2$. The best known classical algorithm for this problem is due to Delfs and Galbraith [32] and requires $\tilde{O}(p^{1/2})$ bit operations. The best known quantum algorithm is due to Biasse et al [7] and requires $\tilde{O}(p^{1/4})$ bit operations. However, SSCDH problem can be regarded as an instance of the *claw problem* for which the best known classical and quantum attacks requires $O(p^{1/4})$ and $O(p^{1/6})$ bit operations, respectively [98, 92].

Note that the number of possible secret isogenies that A can create is equal to the number of possible distinct kernels, which is $\ell_A^{e_A-1}(\ell_A+1)$ and the number of possible isogeny choices for B is $\ell_B^{e_B-1}(\ell_B+1)$.

Further note that the time needed to compute an isogeny grows linearly with the degree of the isogeny. Representing a large degree isogeny as a composition of small prime degree isogenies makes isogeny crypto feasible. For example, SIDH decomposes a degree $\ell_A^{e_A}$ isogeny into a sequence of e_A isogenies of degree ℓ_A instead of computing the isogeny in a single step using Vélu's formulas. while the computation cost of the latter one is $O(\ell_A^{e_A})$, the cost to compute the former one is proportional to ℓ_A. To reduce the cost of the computation of sequences of isogenies and speed the computation up, Jao and De Feo proposed a new method. For further details, we refer the readers to [52].

The selection of the primes, the selection of the curve equation and the elliptic curve point representation (affine vs projective) together yield efficient implementations of the SIDH algorithm. For the fast arithmetic computation inside the SIDH protocol, it is more convenient to use the primes of the form $p = 2^{e_A}3^{e_B} \pm 1$. For an initial curve $\mathscr{E}/\mathbb{F}_{p^2} : y^2 = x^3 + x$ where $p = 2^{e_A}3^{e_B} \pm 1$, the 751-bit prime $p = 2^{372}3^{239} - 1$ provides 125-bit post-quantum security level matching security of AES-192 and the 964-bit prime $p = 2^{486}3^{301} - 1$ provides 161-bit post-quantum security level matching AES-256.

1.5 HOMOMORPHIC ENCRYPTION

Cloud computing offers many services to users, including storage of and computation with large amounts of data. To take the advantage of the cloud computing, users must share their data with the service provider. These data might be sensitive (for example, financial data or patients' medical records). A simple solution to ensure data privacy is to encrypt the data that is sent to the cloud. However, a user cannot compute on the encrypted data in the cloud. To perform computations, the data must be downloaded and decrypted, or the secret key must be shared with the service provider. The former process nullifies the major advantage of using cloud services while the later process sacrifices privacy. This is where *homomorphic encryption* (HE) comes into play. While the conventional encryption schemes does not allow operations to be performed on the encrypted data without decrypting it first, HE allows the cloud servers to compute on encrypted data without decrypting it in advance. This concept was first introduced in 1978, shortly after the invention of RSA cryptosystem [79], by Rivest, Adleman and Dertouzous [78], in their work entitled "On data banks and privacy homomorphism".

A homomorphic (public-key) encryption scheme $\mathscr{E}$ consists of four efficient algorithms: $\mathbf{KeyGen}_{\mathscr{E}}$, $\mathbf{Enc}_{\mathscr{E}}$, $\mathbf{Dec}_{\mathscr{E}}$ and $\mathbf{Eval}_{\mathscr{E}}$, where the first three algorithms are the usual 3-tuples of any conventional public-key encrption scheme whereas the fourth one is an HE-specific algorithm which is associated to a set of permitted functions $\mathscr{F}_{\mathscr{E}}$. These algorithms are efficient in the sense that their computational complexity must be polynomial in security parameter λ that specifies the bit-length of the keys. $\mathbf{KeyGen}_{\mathscr{E}}$ takes a security parameter λ as input, and outputs a pair of keys (pk, sk), where pk denotes the public key and sk denotes the secret key. $\mathbf{Enc}_{\mathscr{E}}$ takes the public key pk, a plaintext m from the underlying plaintext space $\mathscr{M}$ and some randomness as inputs, and outputs a ciphertext $c \in \mathscr{C}$ where $\mathscr{C}$ is the ciphertext space. $\mathbf{Dec}_{\mathscr{E}}$ takes the secret key sk and a ciphertext c as inputs, and outputs a plaintext m. *Correct decryption* is required to be able to call $\mathscr{E}$ an encryption scheme, i.e., the equality

$$\mathbf{Dec}_{\mathscr{E}}(sk, \mathbf{Enc}_{\mathscr{E}}(pk, m)) = m$$

should be satisfied. $\mathbf{Eval}_{\mathscr{E}}$ takes the public key pk, any ciphertexts $c_1, \ldots, c_t \in \mathscr{C}$ with $\mathbf{Enc}_{\mathscr{E}}(pk, m_i) = c_i$ and any permitted function f in $\mathscr{F}_{\mathscr{E}}$ as inputs. It outputs an evaluated ciphertext that encrypts $f(m_1, \ldots, m_t)$. *Correct evaluation* is satisfied if the following holds:

$$\mathbf{Dec}_{\mathscr{E}}(sk, \mathbf{Eval}_{\mathscr{E}}(pk, f, c_1, \ldots, c_t)) = f(m_1, \ldots, m_t),$$

i.e., the evaluated ciphertext decrypts to the computation of the plaintexts through $f \in \mathscr{F}_{\mathscr{E}}$. If f is not in $\mathscr{F}_{\mathscr{E}}$, with an overwhelming probability, $\mathbf{Eval}_{\mathscr{E}}$ algorithm will not produce a meaningful output.

If $\mathscr{E}$ has the properties of both correct decryption and correct evaluation for the functions in $\mathscr{F}_{\mathscr{E}}$, then it is called an **$\mathscr{F}_{\mathscr{E}}$-homomorphic scheme**. However, mere correctness does not rule out trivial schemes where the evaluation algorithm just output $(f, c_1 \ldots, c_t)$ without processing the ciphertexts at all, and the decryption function decrypts the ciphertexts $c_1, \ldots, c_t$ and then apply f to the resulting plaintexts. Further important attribute of an homomorphic encryption scheme, which is referred as *compactness* (or *compact ciphertext requirement*), excludes this trivial case. Compactness property requires the ciphertext size and decryption time to be completely independent of the homomorphically evaluated function f but only dependent on the security parameter λ. For example, decryption of an evaluated ciphertext takes the same amount of computation as decryption of a fresh ciphertext $c = \mathbf{Enc}_{\mathscr{E}}(pk, m)$. More formally, $\mathscr{E}$ is *compact* if there exists a polynomial g such that, for every value of the security parameter λ, $\mathbf{Dec}_{\mathscr{E}}$ can be expressed as a circuit of size at most $g(\lambda)$. Note that an $\mathscr{F}_{\mathscr{E}}$-homomorphic scheme is not necessarily compact.

An arithmetic function f can also be represented as a circuit which breaks the computation of f down into AND, OR and NOT gates. Addition, subtraction and multiplication operations (in fact, these operations *modulo* 2) are enough to evaluate these gates. For $x, y \in \{0,1\}$, we have $\text{AND}(x,y) = xy$, $\text{NOT}(x) = 1 - x$ and $\text{OR}(x,y) = 1 - (1-x)(1-y)$.

Homomorphic encryption schemes are categorized into three classes according to the set of permitted functions. If an encryption scheme permits only one type of operation (either addition or multiplication) with an unlimited number of times, then it is called a **partially homomorphic encryption (PHE)** scheme; if it allows one type of operation with a limited number of times while allowing another infinitely many times, it is called a **somewhat homomorphic encryption (SWHE)** scheme. In PHE and SWHE schemes, there is no compactness requirement, i.e., the ciphertexts can get quite larger with each homomorphic operation. If an encryption scheme can handle all functions (i.e., allows both addition and multiplication infinitely many times) and fulfill the compactness requirement then it is called as **fully homomorphic encryption (FHE)** scheme.

PHE schemes are deployed in some particular real-life applications like electronic voting [3] and Private Information Retrieval (PIR) [57] whose algorithms support only addition operation. Although, SWHE schemes support both addition and multiplication, the maximum number of operations performed homomorphically is limited since each operation contributes "noise" to the ciphertext and after a threshold decryption fails. However, explosion in demand for cloud computing platforms accelerated the construction of FHE schemes which enables arbitrary computation on encrypted data.

1.5.1 PARTIALLY HOMOMORPHIC ENCRYPTION

There are several PHE schemes [79, 46, 36, 2, 64, 72, 74, 31], supporting either addition or multiplication operation, in the literature . In this section, we focus on three of them. It is worth noting that Paillier and ElGamal algorithms are standardized in ISO (ISO/IEC 18033-6:2019).

1.5.1.1 Goldwasser-Micali Algorithm

The Goldwasser–Micali (GM) algorithm [46], developed by Shafi Goldwasser and Silvio Micali in 1982, has the distinction of being the first probabilistic public-key encryption scheme, where each plaintext has several corresponding ciphertexts. The security of the GM algorithm is based on the Quadratic Residuosity Problem modulo $n = p \cdot q$.

An integer $a \in \mathbb{Z}_n^*$ is called a *quadratic residue* modulo n if there exists an integer $x \in \mathbb{Z}_n^*$ such that $a \equiv x^2 \pmod{n}$. If there is no solution to this congruence, then a is called a *quadratic non-residue* modulo n.

There is a special number-theoretic tool associated with quadratic residues, the *Jacobi symbol*, denoted by $\left(\frac{a}{n}\right)$, which is defined for all $a \geq 0$ and all odd positive integers n. To understand the GM algorithm in detail, we refer the reader to Section 2.1.10 (Quadratic Residues) in [73]. If $n > 3$ is an odd composite integer, the problem of determining whether a non-negative integer a with Jacobi symbol 1 is a quadratic residue modulo n is called the *Quadratic Residuosity Problem.*

KeyGen: Two random large primes p and q are chosen, and $n = p \cdot q$ is computed. Then a quadratic non-residue $x \in \mathbb{Z}_n^*$ with Jacobi symbol $\left(\frac{x}{n}\right) = 1$ is chosen. This choice is accomplished by finding $x \in \mathbb{Z}_n^*$ such that $\left(\frac{x}{p}\right) = \left(\frac{x}{q}\right) = -1$. By choosing p and q as Blum integers, i.e. $p \equiv 3 \pmod{4}$ and $q \equiv 3 \pmod{4}$, the integer $n-1$ is guaranteed to be a quadratic non-residue with $\left(\frac{n-1}{n}\right) = \left(\frac{-1}{n}\right) = 1$. The public key pair is (n,x) and the private key pair is (p,q).

Enc: After converting the message into a plaintext which is a string of bits $(m_1, m_2, \ldots, m_k)$, the sender picks uniformly at random $y_i \in \mathbb{Z}_n^*$ for each bit m_i and encrypts each bit by computing

$$c_i = E(m_i) \equiv y_i^2 \cdot x^{m_i} \pmod{n}.$$

via the encryption function

$$\begin{aligned} E: (\{0,1\}, \oplus) &\rightarrow (\mathbb{Z}_n^*, \cdot) \\ m &\mapsto y^2 \cdot x^m, \end{aligned}$$

where $\oplus$ denotes addition modulo 2, $\cdot$ denotes modular multiplication and $\mathbb{Z}_n^*$ denotes the set of positive integers that are less than n and relatively prime to n. The ciphertext generated is $(c_1, c_2, \ldots, c_k)$, such that $c_i \in \mathbb{Z}_n$ for $i = 1, 2, \ldots, k$.

Dec: The legitimate receiver knows the private key pair (p,q) and can decide the quadratic residuosity of c_i modulo p and modulo q. To decrypt the message and get the plaintext back, she determines whether c_i is a quadratic residue modulo n for $i = 1, \ldots, k$. If c_i is a quadratic residue modulo both

p and q, then c_i is a quadratic residue modulo n, which necessarily yields $m_i = 0$. Otherwise, c_i is a quadratic non-residue modulo n which implies $m_i = 1$. **Eval:** Let y_1 and y_2 be randomly selected integers in $\mathbb{Z}_n^*$. For bits m_1 and m_2,

$$\begin{aligned} E(m_1) \cdot E(m_2) &\equiv (y_1{}^2 \cdot x^{m_1}) \cdot (y_2{}^2 \cdot x^{m_2}) \pmod{n} \\ &\equiv (y_1 \cdot y_2)^2 \cdot x^{m_1 \oplus m_2} \pmod{n} \\ &\equiv E(m_1 \oplus m_2), \end{aligned}$$

which yields

$$D(E(m_1) \cdot E(m_2)) = m_1 \oplus m_2.$$

The randomness in the encryption of $m_1 \oplus m_2$ is $y_1 \cdot y_2$, which is neither uniformly distributed in $\mathbb{Z}_n^*$ nor independent of the randomness in $E(m_1)$ and $E(m_2)$. However this can be addressed by the re-randomization property of GM algorithm. Let $r \in \mathbb{Z}_n^*$ be a random number. Then,

$$r^2 \cdot E(m) \equiv r^2 \cdot y^2 \cdot x^m \pmod{n} \equiv (r \cdot y)^2 \cdot x^m \pmod{n},$$

which is a valid encryption of m with the randomness $r \cdot y \in \mathbb{Z}_n^*$.

1.5.1.2 ElGamal Algorithm

The ElGamal cryptosystem [36], which is a public-key encryption scheme proposed by Taher ElGamal in 1985, improves the Diffie-Hellman key exchange method [33] into an encryption algorithm. There are two number theoretic versions of this algorithm; one is multiplicatively homomorphic and the other is additively homomorphic. Additively homomorphic version is not practical in use since it forces the legitimate receiver to solve a discrete logarithm problem, which is intractable, to decrpyt a ciphertext. Therefore, we focus our attention on the multiplicatively homomorphic version of ElGamal. Its security is based on the hardness of both the Computational Diffie-Hellman Problem and the Decisional Diffie-Hellman Problem in the underlying group G_q.
KeyGen: Two random large primes p and q satisfying $q \mid (p-1)$ are chosen. Next, a cyclic subgroup G_q of $\mathbb{Z}_p^*$ of order q with generator g is chosen. This choice is accomplished by selecting some $y \in \mathbb{Z}_p^*$ and computing $g \equiv y^{(p-1)/q} \pmod{p}$. Finally a random $x \in \mathbb{Z}_q$ is selected an $h = g^x \pmod{p}$ is computed. The public key quadruple is (p, q, g, h) and the private key is x.
Enc: The plaintext is $m \in G_q$. The sender generates a random number $r \in \mathbb{Z}_q$ and computes the ciphertext pair

$$E(m) = (c_1, c_2) = (g^r \pmod{p}, m \cdot h^r \pmod{p})$$

via the encryption function

$$\begin{aligned} E : (G_q, \cdot) &\rightarrow (G_q \times G_q, \cdot) \\ m &\mapsto (g^r, m \cdot h^r), \end{aligned}$$

For encryption of each message, a new r is chosen to be a uniformly random integer in order to ensure security.

Dec: The legitimate receiver who holds the private key x can decrypt the ciphertext (c_1, c_2), without knowing the value of r, by computing u_1 and u_2 as

$$\begin{aligned} u_1 &= (g^r)^x = (g^x)^r \equiv h^r \pmod{p} \\ u_2 &= {u_1}^{-1} \cdot c_2 \equiv h^{-r} \cdot (m \cdot h^r) \equiv m \pmod{p}, \end{aligned}$$

where ${u_1}^{-1}$ is the multiplicative inverse of u_1 in the group G_q. This inverse can be computed using the Extended Euclidean Algorithm in number theory.

Eval: Let m_1 and m_2 be two plaintexts with accompanying random numbers r and r', respectively. Then the pairwise products of the ciphertext pairs are

$$\begin{aligned} E(m_1) \cdot E(m_2) &= (c_1 \cdot c_1', c_2 \cdot c_2') \\ &= (g^r \cdot g^{r'} \pmod{p}, (m_1 \cdot h^r) \cdot (m_2 \cdot h^{r'}) \pmod{p}) \\ &= (g^{r+r'} \pmod{p}, m_1 \cdot m_2 \cdot h^{r+r'} \pmod{p}) \\ &= E(m_1 \cdot m_2) \end{aligned}$$

where the randomness in the encryption of $m_1 \cdot m_2$ is $r + r'$, which is neither uniformly distributed in $\mathbb{Z}_q$ nor independent of the randomness in $E(m_1)$ and $E(m_2)$. However this can be addressed by the re-randomization property of the multiplicative ElGamal algorithm.

Let $E(m) = c = (c_1, c_2) \equiv (g^r, m \cdot h^r) \pmod{p}$ for random $r \in \mathbb{Z}_q$, and $r' \in \mathbb{Z}_q$ be another chosen random number. Then

$$\begin{aligned} (c_1 \cdot g^{r'}, c_2 \cdot h^{r'}) &\equiv (g^r \cdot g^{r'}, m \cdot h^r \cdot h^{r'}) \pmod{p} \\ &\equiv (g^{r+r'}, m \cdot h^{r+r'}) \pmod{p}, \end{aligned}$$

which is a re-randomized ciphertext of the original message m where $r + r' \in \mathbb{Z}_q$.

1.5.1.3 Paillier Algorithm

The Paillier algorithm was developed by Pascal Paillier [74] in 1999. Its security is based on the Composite Residuosity Class Problem.

KeyGen: Two distinct large primes p and q are chosen such that $\gcd(p, q-1) = \gcd(q, p-1) = 1$. Then $\lambda(n) = \text{lcm}(p-1, q-1)$ is computed, where $n = p \cdot q$. Next, a semi-random base $g \in \mathbb{Z}_{n^2}^*$ with $n | \text{ord}(g)$ is chosen, where $\text{ord}(g)$ denotes the multiplicative order of g in the cyclic group $\mathbb{Z}_{n^2}^*$. The public key pair is (n, g) and private key is $\lambda(n)$.
Enc: After converting the message into a plaintext m, where $0 \leq m < n$, the sender chooses a random $u \in \mathbb{Z}_{n^2}^*$ such that $0 < u < n$. Then the ciphertext is computed as

$$c \equiv g^m \cdot u^n \pmod{n^2}$$

via the encryption function

$$\begin{aligned} E_g : \quad \mathbb{Z}_n \times \mathbb{Z}_n^* \quad & \rightarrow \mathbb{Z}_{n^2}^* \\ (m, u) \quad & \mapsto g^m \cdot u^n. \end{aligned}$$

Note that during key generation process, the condition on g (namely, $n | \text{ord}(g)$) ensures that the encryption function E_g is bijective. More clearly, given any $w \in \mathbb{Z}_{n^2}^*$, with n fixed, the pair (m, u) is the unique pair satisfying the equation $E_g(m, u) \equiv w \pmod{n^2}$. For the proof, see Lemma 14, Chapter 9 in [73].

The nth residuosity class of $w \in \mathbb{Z}_{n^2}^*$ with respect to g, denoted by $[\![w]\!]_g$, is the unique integer $m \in \mathbb{Z}_n$ for which there exists $u \in \mathbb{Z}_n^*$ such that

$$E_g(m, u) = g^m \cdot u^n = w.$$

Given $g, w \in \mathbb{Z}_{n^2}^*$ such that $n | \text{ord}(n)$, computing the class $[\![w]\!]_g$ is called *Composite Residuosity Class Problem* of base g on which the security of Paillier's algorithm is based. For more details about this problem, see Section 2.1.12 in [73].
Dec: The legitimate receiver decrypts the encrypted message c by using the private key $\lambda(n)$ as follows:

$$m \equiv L(c^{\lambda(n)} \pmod{n^2}) \cdot (L(g^{\lambda(n)} \pmod{n^2}))^{-1} \pmod{n}.$$

Here, $L(x) = (x-1)/n \pmod{n}$. To see the correctness proof of the decryption algorithm, the reader is referred to Chapter 9 in [73].
Eval: Consider the messages m_1 and m_2 with their accompanying random u-values u_1 and u_2, respectively. Then the product of corresponding ciphertexts is

$$\begin{aligned} E_g(m_1, u_1) \cdot E_g(m_2, u_2) &= (g^{m_1} \cdot (u_1)^n) \cdot (g^{m_2} \cdot (u_2)^n) \pmod{n^2} \\ &\equiv g^{m_1+m_2} (u_1 \cdot u_2)^n \pmod{n^2} \\ &= E_g(m_1 + m_2, u_1 \cdot u_2), \end{aligned}$$

where the randomness in the encryption of $m_1 + m_2$ is $u_1 \cdot u_2$, which is neither uniformly distributed in $\mathbb{Z}^*_{n^2}$ nor independent of the randomness in $E_g(m_1, u_1)$ and $E_g(m_2, u_2)$. However, this can be addressed by re-randomization property of the additively homomorphic Paillier algorithm.

For randomly chosen $u, u' \in \mathbb{Z}^*_n$, the ciphertext $c = E_g(m, u)$ can be re-randomized by calculating

$$\begin{aligned} E_g(m,u) \cdot (u')^n &\equiv c \cdot (u')^n \pmod{n^2} \\ &\equiv (g^m \cdot u^n) \cdot (u')^n \pmod{n^2} \\ &\equiv g^m \cdot (u \cdot u')^n \pmod{n^2} \\ &= E_g(m, u \cdot u'). \end{aligned}$$

1.5.2 SOMEWHAT HOMOMORPHIC ENCRYPTION

Until 2005, all proposed encryption schemes had partial (either additive or multiplicative) homomorphic property. In 2005, Boneh, Goh and Nissim constructed BGN cryptosystem based on bilinear pairings on elliptic curves that can support arbitrarily many additions and a single multiplication by keeping the ciphertext size constant. While BGN scheme meets the compactness requirement, allowing only one multiplication makes it somewhat homomorphic. After the first plausible FHE published in 2009 [42], some SWHE versions of FHE schemes were also proposed due to the performance issues associated with FHE schemes.

1.5.3 FULLY HOMOMORPHIC ENCRYPTION

Fully Homomorphic Encryption (FHE) is a special type of encryption which is both additively and multiplicatively homomorphic. Since addition and multiplication form a complete set of operations, an FHE scheme allows any polynomial-time computation on encrypted data. In 1978, Rivest, Adleman and Dertouzos [78] first proposed theoretic possibility of a scheme supporting arbitrarily complex computation in their paper titled "On Data Banks and Privacy Homomorphisms". However, for more than 30 years, this theoretic possibility could not be put into practice and so it has been regarded as a "holy grail" of cryptography. Craig Gentry proposed the first plausible way of obtaining an FHE scheme based on ideal lattices in his seminal Stanford PhD thesis [42].

Gentry's scheme is not only an FHE scheme but also a blueprint to obtain an FHE scheme from an SWHE scheme. Although this scheme was considered as a major breakthrough, it was not efficient and hard to implement. Since the release of this blueprint, significant progress has been made in the direction of finding efficient and simpler FHE schemes ([88, 94, 89, 15, 14, 12]). His construction has three components: an SWHE scheme that can support a limited number of operations (a few multiplications and arbitrarily many additions), *squashing* method which converts the SWHE scheme into a bootstrappable one and finally a method of *bootstrapping* which turns the (bootstrappable) SWHE scheme into an FHE scheme.

Encryption functions of all existing SWHE schemes works by adding a small amount of noise to the plaintext. Homomorphic evaluations on ciphertexts increase this noise and once it exceeds a certain threshold, the decryption fails. *Bootstrapping* refreshes a ciphertext by running the decryption function on it homomorphically. An SWHE scheme $\mathscr{E}$ is called *bootstrappable* if it can evaluate its own decryption function, plus one addition or multiplication gate modulo 2. When these augmented circuits are in the permitted set of functions (or circuits) $\mathscr{F}_{\mathscr{E}}$, one can construct a fully homomorphic encryption scheme from $\mathscr{E}$. A bootstrappable scheme refreshes the evaluated ciphertext for more homomorphic computations by reducing the noise in the ciphertext via the following $\textbf{Recrypt}_{\mathscr{E}}$ algorithm.

$\textbf{Recrypt}_{\mathscr{E}}(pk_2, D_{\mathscr{E}}, \overline{sk_1}, c_1)$

- Generate $\overline{c_1}$ via $\textbf{Encrypt}_{\mathscr{E}}(pk_2, c_{1j})$ over the bits of c_1
- Output $c = \textbf{Eval}_{\mathscr{E}}(pk_2, D_{\mathscr{E}}, \overline{sk_1}, \overline{c_1})$

First, it is supposed that two different public and secret key pairs are generated, (pk_1, sk_1) and (pk_2, sk_2). Let c_1 be the encryption of the message bit m with pk_1 and let $\overline{sk_1}$ be a vector of ciphertexts encrypted with pk_2 over the bits of sk_1. The public key pk_2, the decryption circuit $D_{\mathscr{E}}$, $\overline{sk_1}$ and c_1 are taken as inputs by the $\textbf{Recrypt}_{\mathscr{E}}$ function. First, $\overline{c_1}$ is generated as a bitwise encryption of c_1 with the key pk_2 using the encryption function. It is easy to recognize that $\overline{c_1}$ is doubly-encrypted. Since the SWHE scheme $\mathscr{E}$ can evaluate its own decryption function homomorphically, the noisy inner ciphertext is decrypted homomorphically with $\overline{sk_1}$. After the evaluation, a new encryption of m but under pk_2 is obtained. While the noise is decreased by eliminating the noise from the inner ciphertext, additional noise is added during the homomorphic evaluation of the decryption function. As long as the new noise added is less than the old noise removed, there is a progress. Further homomorphic operations can be done repeatedly on the obtained "fresh" ciphertext until reaching again a threshold point.

Gentry's bootstrapping technique can be applied only if the decryption function is simple enough. Otherwise, first *squashing* method should be applied in order to reduce the complexity of the decryption function so that it is in the set of permitted functions. In brief, squashing converts an SWHE scheme into a bootstrappable one.

The development of FHE since the release of Gentry's work [42] can be roughly divided into four generations according to the techniques used in constructing the FHE schemes.

1.5.3.1 First-Generation FHE

This starts with Gentry's original scheme using ideal lattices [42]. The security of the underlying SWHE scheme is based on the hardness of an average-case decision problem over ideal lattices, namely a variant of the "bounded distance decoding problem (BDDP)" on ideal lattices. The semantic security of the achieved FHE scheme is based on an additional assumption called "sparse subset sum assumption". Subsequently, Gentry [43] showed a worst-case to average-case reduction for BDDP over ideal lattices. In the same year, van Dijk et al. [94] presented the second FHE scheme based on the Gentry's idea, but the ideal lattice computations were replaced by simple integer aritmetic operations. The security of this fully homomorphic DGHV scheme is based on the "approximate gcd (AGCD) problem" and "sparse subset sum problem (SSSP)". Then, Smart and Vercauteren [88] introduced a third variant of Gentry's scheme which uses both relatively small key and ciphertext size. Afterward, a series of articles [71, 45, 82] presented optimized the key generation algorithms in order to implement Gentry's FHE scheme efficiently.

These first-generation schemes have several bottlenecks in terms of applicability in real life. Firstly, they have limited homomorphic capacity due to very rapid noise growth. Squashing the decryption circuit to make the underlying SWHE schemes bootstrappable comes at the expense of additional and fairly strong security assumption namely the sparse subset sum assumption. Moreover, the schemes that follow Gentry's blueprint have inherent efficiency limitations. The efficiency of an FHE scheme is measured by the ciphertext and key size, the time it takes to encrypt and decrypt, and more importantly per-gate computation overhead. The per-gate computation overhead is defined as the ratio between the time it takes to compute a circuit homomorphically on encrypted inputs to the time it takes to compute it on clear inputs. The first-generation FHE schemes that follow Gentry's blueprint have a quite poor performance so that their per-gate computation overhead is $p(\lambda)$, a large polynomial in the security parameter.

In 2011, Gentry and Halevi [44] constructed a new approach which is one of the first major deviations from Gentry's blueprint. Their construction still relies on ideal lattices and on bootstrapping but eliminates the need for squashing and thereby does not rely on the hardness of the SSSP. However, there is no noteworthy improvement on the efficiency aside from the optimization that reduces the ciphertext length.

1.5.3.2 Second-Generation FHE

The second generation began in 2011 with the work of Brakerski and Vaikuntanathan [14]. They introduced *re-linearization technique* to control ciphertext dimension in homomorphic multiplications. Further, they showed how to construct a bootstrappable scheme without using squashing, instead using a new technique to simplify the decryption algorithm. This technique, named as *dimension-modulus reduction*, does not require sparse subset sum assumption for security. The security of BV scheme is based solely on the hardness of much more standard "learning with error (LWE)" problem introduced by Regev [77] as a generalization of "learning parity with noise" problem. Compared with the previous schemes using squashing method, BV scheme [14] (as well as GH scheme [44]) has no noteworthy efficiency improvement because of costly bootstrapping operation. The real cost of bootstrapping for FHE schemes that follow Gentry's blueprint is much worse than quadratic (see [12] for a detailed analysis). Brakerski, Gentry and Vaikuntanathan leveraged the techniques in [14] and constructed a *leveled*-FHE scheme [12]. Leveled-FHE is a relaxation of FHE, in which the parameters depend (polynomially) on the depth of the circuits that the scheme is capable of evaluating. The depth here referred to the multiplicative depth which is the maximal number of sequential multiplications that can be performed on ciphertexts. The re-linearization and dimension-modulus reduction techniques in [14] were enhanced as the *key switching* and *modulus switching* techniques in BGV scheme. Modulus switching is a powerful noise management technique that control the noise without bootstrapping and it is computationally cheaper than bootstrapping. This technique sacrifices modulus size without jeopardizing the correctness of decryption. In other words, a ciphertext modulo q is replaced with a ciphertext modulo a smaller modulus p which decrypts to the same plaintext. Although BGV scheme does not requires bootstrapping, they used it as an optimization to reduce the per-gate computation overhead. The security of BGV scheme is based on RLWE (ring learning with error) problem [62] with quasi-polynomial approximation factors whereas all the previous schemes relies on the hardness of problems with sub-exponential approximation factors. BGV scheme can also be instantiated with LWE rather than RLWE, albeit with worse performance. After BGV scheme, Brakerski [11], introduced a new scale-invariant FHE without modulus switching. In this scheme, the same modulus is used throughout the homomorphic evaluation process. Compared with previous LWE-based FHE schemes, in [11] the ciphertext noise grows only linearly with the homomorphic operations rather than exponentially. Then, Fan and Vercauteren [37] optimized the Brakerski's scheme by changing the based assumption from LWE problem to RLWE problem. Another improvements of Brakerski's scheme was reducing the computational overhead of key switching, faster execution of homomorphic operations and efficiency improvement [96]. Later Zhang et al. modified and improved Brakerski's scheme [99].

It is also worth noting that in 2012 a NTRU-based multikey FHE scheme was proposed by Lopez-Alt, Tromer and Vaikuntanathan (LTV) [61] for its promising efficiency and standardization properties. However, to allow homomorphic operations and prove its security, a non-standard assumption is required in LTV scheme. In the following year, Bos, Lauter, Loftus, and Naehrig [9] showed how to remove this non-standard assumption via Brakerski's scale invariant technique [11].

In second-generation FHE schemes, noise growth is slower during homomorphic evaluations compared with first-generation FHE schemes. Moreover, although second generation follows Gentry's blueprint in the sense that they first construct a SWHE scheme and then transform it into a FHE scheme using bootstrapping, they can even be operated in the leveled-FHE mode without bootstrapping and this makes them more efficient. However, the complex process of key-switching (or re-linearization) still introduces a huge computational cost which is a main bottleneck for practicality.

1.5.3.3 Third-Generation FHE

In 2013, Gentry, Sahai and Waters proposed a new LWE-based FHE scheme, known as GSW, which uses *approximate eigenvector* method instead of the expensive relinearization (or key switching) technique. Since the ciphertexts of GSW scheme are matrices that are added and multiplied homomorphically in a natural way, the ciphertext dimension is kept constant. GSW scheme is simpler and asymptotically faster than the previous LWE-based FHE schemes. In the following years, two efficient ring variants of the GSW cryptosystem known as FHEW [35] and TFHE [25] were introduced by Ducas and Micciancio and by Chillotti et al, respectively.

1.5.3.4 Fourth-Generation FHE

All three generation FHE schemes mentioned above support the exact arithmetic operations over some discrete spaces like rings or finite fields. However, majority of real-world applications require computations in a continuous space such as $\mathbb{R}$ or $\mathbb{C}$. To address this issue, Cheon et al. proposed CKKS algorithm [22] which provides a natural setting for performing operations on approximate numbers. The CKKS algorithm is particularly suitable for implementing prediction and machine learning methods. The name of the algorithm originally went by the name HEAAN, but later the authors changed it to CKKS in order to distinguish it from the homomorphic encryption library HEAAN (https://github.com/snucrypto/HEAAN) which implements CKKS. After the release of the CKKS scheme, a full residue number system (RNS) variant was introduced in [21].

Bootstrapping to extend the original leveled encryption scheme CKKS to a fully homomorphic encryption was first proposed by Cheon et al [20]. Subsequently, several newer and better algorithms have been presented for bootstrapping CKKS and its full-RNS variants [18, 48, 10, 53, 59].

1.5.4 IMPLEMENTATION ISSUES

Several open-source FHE libraries exist today. Below we list the most popular ones with the authors (developers) created them, the schemes they support and the languages they are implemented in.

SEAL : Authored by Microsoft; includes BFV, CKKS and written in C++ (https://github.com/Microsoft/SEAL)

PALISADE : Authored by a consortium of DARPA-funded defense contractors; includes BGV, BFV, CKKS, TFHE, FHEW and written in C++ (https://palisade-crypto.org/)

HELib : Authored by Halevi and Shoup; includes BGV, CKKS and written in C++ (https://github.com/homenc/HElib)

HEAAN : Authored by Cheon, Kim, Kim, and Song; includes CKKS and written in C++ (https://github.com/snucrypto/HEAAN)

FHEW : Authored by Ducas and Micciancio; includes FHEW and written in C++ (https://github.com/lducas/FHEW)

TFHE : Authored by Chillotti et al; includes TFHE and written in C++ (https://github.com/tfhe/tfhe)

FV-NFLlib : Authored by CryptoExperts; includes BFV and written in C++ (https://github.com/CryptoExperts/FV-NFLlib)

Lattigo : Authored by EPFL-LDS; includes BFV, CKKS and written in Go (https://github.com/tuneinsight/lattigo)

With the rapid development of FHE schemes and libraries, and frameworks, it is important that the cryptography community has a standard for how to safely set the security parameters. In order for homomorphic encryption to be adopted in medical, health, and financial sectors, an important part of the standardization process is the agreement on security levels for varying parameter sets. HomomorphicEncryption.org has undertaken the task of this standardization [1].

In the remainder of this section, we describe the ideas behind Gentry's lattice-based original construction forming the first generation with the conceptually simpler DGHV scheme [94]. Then, BGV scheme will be presented to describe the ideas behind the second-generation FHE. Finally, the fourth-generation CKKS scheme will be explained.

1.5.5 THE DGHV SCHEME

In [94], van Dijk et al. described a remarkably simple SWHE scheme using only modular arithmetic and used Gentry's techniques to convert it into a fully homomorphic scheme. The construction is based on the hardness of the Approximate Greatest Common Divisor (AGCD) problem formulated by Howgrave-Graham [50]. It is easy to compute the greatest common divisor of a given set of integers by Euclidean Algorithm. However, given polynomially many near-multiples $x_i = s_i + p \cdot q_i$ of a number p, where s_i is much smaller than $p \cdot q_i$, it is hard to compute p. In fact, AGCD assumption states that when the multiples are "noisy", it is not possible to compute p efficiently. AGCD problem can be reduced to the security of the scheme of van Dijk et al.

A secret-key SWHE scheme will be described first. Then a public-key version will be obtained by invoking the result of Ron Rothblum [81] that shows how to transform any secret-key homomorphic encryption scheme into a public-key one.

DGHV construction uses a number of parameters (all polynomial in the security parameter λ) adapted from AGCD problem and they are set under some constraints. As a convenient parameter setting, set $N = \lambda, P = \lambda^2$ and $Q = \lambda^5$.

KeyGen$_{\mathcal{E}}$: A random P-bit odd integer p (not necessarily prime) is generated.

Enc$_{\mathcal{E}}$: To encrypt a bit $m \in \{0,1\}$, a random N-bit number μ is chosen such that $\mu = m \bmod 2$ and a random Q-bit number q is chosen. Write $\mu = m + 2r$ for $r \ll p$. The output is a fresh ciphertext $c = E(m) = \mu + pq = m + 2r + pq$ with a small "noise" μ which masks the actual message.

Dec$_{\mathcal{E}}$: The ciphertext is decrypted as $m = D(c) = (c \bmod p) \bmod 2$. Decryption works properly as long as the noise $c \bmod p$ is in the range

$(-p/2, p/2)$ such that p divides $c - c'$. This condition put a limit on the number of homomorphic operations performed on the ciphertexts. As the noise of the system grows over $p/2$, the decryption no longer returns the correct result.

Eval$_{\mathscr{E}}$: Consider the ciphertexts $c_1 = m_1 + 2r_1 + pq_1$ and $c_2 = m_2 + 2r_2 + pq_2$, where c_i's noise is $m_i + 2r_i$. Then homomorphic addition computes

$$E(m_1) + E(m_2) = (m_1 + m_2) + 2(r_1 + r_2) + p(q_1 + q_2)$$

which is a valid ciphertext of $m_1 + m_2$ as long as the noises are small enough so that $|(m_1 + m_2) + 2(r_1 + r_2)| < p/2$. It is possible to perform various number of homomorphic additions before noise goes beyond $p/2$.

Homomorphic multiplication computes

$$E(m_1)E(m_2) = m_1 m_2 + 2(2r_1 r_2 + r_1 m_2 + r_2 m_1) + pq'$$

for some integer q'. This is a valid ciphertext of $m_1 m_2$ and can be decryted as long as the noises are small enough so that $| = m_1 m_2 + 2(2r_1 r_2 + r_1 m_2 + r_2 m_1)| < p/2$. It is clear that multiplication increases the noise faster than addition.

After performing many multiplications and additions, the noise can go beyond $p/2$ and the decryption function of the scheme $\mathscr{E}$ no longer outputs the correct plaintext. Hence, this somewhat homomorphic encryption scheme is not fully homomorphic. But still $\mathscr{E}$ is homomorphic enough. It can handle an elementary symmetric polynomial in t variables of degree (roughly) $d < P/(N \cdot \log t)$ as long as $2^{Nd} \cdot (\frac{t}{d}) < p/2$.

The scheme described so far was the secret-key version of the homomorphic encryption. A public-key version is presented in [94]. The secret key of the scheme is p as before. The public key is a list of encryptions of zero under the secret-key version: $\{x_i = 2r_i + pq_i\}_{i=0}^{k}$ where r_i and q_i are chosen as before. Here the x_i are sampled so that x_0 is the largest, x_0 is odd and x_0 mod p is even. To encrypt a bit m, a random subset $S \subset \{1, 2, \ldots, k\}$ and a random integer in a certain range are chosen. The encryption is

$$c = m + 2r + 2\sum_{i \in S} x_i \bmod x_0$$

The ciphertext is decrypted as $(c \bmod p) \bmod 2$ as long as c has a small noise (which is possible only if the encyptions of zero in the public key have small noises).

Now it is time to ask this question: Is the somewhat homomorphic scheme $\mathscr{E}$ described above "bootstrappable" ? The answer is "yes" only if $\mathscr{E}$ is capable of evaluating its own decryption circuit (plus some)

For the bootsrapping analysis, consider the decryption function

$$m = (c \bmod p) \bmod 2.$$

Since $c \bmod p = c - p \cdot \lfloor c/p \rceil$, where $\lfloor \cdot \rceil$ rounds to the neraest integer, and also p is odd, decryption function can be written more simply as

$$c - \lfloor c/p \rceil \bmod 2 = (c \bmod 2) \oplus (\lfloor c/p \rceil \bmod 2).$$

This is just the XOR of the least significant bits of c and $\lfloor c/p \rceil$.

Computing the least significant bit and XOR is immediate. However, computing $\lfloor c/p \rceil$ is complicated. Because, each large integer c and $1/p$ need to be expressed with at least $P \approx \log p$ bits of precision to guarantee that $\lfloor c/p \rceil$ is computed correctly. As two P-bit numbers are multiplied, a bit of the result may be a high-degree polynomial of the input bits. This degree is also roughly P. Since $\mathscr{E}$ can handle an elementary symmetric polynomial in t variables of approximate degree $d < P/(N \cdot \log t)$, it is not possible for $\mathscr{E}$ to handle even a single monomial of degree P, where the noise of output ciphertext is upper-bounded by $(2^N)^P \approx p^N \gg p/2$. It turns out that $\mathscr{E}$ cannot handle its decryption function, which means it is not bootstrappable.

However, it is possible to transform the scheme, by using Gentry's ingenious *squashing* technique, into a bootstrappable one with the same homomorphic capacity but a decryption function that is simple enough. This transformation is accomplished by augmenting the public key with a "hint" about the secret key. The hint is a large set of rational numbers that has a secret sparse subset which sums to the original secret key. The "post-processed" ciphertext via this hint, which contains a sum of a small set of nonzero terms instead of the multiplication of large integers c and $1/p$, is decrypted more efficiently than the original ciphertext. In order to guarantee that the hint in the public key does not reveal any adversary information about the secret key, an additional security assumption is required, namely "sparse subset sum" assumption. This assumption is based on the difficulty of sparse subset sum problem (SSSP) used by Gentry [42] and studied previously in the context of server-aided cryptography [68]. For more details on this, we refer the reader to [94].

DGHV scheme is conceptually very simple but less efficient than the lattice-based scheme. Several optimizations and new variants over integers was introduced to address the efficiency problem [27, 28, 97, 19, 26, 76, 69, 23, 4].

1.5.6 THE BGV SCHEME

We will describe here the RLWE instantiation of the BGV scheme [12] which has a considerably better performance compared to the LWE instantiation. Let λ and μ be security parameters. In the setup procedure, a 4-tuple of parameters $params = (q, d, N, \chi)$ is chosen, where $q = q(\lambda)$ is a μ-bit odd modulus, $d = d(\lambda)$ is a power of 2, $N > \lceil 3\log q \rceil$ and χ is a discrete Gaussian distribution over $\mathbb{Z}$. The underlying ring of this scheme is the ring of polynomials of degree less than d with integer coefficients denoted as $R = \mathbb{Z}[x]/(x^d + 1)$. R_q is used to denote the quotient ring $R/qR = \mathbb{Z}_q[x]/(x^d + 1)$ where the coefficients of polynomials are integers modulo q.

Vectors will be written in bold lowercase letters.

SecretKeyGen: The secret key $\mathbf{s}$ is generated by drawing $\mathbf{s}' \leftarrow \chi$ and setting $\mathbf{s} = (1, \mathbf{s}') \in {R_q}^2$.

PublicKeyGen: The public key is obtained by generating a column matrix $\mathbf{A}' \leftarrow \mathbb{R}_q^{N\times 1}$ uniformly and an error vector $\mathbf{e} \leftarrow \chi^N$, and then setting $\mathbf{b} \leftarrow \mathbf{A}'\mathbf{s}' + 2\mathbf{e}$. The public key $\mathbf{A}$ is an $N \times 2$ matrix over R_q whose first column is $\mathbf{b}$ and the second column is **-A**$'$.

Enc: A message $m \in R_2$ is encrypted by setting $\mathbf{m} = (m, 0) \in {R_q}^2$, generating $\mathbf{r} \leftarrow {R_2}^N$ uniformly at random and computing the ciphertext $\mathbf{c} \leftarrow \mathbf{m} + \mathbf{A}^T\mathbf{r} \in {R_q}^2$.

Dec: A ciphertext $\mathbf{c}$ is decrypted as $m \leftarrow [[\langle \mathbf{c}, \mathbf{s} \rangle]_q]_2$ which is the reduction of the dot product of $\mathbf{c}$ and $\mathbf{s}$ first modulo q (into the interval $(-q/2, q/2)$) and then modulo 2.

In order to construct a leveled homomorphic encryption scheme from the encryption scheme defined above, some operations must be defined, namely *BitDecomp*, *Powersof2*, *SwitchKeyGen*, *SwitchKey* and *Scale*.

BitDecomp($\mathbf{x} \in R_q^n$) operation decomposes $\mathbf{x}$ into its bit representation

$$(\mathbf{u}_0, \mathbf{u}_1, \ldots, \mathbf{u}_{\lfloor \log q \rfloor}) \in {R_2}^{n\cdot\lceil \log q \rceil},$$

where $\mathbf{x} = \sum_{j=0}^{\lfloor \log q \rfloor} 2^j \cdot \mathbf{u}_j$ with all $\mathbf{u}_j \in R_2^n$.

Powersof2($\mathbf{x} \in R_q^n$) operation outputs the vector

$$(\mathbf{x}, 2\cdot\mathbf{x}, \ldots, 2^{\lfloor \log q \rfloor}\cdot\mathbf{x}) \in R_q^{n\cdot\lceil \log q \rceil}.$$

For vectors $\mathbf{c}$ and $\mathbf{s}$ of equal length, it is easy to observe that

$$\langle BitDecomp(\mathbf{c}, q), Powersof2(\mathbf{s}, q) \rangle = \langle \mathbf{c}, \mathbf{s} \rangle \pmod{q}.$$

Key switching method consists of two procedures described below.

SwitchKeyGen($\mathbf{s}_1 \in R_q^{n_1}, \mathbf{s}_2 \in R_q^{n_2}$) operation starts by generating a public key $\mathbf{A}$ from the secret key $\mathbf{s}_2$ for $N = n_1 \cdot \lceil \log q \rceil$ as described above.

Then it ouputs a (public key) matrix $\mathbf{B}$ by adding $Powersof2(\mathbf{s}_1) \in R_q^N$ to the first column of the matrix A.

SwitchKey$(\mathbf{B}, \mathbf{c}_1)$ takes the ciphertext $\mathbf{c}_1$ encrypted under the secret key $\mathbf{s}_1$ and the output $\mathbf{B}$ of *SwitchKeyGen*, then outputs a new ciphertext $\mathbf{c}_2$ that encrypts the same message under $\mathbf{s}_2$, namely

$$c_2 = BitDecomp(\mathbf{c}_1)^T \cdot \mathbf{B} \in R_q^{n_2},$$

where n_2 is the dimension of $\mathbf{s}_2$.

Finally, for the sake of completeness, the *Scale* operation must be defined.

Scale$(\mathbf{x}, q, p, r)$ outputs $\mathbf{x}'$ defined as the R-vector closest to $(p/q) \cdot \mathbf{x}$ that satisfies $\mathbf{x}' = \mathbf{x} \pmod{r}$, where $q > p$.

Let $\mathbf{c}$ be a valid encryption of m under the secret key $\mathbf{s}$ modulo q (i.e., $m = [[\langle \mathbf{c}, \mathbf{s} \rangle]_q]_2$) and let $\mathbf{s}$ be a short vector. Further let $\mathbf{c}'$ be a simple scaling of $\mathbf{c}$, that is the R-vector closest to $(p/q) \cdot \mathbf{c}$ such that $\mathbf{c}' = \mathbf{c} \bmod 2$. It turns out that $\mathbf{c}'$ is a valid encryption of m under $\mathbf{s}$ modulo $p < q$ using the usual decryption equation (i.e., $m = [[\langle \mathbf{c}', \mathbf{s} \rangle]_p]_2$). In a nutshell, it is possible to change the inner modulus in the decryption equation to a smaller number while preserving the correctness of decrption under the same secret key. An evaluator, who does not know the secret key but only knows a bound on its length, can transform a ciphertext $\mathbf{c}$ satisfying $m = [[\langle \mathbf{c}, \mathbf{s} \rangle]_q]_2$ into a ciphertext $\mathbf{c}'$ satisfying $m = [[\langle \mathbf{c}', \mathbf{s} \rangle]_p]_2$ (see Lemma 5 in [12]). Most interestingly, if $\mathbf{s}$ has coefficients that are small in relation to q and p is sufficiently smaller than q, then the magnitude of the noise in the ciphertext essentially decreases (Corollary 1 in [12])

$$|[\langle \mathbf{c}', \mathbf{s} \rangle]_p| < |[\langle \mathbf{c}, \mathbf{s} \rangle]_q|.$$

Given the scheme and operations described above, it is now possible to define a leveled FHE scheme which can be transfromed into a FHE scheme by using Gentry's bootstrapping technique.

Let L be a parameter indicating the number of levels of arithmetic circuit that the FHE scheme is capable of evaluating. Further let $\mu = \mu(\lambda, L)$, where λ is the security parameter. The setup procedure defined previously must be called from L(input level of circuit) to 0 (output level) in order to obtain a ladder of parameters. Namely, $params_j = (q_j, d, N_j, \chi)$ where $q_l > q_{L-1} > \cdots > q_1 > q_0$ has size $(j+1)\mu$ bits and $N_j > \lceil 3 \log q_j \rceil$ for $j = 0, 1, \ldots, L$. The parameter sets $params_j$ is used to generate the secret key $\mathbf{s}_j$, by executing the SecretKeyGen procedure, and the public key $\mathbf{A}_j$, by executing the PublicKeyGen procedure described earlier for ecah level $j = L, L-1, \ldots, 1, 0$. Then by tensoring $\mathbf{s}_j$ with itself, set $\mathbf{s}_j' = \mathbf{s}_j \otimes \mathbf{s}_j$ whose coefficients are each of the product of two coefficients of $\mathbf{s}_j$ in R_{q_j}. Afterward, set $\mathbf{s}_j'' = BitDecomp(\mathbf{s}_j')$ and perform $\mathbf{B}_j = SwitchKeyGen(\mathbf{s}_j'', s_{j-1})$.

Encryption is done by carrying out the encryption operation defined before using the public keys A_j and decryption is done by executing the decryption operation defined before using the secret key $\mathbf{s}_j$. The ciphertexts in depth j of the circuit are assumed to be encrypted under $\mathbf{s}_j$ using the modulus q_j. Homomorphic addition and multiplication operations are executed on the ciphertexts, and after performing each operation, a function named *Refresh* is called. *Refresh* calls the *Scale* function to switch the moduli and then invokes the *SwitchKey* function to switch the key under which the resulting ciphertext is encrypted. Indeed, since addition increases the noise much more slowly than multiplication, it is not necessarily required to refresh after additions.

1.5.7 THE CKKS SCHEME

Cheon et al. [22] proposed CKKS scheme in 2017 for efficient approximate computation on encrypted data. The CKKS algorithm works in the ring of polynomials with integer coefficients modulo the mth cyclotomic polynomial $\Phi_m(x)$ that is $R = \mathbb{Z}[x]/(\Phi_m(x))$. The degree of $\Phi_m(x)$ is $n = \phi(m)$, where ϕ is the Euler's totient function. In the ring $R_q = \mathbb{Z}_q[x]/(\Phi_m(x))$, the elements are polynomials whose degree is up to $n-1$ with coefficients in the range $(-q/2, q/2]$. If $\zeta = e^{\frac{2\pi i}{m}}$ is a primitive mth root of unity, then the mth cyclotomic polynomial is

$$\Phi_m(x) = \prod_{\substack{1 \leq j \leq m \\ \gcd(j,m)=1}} (x - \zeta^j).$$

In CKKS, $m \geq 2$ is taken as a power of 2. Then $\Phi_m(x) = x^{m/2} + 1 = x^n + 1$. Before encryption and after decryption of CKKS scheme, encoding and decoding functions are called, respectively. Consider the canonical embedding map

$$\begin{aligned} \sigma : R &\rightarrow \mathbb{C}^n \\ a(x) &\mapsto (a(\zeta^j))_{j \in \mathbb{Z}_m^*}, \end{aligned}$$

where the second half of the complex values in the image vector $\sigma(a)$ are the symmetric complex conjugates of the first half. So we can project the image vectors onto their first half via the natural projction $\pi : \mathbb{C}^n \rightarrow \mathbb{C}^{n/2}$. Then the decoding function transforms an arbitrary polynomial $a(x) \in R$ into a complex vector $\mathbf{z}$ such that $\mathbf{z} = \pi \circ \sigma(a) \in \mathbb{C}^{n/2}$. The encoding function is defined as the inverse of this decoding function. Specifically, it encodes an input vector $\mathbf{z} \in \mathbb{C}^{n/2}$ into a polynomial $a(x) = \sigma^{-1} \circ \pi^{-1}(\mathbf{z})$.

The L-infinity norm of $\sigma(a)$ for $a \in R$ is denoted by $||a||_\infty = ||\sigma(a)||_\infty$, which is equal to the largest of the absolute value of the complex components of the vector $\sigma(a)$. Following notations in [22], we define three distributions as follows. Given a real $\gamma > 0$, $\mathcal{DG}(\gamma^2)$ denotes a distribution over $\mathbb{Z}^n$ which samples its components independently from the discrete Gaussian distribution of variance γ^2. For a positive integer h, $\mathcal{HWT}(h)$ denotes uniform distribution over the set of vectors in $\{0,+1,-1\}^n$ whose Hamming weight is exactly h. For a real $o \le \rho \le 1$, the distribution $\mathcal{ZO}(\rho)$ draws each vector from $\{0,+1,-1\}^n$ with probability $\rho/2$ for each of $+1$ and -1, and probability of being zero is $1-\rho$.

The aim is to construct a leveled HE scheme for approximate arithmetic. Let the integer L be the depth of the arithmetic circuit to be evaluated homomorphically and $p > 0$ be a base. The ciphertext modulus is $q_k = p^k$ for each level $k = 1,\ldots,L$. Parameters for level k come from $\mathbb{Z}_{q^k}[x]/(x^n+1)$ for each $k = 1,\ldots,L$. The input level of the arithmetic circuit uses the modulus $q_L = p^L$, and the next level uses $q_{L-1} = p^{L-1}$ and so on. The output level uses the modulus $q_1 = p$.
SecretKeyGen: For $O(2^\lambda)$ security, we choose the parameters of the scheme as a power of two $m = 2n$, a real value γ, an integer h, an integer P, and the base p.

Then we sample $\mathbf{s} \leftarrow \mathcal{HWT}(h)$, $\mathbf{a} \leftarrow R_{q_L}$, $\mathbf{a}' \leftarrow R_{P\cdot q_L}$, $\mathbf{e} \leftarrow \mathcal{DG}(\gamma^2)$ and $\mathbf{e}' \leftarrow \mathcal{DG}(\gamma^2)$ to generate the following secret key **sk**, the public key **pk** and the evaluation key **evk**, respectively.

$$\begin{aligned}
\mathbf{sk} &= (1,\mathbf{s}) \\
\mathbf{pk} &= (\mathbf{b},\mathbf{a}) \in R_{q_L}^2 \text{ where } \mathbf{b} = -\mathbf{a}\cdot\mathbf{s}+\mathbf{e} \pmod{q_L} \\
\mathbf{evk} &= (\mathbf{b}',\mathbf{a}') \in R_{Pq_L}^2 \text{ where } \mathbf{b}' = -\mathbf{a}'\cdot\mathbf{s}+\mathbf{e}'+P\mathbf{s}^2 \pmod{P\cdot q_L}.
\end{aligned}$$

Note that vectors above also represents polynomials whose coefficients are the components of the corresponding vector. So the vectors are multiplied as polynomials in the corresponding polynomial ring and then written back as a vector.
Enc: After encoding an input message $\mathbf{z} \in \mathbb{C}^{n/2}$ into the plaintext $\mathbf{m} \in R$ using the procedure described previously, and sampling $\mathbf{v} \leftarrow \mathcal{ZO}(0.5)$ and $\mathbf{e}_0, \mathbf{e}_1 \leftarrow \mathcal{DG}(\gamma^2)$, we compute the ciphertext via the encryption function $E_{\mathbf{pk}}$ as

$$\mathbf{c} = E_{\mathbf{pk}}(\mathbf{m}) = \mathbf{v}\cdot\mathbf{pk} + (\mathbf{m}+\mathbf{e}_0, \mathbf{e}_1) \pmod{q_L}.$$

Dec: The plaintext polynomial $\mathbf{m}$ is computed from a ciphertext $\mathbf{c}$ in level k via the decryption function $D_{\mathbf{sk}}$ as

$$\mathbf{m} = D_{\mathbf{sk}}(\mathbf{c}) = \langle \mathbf{c}, \mathbf{sk} \rangle \pmod{q_k}.$$

The CKKS algorithm introduces an error so that the decrypted value is not exactly the same as the input value, indeed we have

$$D_{\mathbf{sk}}(E_{\mathbf{pk}}(\mathbf{m})) \approx \mathbf{m}.$$

During the evaluation of the arithmetic circuit, the CKKS algorithm performs homomorphic addition, homomorphic multiplication, and rescale operations.

The homomorphic addition of two ciphertexts $\mathbf{c} = (\mathbf{c}_0, \mathbf{c}_1)$ and $\mathbf{c}' = (\mathbf{c}'_0, \mathbf{c}'_1)$ in the same circuit level k is performed using

$$\begin{aligned} \mathbf{c}_{add} &= \mathbf{c} + \mathbf{c}' \pmod{q_k} \\ &= (\mathbf{c}_0, \mathbf{c}_1) + (\mathbf{c}'_0, \mathbf{c}'_1) \pmod{q_k} \\ (\mathbf{d}_0, \mathbf{d}_1) &= (\mathbf{c}_0 + \mathbf{c}'_0, \mathbf{c}_1 + \mathbf{c}'_1) \pmod{q_k}. \end{aligned}$$

Here the input values $\mathbf{c}_0, \mathbf{c}'_0, \mathbf{c}_1, \mathbf{c}'_1$ and the output values $\mathbf{d}_0, \mathbf{d}_1$ are the elements of the ring R_{q_k} and the arithmetic is performed in this polynomial ring.

The homomorphic multiplication of two ciphertexts $\mathbf{c} = (\mathbf{c}_0, \mathbf{c}_1)$ and $\mathbf{c}' = (\mathbf{c}'_0, \mathbf{c}'_1)$ in the same circuit level k is performed using

$$\begin{aligned} \mathbf{c}_{mult} &= \mathbf{c} \odot \mathbf{c}' \pmod{q_k} \\ &= (\mathbf{d}_0, \mathbf{d}_1) + \lfloor P^{-1} \cdot \mathbf{d}_2 \cdot \mathbf{evk} \rceil \pmod{q_k} \end{aligned}$$

where $(\mathbf{d}_0, \mathbf{d}_1, \mathbf{d}_2) = (\mathbf{c}_0 \cdot \mathbf{c}'_0, \mathbf{c}_0 \cdot \mathbf{c}'_1 + \mathbf{c}'_0 \cdot \mathbf{c}_1, \mathbf{c}_1 \cdot \mathbf{c}'_1) \pmod{q_k}$ and $\lfloor \cdot \rceil$ stands for rounding to the nearest integer. The output components of $\mathbf{c}_{mult}$ are also the elements of the ring R_{q_k} and the arithmetic is performed in this ring.

Rescale operation $Rescale_{k \to k'}(\mathbf{c})$ transfroms the ciphertext $\mathbf{c}$ from level k to level k' by computing

$$\begin{aligned} \mathbf{c}' &= \lfloor p^{k'-k} \cdot \mathbf{c} \rceil \pmod{q_{k'}} \\ (\mathbf{c}'_0, \mathbf{c}'_1) &= \lfloor p^{k'-k} \cdot (\mathbf{c}_0, \mathbf{c}_1) \rceil \pmod{q_{k'}} \\ &= (\lfloor p^{k'-k} \cdot \mathbf{c}_0 \rceil, \lfloor p^{k'-k} \cdot \mathbf{c}_1 \rceil) \pmod{q_{k'}} \end{aligned}$$

Generally, $k' = k - 1$, and therefore, the resclae transforms $\mathbf{c}$ from k to $k - 1$ (one level closer to the output level)

$$\begin{aligned} \mathbf{c}' &= \lfloor p^{-1} \cdot \mathbf{c} \rceil \pmod{q_{k-1}} \\ (\mathbf{c}'_0, \mathbf{c}'_1) &= \lfloor p^{-1} \cdot (\mathbf{c}_0, \mathbf{c}_1) \rceil \pmod{q_{k-1}} \\ &= (\lfloor \mathbf{c}_0 / p \rceil, \lfloor \mathbf{c}_1 / p \rceil) \pmod{q_{k-1}} \end{aligned}$$

1.6 CONCLUSIONS

This paper presented an extensive summary of the evolution of cryptography since Shannon's seminal paper "Communication Theory of Secrecy Systems" [83]. The first milestone point is the development of secret-key cryptographic methods LUCIFER, DES, and AES [39, 65, 67], that started in 1958 and continue to-day. The second milestone was the invention of public-key cryptography, starting with Diffie-Hellman key exchange [33] and Rivest-Shamir-Adleman [79] between 1976-1978. Followed up public-key cryptography, a variety of post-quantum cryptographic (PQC) algorithms [60] have been developed, that are expected to make us safe with the advent of quantum computers. Then, we have partially homomorphic encryption (HE) methods [73] that have been flourishing since the day public-key cryptography was invented, and finally fully-homomorphic encryption methods which are based on the ideas of Craig Gentry [42]. The PQC and HE methods are the two directions cryptographic research and development will move on in the next two decades.

Our interest in cryptography is as old as the invention of writing, and it is doubtful this fascination will vane. There will be many information security challenges ahead, and we will attempt to understand and bring solutions for them using cryptographic ideas and tools.

REFERENCES

1. M. Albrecht, M. Chase, H. Chen, J. Ding, S. Goldwasser, S. Gorbunov, S. Halevi, J. Hoffstein, K. Laine, K. Lauter, S. Lokam, D. Micciancio, D. Moody, T. Morrison, A. Sahai, and V. Vaikuntanathan. Homomorphic encryption security standard, 2018.

2. J. Benaloh. Dense probabilistic encryption. In *Proceedings of the Workshop on Selected Areas of Cryptography*, pages 120–128, 1994.

3. J. C. Benaloh. *Verifiable Secret-Ballot Elections*. PhD thesis, Yale University, 1988.

4. D. Benarroch, Z. Brakerski, and T. Lepoint. Fhe over the integers: Decomposed and batched in the post-quantum regime. In S. Fehr, editor, *Public-Key Cryptography – PKC 2017*, pages 271–301. Springer, LNCS N. 10175, 2017.

5. E. Berlekamp, R. McEliece, and H. van Tilborg. On the inherent intractability of certain coding problems. *IEEE Transactions on Information Theory*, 24(3):384–386, 1978.

6. D. J. Bernstein, T. Lange, and C. Peters. Attacking and defending the McEliece cryptosystem. In J. Buchmann and J. Ding, editors, *Post-Quantum*

Cryptography-PQCrypto 2008, pages 31–46. Springer, LNCS Nr. 5299, 2008.

7. J. Biasse, D. Jao, and A. Sankar. A quantum algorithm for computing isogenies between supersingular elliptic curves. In W. Meier and D. Mukhopadhyay, editors, *INDOCRYPT 2014*, pages 428–442. Springer, LNCS Nr. 8885, 2014.

8. E. Biham and A. Shamir. *Differential Cryptanalysis of the Data Encryption Standard*. Springer Verlag, 1993.

9. J. W. Bos, K. Lauter, J. Loftus, and M. Naehrig. Improved security for a ring-based fully homomorphic encryption scheme. In M. Stam, editor, *IMACC 2013*, pages 45–64. Springer, LNCS Nr. 8308, 2013.

10. J.-P. Bossuat, C. Mouchet, J. Troncoso-Pastoriza, and J.-P. Hubaux. Efficient bootstrapping for approximate homomorphic encryption with non-sparse keys. In A. Canteaut and F.-X. Standaert, editors, *Advances in Cryptology – EUROCRYPT 2021*, pages 587–617. Springer, LNCS Nr. 12696, 2021.

11. Z. Brakerski. Fully homomorphic encryption without modulus switching from classical GapSVP. In R. Safavi-Naini and R. Canetti, editors, *Advances in Cryptology - CRYPTO 2012*, pages 868–886. Springer, LNCS Nr. 7417, 2012.

12. Z. Brakerski, C. Gentry, and V. Vaikuntanathan. (Leveled) Fully homomorphic encryption without bootstrapping. *ACM Transactions on Computation Theory*, 6(3), 2014.

13. Z. Brakerski and V. Vaikuntanathan. Efficient fully homomorphic encryption from (standard)-LWE. In *2011 IEEE 52nd Annual Symposium on Foundations of Computer Science*, pages 97–106. IEEE, 2011.

14. Z. Brakerski and V. Vaikuntanathan. Efficient fully homomorphic encryption from (standard)-LWE. In *2011 IEEE 52nd Annual Symposium on Foundations of Computer Science*, pages 97–106. IEEE, 2011.

15. Z. Brakerski and V. Vaikuntanathan. Fully homomorphic encryption from ring-LWE and security for key dependent messages. In P. Rogaway, editor, *CRYPTO*, pages 505–524. Springer, LNCS Nr. 6841, 2011.

16. R. Broker. Constructing supersingular elliptic curves. *Journal of Combinatorics and Number Theory 1*, 1(3):269 –273, 2009.

17. W. Castryck, T. Lange, C. Martindale, L. Panny, and J. Renes. Csidh: An efficient post-quantum commutative group action. In T. Peyrin and S. Galbratith, editors, *Advances in Cryptology – ASIACRYPT 2018*, pages 395–427. Springer, LNCS Nr. 11274, 2018.

18. H. Chen, I. Chillotti, and Y. Song. Improved bootstrapping for approximate homomorphic encryption. In J. B. Nielsen and V. Rijmen, editors, *EUROCRYPT*, pages 34–54. Springer, LNCS Nr. 11477, 2019.

19. J. H. Cheon, J. S. Coron, J. Kim, M. S. Lee, T. Lepoint, M. Tibouchi, and A. Yun. Batch fully homomorphic encryption over the integers. In T. Johansson and P. Q. Nguyen, editors, *Advances in Cryptology – EUROCRYPT 2013*, pages 315–335. Springer, LNCS Nr. 7881, 2013.

20. J. H. Cheon, K. Han, A. Kim, M. Kim, and Y. Song. Bootstrapping for approximate homomorphic encryption. In J. B. Nielsen and V. Rijmen, editors, *EUROCRYPT*, pages 360–384. Springer, LNCS Nr. 10820, 2018.

21. J. H. Cheon, K. Han, A. Kim, M. Kim, and Y. Song. A full RNS variant of approximate homomorphic encryption. In C. Cid and M. J. Jacobson, editors, *Selected Areas in Cryptography – SAC 2018*, pages 347–368. Springer, LNCS Nr. 11349, 2018.

22. J. H. Cheon, A. Kim, M. Kim, and Y. Song. Homomorphic encryption for arithmetic of approximate numbers. In T. Takagi and T. Peyrin, editors, *ASIACRYPT*, pages 409–437. Springer, LNCS Nr. 10624, 2017.

23. J. H. Cheon and D. Stehlé. Fully homomophic encryption over the integers revisited. In E. Oswald and M. Fischlin, editors, *Advances in Cryptology – EUROCRYPT 2015*, pages 513–536. Springer, LNCS Nr. 9056, 2015.

24. A. M. Childs, D. Jao, and V. Soukharev. Constructing elliptic curve isogenies in quantum subexponential time. *Journal of Mathematical Cryptology*, 8(1):1–29, 2014.

25. I. Chillotti, N. Gama, M. Georgieva, and M. Izabachène. Faster fully homomorphic encryption: Bootstrapping in less than 0.1 seconds. In J. Cheon and T. Takagi, editors, *Advances in Cryptology – ASIACRYPT 2016*, pages 3–33. Springer, LNCS Nr. 10031, 2016.

26. J. S. Coron, T. Lepoint, and M. Tibouchi. Scale-invariant fully homomorphic encryption over the integers. In H. Krawczyk, editor, *Public-Key Cryptography – PKC 2014*, pages 311–328. Springer, LNCS Nr. 8383, 2014.

27. J. S. Coron, A. Mandal, D. Naccache, and M. Tibouchi. Fully homomorphic encryption over the integers with shorter public keys. In P. Rogaway, editor, *Advances in Cryptology – CRYPTO 2011*, pages 487–504. Springer, 2011.

28. J. S. Coron, D. Naccache, and M. Tibouchi. Public key compression and modulus switching for fully homomorphic encryption over the integers. In D. Pointcheval and T. Johansson, editors, *Advances in Cryptology – EUROCRYPT 2012*, pages 446–464. Springer, LNCS Nr. 7237, 2012.

29. J. Couveignes. Hard homogeneous spaces. IACR Cryptology ePrint Archive, Report 2006/291 https://eprint.iacr.org/2006/291, 2006.

30. J. Daemen and V. Rijmen. *The Design of Rijndael*. Springer Verlag, 2002.

31. I. Damgard and M. Jurik. A generalisation, a simplification and some applications of paillier's probabilistic public-key system. In K. Kim, editor, *Public Key Cryptography*, pages 119–136. Springer, 2001.

32. C. Delfs and S. D. Galbraith. Computing isogenies between supersingular elliptic curves over $\mathbb{F}_p$. *Designs, Codes and Cryptography*, 78(2):425–440, 2016.

33. W. Diffie and M. E. Hellman. New directions in cryptography. *IEEE Transactions on Information Theory*, 22:644–654, November 1976.

34. J. F. Dooley. *History of Cryptography and Cryptanalysis*. Springer, 2018.

35. L. Ducas and D. Micciancio. Fhew: Bootstrapping homomorphic encryption in less than a second. In E. Oswald and M. Fischlin, editors, *Advances in Cryptology - EUROCRYPT 2015*, pages 617–640. Springer, LNCS Nr. 9056, 2005.

36. T. ElGamal. A public key cryptosystem and a signature scheme based on discrete logarithms. In G. R. Blakley and D. Chaum, editors, *Crypto – Advances in Cryptology*, pages 10–18. Springer, 1985.

37. J. Fan and F. Vercauteren. Somewhat practical fully homomorphic encryption. IACR ePrint Archive, 144, http://eprint.iacr.org/2012/144, 2012.

38. H. Feistel, W. A. Notz, and J. L. Smith. Some cryptographic technique for machine-to-machine data communications. *Proceedings of the IEEE*, 63(11):1545–1554, 1975.

39. H. A. Feistel. A survey of problems inauthenticated communication and control. Report, MIT Lincoln Laboratory, 20 May 1958.

40. L. De Feo, D. Jao, and J. Plût. Towards quantum-resistant cryptosystems from supersingular elliptic curve isogenies. *Journal of Mathematical Cryptology*, 8(3):209–247, 2014.

41. S. D. Galbraith. *Mathematics of Public Key Cryptography*. Cambridge University Press, 2012.

42. C. Gentry. *A Fully Homomorphic Encryption Scheme*. PhD thesis, Stanford University, 2009.

43. C. Gentry. Toward basing fully homomorphic encryption on worst-case hardness. In T. Rabin, editor, *Advances in Cryptology – CRYPTO 2010*, pages 116–137. Springer, LNCS Nr. 6223, 2010.

44. C. Gentry and S. Halevi. Fully homomorphic encryption without squashing using depth-3 arithmetic circuits. In *2011 IEEE 52nd Annual Symposium on Foundations of Computer Science*, pages 107–109, 2011.

45. C. Gentry and S. Halevi. Implementing gentry's fully-homomorphic encryption scheme. In K. G. Paterson, editor, *Advances in Cryptology – EUROCRYPT 2011*, pages 129–148. Springer, LNCS Nr. 6632, 2011.

46. S. Goldwasser and S. Micali. Probabilistic encryption. *Journal of Computer and System Sciences*, 28(2):270 – 299, 1984.

47. V. D. Goppa. A new class of linear error-correcting codes. *Probl. Peredachi Inf.*, 6(3):24–30, 1970.

48. K. Han and D. Ki. Better bootstrapping for approximate homomorphic encryption. In S. Jarecki, editor, *Topics in Cryptology - CT-RSA 2020*, pages 364–390. Springer, LNCS Nr. 12006, 2020.

49. M. E. Hellman. DES will be totally insecure within ten years. *IEEE Spectrum*, 16(7):32–40, 1979.

50. N. Howgrave-Graham. Approximate integer common divisors. In Joseph H. Silverman, editor, *Cryptography and Lattices - CaLC 2001*, pages 51–66. Springer, LNCS Nr. 2146, 2001.

51. D. Jao, R. Azarderakhsh, M. Campagna, C. Costello, L. De Feo, B. Hess, A. Jalali, B. Koziel, B. LaMacchia, P. Longa, M. Naehrig, J. Renes, V. Soukharev, D. Urbanik, G. Pereira, K. Karabina, and A. Hutchinson. Sike. National Institute of Standards and Technology Report 2020 https://csrc.nist.gov/projects/post-quantum-cryptography/round-3-submissions, 2020.

52. D. Jao and L. De Feo. Towards quantum-resistant cryptosystems from supersingular elliptic curve isogenies. In B. Y. Yang, editor, *PQCrypto 2011*, pages 19–34. Springer, LNCS Nr. 7071, 2011.

53. C. S. Jutla and N. Manohar. Modular Lagrange interpolation of the Mod function for bootstrapping for approximate HE. IACR ePrint Archive, 1355, 2020.

54. A. G. Konheim. *Cryptography, A Primer*. John Wiley & Sons, 1981.

55. A. G. Konheim. Automated teller machines: their history and authentication protocols. *Journal of Cryptographic Engineering*, 6(1):1–29, April 2016.

56. A. G. Konheim. Horst Feistel: the inventor of LUCIFER, the cryptographic algorithm that changed cryptology. *Journal of Cryptographic Engineering*, 9(1):85–100, April 2019.

57. E. Kushilevitz and R. Ostrovsky. Replication is not needed: Single database, computationally-private information retrieval. In *Proc. of the 38th Annu. IEEE Symp. on Foundations of Computer Science*, pages 364–373, 1997.

58. T. Lange. Sd8 (post-quantum cryptography) - part 4: Code-based cryptography. ISO/IEC JTC 1/SC 27/WG 2 Cryptography and Security Mechanisms Convenorship: JISC (Japan)- Standing Document https://www.din.de/resource/blob/721042/4f1941ac1de9685115cf53bc1a14ac61/sc27wg2-sd8-data.zip, 2020.

59. Y. Lee, J.-W. Lee, Y.-S. Kim, and J.-S. No. Near-optimal polynomial for modulus reduction using L2-norm for approximate homomorphic encryption. *IEEE Access*, 8:144321–144330, 2020.

60. Z. Liu, P. Longa, and Ç. K. Koç. Guest editors' introduction to the special issue on cryptographic engineering in a post-quantum world: State of the art advances. *IEEE Transactions on Computers*, 67(11):1532–1534, 2018.

61. A. Lopez-Alt, E. Tromer, and V. Vaikuntanathan. On-the-fly multiparty computation on the cloud via multikey fully homomorphic encryption. In *STOC '12: Proceedings of the Forty-Fourth Annual ACM Symposium on Theory of Computing*, pages 1219–1234. ACM, 2012.

62. V. Lyubashevsky, C. Peikert, and O. Regev. On ideal lattices and learning with errors over rings. In H. Gilbert, editor, *Advances in Cryptology - EUROCRYPT 2010*, pages 1–23. Springer, LNCS Nr. 6110, 2010.

63. R. J. McEliece. A public-key cryptosystem based on algebraic coding theory. *DSN Progress Report*, 44:114–116, 1978.

64. D. Naccache and J. Stern. A new public key cryptosystem based on higher residues. In *Proceedings of the 5th ACM Conference on Computer and Communications Security*, page 59–66. ACM, 1998.

65. National Institute for Standards and Technology. Data Encryption Standard, FIPS 46 1977.

66. National Institute for Standards and Technology. Data Encryption Standard. https://nvlpubs.nist.gov/nistpubs/sp958-lide/250-253.pdf, 1999.

67. National Institute for Standards and Technology. Advanced Encryption Standard (AES), FIPS 197, November 2001.

68. P. Q. Nguyen and I. Shparlinski. On the insecurity of a server-aided rsa protocol. In C. Boyd, editor, *ASIACRYPT*, pages 21–35. Springer, LNCS Nr. 2248, 2001.

69. K. Nuida and K. Kurosawa. (batch) fully homomorphic encryption over integers for non-binary message spaces. In E. Oswald and M. Fischlin, editors, *Advances in Cryptology – EUROCRYPT 2015*, pages 537–555. Springer, LNCS Nr. 9056, 2015.

70. National Institute of Standards and Technology. Submission requirements and evaluation criteria for the post-quantum cryptography standardization process. https://csrc.nist.gov/CSRC/media/Projects/Post-Quantum-Cryptography/documents/call-for-proposals-final-dec-2016.pdf, 2016.

71. N. Ogura, G. Yamamoto, T. Kobayashi, and S. Uchiyama. Fully homomorphic encryption with relatively small key and ciphertext sizes. In I. Echizen, N. Kunihiro, and R. Sasaki, editors, *Advances in Information and Computer Security – IWSEC 2010*, pages 70–83. Springer, LNCS Nr. 6434, 2010.

72. T. Okamoto and S. Uchiyama. A new public-key cryptosystem as secure as factoring. In K. Nyberg, editor, *Advances in Cryptology — EUROCRYPT'98*, pages 308–318. Springer, 1998.

73. F. Özdemir, Z. Ö. Özger, and Ç. K. Koç. *Partially Homomorphic Encryption*. Springer, 2021.

74. P. Paillier. Public-key cryptosystems based on composite degree residuosity classes. In Jacques Stern, editor, *Advances in Cryptology — EUROCRYPT '99*, pages 223–238. Springer, 1999.

75. N. Patterson. The algebraic decoding of Goppa codes. *IEEE Transactions on Information Theory*, 21(2):203–207, 1975.

76. Y. G. Ramaiah and G. V. Kumari. Efficient public key generation for homomorphic encryption over the integers. In V. V. Das and J. Stephen, editors, *Advances in Communication, Network, and Computing- CNC 2012*, pages 262–268. Springer, LNICST Nr. 108, 2012.

77. O. Regev. On lattices, learning with errors, random linear codes, and cryptography. *Journal of ACM*, 56(6), 2009.

78. R. Rivest, L. Adleman, and M. L. Dertouzos. On data banks and privacy homomorphisms. In R. DeMillo, D. Dobkin, A. Jones, and R. Lipton, editors, *Foundations of Secure Computation*, pages 169–177. Academic Press, 1978.

79. R. L. Rivest, A. Shamir, and L. Adleman. A method for obtaining digital signatures and public-key cryptosystems. *Communications of the ACM*, 21(2):120–126, February 1978.

80. A. Rostovtsev and A. Stolbunov. Public-key cryptosystem based on isogenies. IACR Cryptology ePrint Archive, Report 2006/145 https://eprint.iacr.org/2006/145, 2006.

81. R. Rothblum. Homomorphic encryption: From private-key to public-key. In Y. Ishai, editor, *Theory of Cryptography - TCC 2011*, pages 219–234. Springer, 2011.

82. P. Scholl and N. P. Smart. Improved key generation for gentry's fully homomorphic encryption scheme. In L. Chen, editor, *Cryptography and Coding – IMACC 2011*, pages 10–22. Springer, LNCS Nr. 7089, 2011.

83. C. E. Shannon. Communication theory of secrecy systems. *Bell System Technical Journal*, 28:656–715, October 1949.

84. P. W. Shor. Polynominal time algorithms for discrete logarithms and factoring on a quantum computer. In Leonard M. Adleman and Ming-Deh A. Huang, editors, *Proceedings of Algorithmic Number Theory, LNCS 877*. Springer, 1994.

85. P. W. Shor. Polynomial-time algorithms for prime factorization and discrete logarithms on a quantum computer. *SIAM Journal on Computing*, 26(5):1484–1509, 1997.

86. P. W. Shor. Polynomial-time algorithms for prime factorization and discrete logarithms on a quantum computer. *SIAM Review*, 41(2):303–332, 1999.

87. J. H. Silverman. *The Arithmetic of Elliptic Curves*. Springer Verlag, 2009.

88. N. P. Smart and F. Vercauteren. Fully homomorphic encrypton with relatively small key and ciphertext sizes. In P. Q. Nguyen and D. Pointcheval, editors, *Public Key Cryptography - PKC 2010*, pages 420–443. Springer, LNCS Nr. 5056, 2010.

89. D. Stehlé and R. Steinfeld. Faster fully homomorphic encryption. In *ASIACRYPT 2010*, pages 377–394. Springer, LNCS Nr. 6477, 2010.

90. A. Stolbunov. Constructing public-key cryptographic schemes based on class group action on a set of isogenous elliptic curves. *Advances in Mathematics of Communications*, 4(2):215–235, 2010.

91. M. Swayne. The world's top 12 quantum computing research universities. https://thequantumdaily.com/2019/11/18/the-worlds-top-12-quantum-computing-research-universities, November 18, 2019.

92. S. Tani. Claw finding algorithms using quantum walk. *Theoretical Computer Science*, 410(50):5285–5297, 2009.

93. J. Tate. Endomorphisms of abelian varieties over finite fields. *Inventiones mathematicae*, 2:134–144, 1966.

94. M. van Dijk, C. Gentry, S. Halevi, and V. Vaikuntanathan. Fully homomorphic encrypton over the integers. In *EUROCRYPT 2010*, pages 24–43. Springer, LNCS Nr. 6110, 2010.

95. J. Vélu. Isogénies entre courbes elliptiques. *C. R. Acad. Sc. Paris, Série A.*, 273:238–241, 1971.

96. T. Wu, H. Wang, and Y.-P. Liu. Optimizations of Brakerski's fully homomorphic encryption scheme. In *Proceedings of 2012 2nd International Conference on Computer Science and Network Technology*, pages 2000–2005, 2012.

97. H. M. Yang, Q. Shia, X. Wang, and D. Tang. A new somewhat homomorphic encryption scheme over integers. In *2012 International Conference on Computer Distributed Control and Intelligent Environmental Monitoring*, pages 61–64, 2012.

98. S. Zhang. Promised and distributed quantum search. In L. Wang, editor, *Computing and Combinatorics - COCOON 2005*, pages 430–439. Springer, LNCS Nr. 3595, 2005.

99. X. Zhang, C. Xu, C. Jin, R. Xie, and J. Zhao. Efficient fully homomorphic encryption from RLWE with an extension to a threshold encryption scheme. *Future Generation Computer Systems*, 36:180–186, 2014.

2 Introduction to Formal Methods for the Analysis and Design of Cryptographic Protocols

Daniele Bringhenti
Politecnico di Torino, Torino, Italy

Riccardo Sisto
Politecnico di Torino, Torino, Italy

Fulvio Valenza
Politecnico di Torino, Torino, Italy

Jalolliddin Yusupov
Turin Polytechnic University, Tashkent, Uzbekistan

CONTENTS

DOI: 10.1201/9781003090052-2

This chapter introduces formal methods, and how they are applied to cryptographic protocols to achieve high security assurance. Cryptographic protocols are communication protocols designed to guarantee security proprieties against the antagonistic actions of attackers. Formal methods are mathematical techniques used to model cryptographic protocols rigorously, and to support their development process in an automated way, so managing their high complexity and reducing the risk of human error. The most important techniques supporting this development process are overviewed, starting from how formal cryptographic protocol models can be built with increasing detail level (belief models, symbolic models, and computational models). Then, this chapter introduces the techniques to formally verify such models at design level, to link design-level abstract models to concrete implementations to reduce the risk of introducing errors in the implementation phase, and the techniques for run-time monitoring of implementations based on formal models. Finally, examples of frameworks combining such techniques are presented.

This chapter introduces formal methods, and how they can be applied to cryptographic protocols, in order to detect their weaknesses early and achieve high assurance about their correctness. Both design and implementation phases will be covered. The chapter starts with an introduction to cryptographic protocols in Section 2.1, and to formal methods in Section 2.2. Then, Section 2.3 presents the various ways formal methods can be used throughout the cryptographic protocols lifecycle. Afterward, in Section 2.4, the different types of formal models that can be used for cryptographic protocols are detailed, while the next two sections focus, respectively, on the main formal verification techniques that can be used for cryptographic protocols at design level (Section 2.5), at implementation level (Section 2.6) and the frameworks for cryptographic protocol engineering (Section 2.7). Finally, Section 2.8 draws conclusions.

2.1 CRYPTOGRAPHIC PROTOCOLS

A communication protocol is a set of rules about communication, designed to achieve some goals in a distributed system. Such rules govern, for instance, the format of exchanged messages, and the procedures that each process in the system should follow to reach the goals, leaving other implementation details free. If each process in the system follows the protocol rules, the goals should be reached, independently of the implementation choices made when implementing each process (this property is called interoperability).

Cryptographic protocols aim to guarantee security proprieties in a distributed system against the antagonistic actions of attackers who threaten the system. Many of these security properties have been defined in recent years. The standard X.800 [27] classifies such proprieties into five macro categories and fourteen sub-categories. Tables 2.1, 2.2 and 2.3, summarize this classification. A precise definition of these and even other properties can be found in the RFC-2828 [75] glossary. In addition to the 5 macro-categories defined in X.800, a sixth one is worth to mention, which is deserving incresing attention. It is the class of privacy-related specific properties, i.e. security properties that aim specifically to guarantee privacy, i.e., "the right of individuals to control what information related to them may be collected and stored and by whom and to whom that information may be disclosed" [75]. The most notable examples of such properties are anonymity (i.e., the concealment of a user's identity) and untraceability (i.e., the concealment of the identity of who made some action).

As their name suggests, cryptographic protocols usually reach their goals by means of cryptography. However, the more general term security protocols is also in use in literature.

We use cryptographic protocols in everyday computing, like log in to a system (e.g., Kerberos authentication), or conduct a secure e-commerce transaction (e.g., TLS). Such real-life protocols are often specified as international standards, but not always: sometimes, custom protocols are defined and used for particular purposes. Real-life cryptographic protocols are characterized by the full complexity of a real protocol, including all the aspects that are necessary for interoperability (e.g., precise defintion of message formats, options, etc.). In the world of research, instead, it is common to focus on more abstract and simpler protocol models, where some details, such as the precise binary message formats, are left out for simplicity (this

Table 2.1
Macro categories of security proprieties (as defined in X.800).

Authentication: guarantee that a communicating entity is who it claims to be (e.g., making sure that the other peer engaged in a point-to-point communication is who it clims to be or that the origin of some data is the claimed one). Several mechanisms/technologies can be used for authentication like passwords, tokens, signatures, challenges.

Access Control: protection against unauthorized use of resources. This is usually achieved by authenticating the user who is trying to use a resource and checking that user's authorization before granting access.

Data confidentiality: protection from unauthorized disclosure of data; (e.g., guarantee that confidential information is only available to the authorized partners of a communication). This is usually achieved by encrypting data.

Data integrity: protection against unauthorized changes to data, including both intentional change or destruction and accidental change or loss. This is usually achieved by making unauthorized changes to data detectable (e.g., by means of digest checking or signature checking).

Non-repudiation: protection against false denial of involvement in an association (e.g., assurance that a signed document cannot be repudiated by the signer). This is usually achieved by employing digital signatures.

is the case, for example, of the Needham-Schroeder Public Key Authentication Protocol that will be presented below as an example). Such models, that we can call theoretical protocols, catch only the essence of the protocol rules, and they can be used to reason about the basic protocol mechanisms and the security properties they can enforce, before defining the full-fledged protocol specification intended for real applications. In 1978, Roger Needham and Michael Schroeder initiated a large body of work on the design and analysis of cryptographic protocols with their proposal of two theoretical authentication protocols [62], one based on symmetric encryption, and the other one based on asymmetric encryption. Later on, it was discovered that both these protocols were flawed, and new theoretical protocols were proposed to fix them. This process of error and fix continued, witnessing how difficult it is to detect flaws in these protocols, finally leading

Table 2.2
Sub-categories of security proprieties (as defined in X.800).

Peer entity authentication: confirmation of the identities of one or more of the entities connected to one or more of the other entities.

Data origin authentication: corroboration of the source of some data.

Connection confidentiality: confidentiality of all user-data exchanged on a connection.

Connectionless confidentiality: confidentiality of all user-data exchanged in a single connectionless message exchange (e.g., confidentiality of a single block of data, like a packet).

Selective field confidentiality: confidentiality of selected fields within the user-data exchanged on a connection or in a single connectionless message exchange.

Traffic flow confidentiality: confidentiality of the information which might be derived from the observation of traffic.

Connection integrity: integrity of all user-data exchanged on a connection or detection of any modification, insertion, deletion or duplication of any data.

Connection integrity with recovery: integrity of all user-data exchanged on a connection or detection of any modification, insertion, deletion or duplication of any data with recovery attempted.

Selective field connection integrity : integrity of selected fields within the user data exchanged on a connection or detection of any modification, insertion, deletion, or duplication of the data in a selected set of fields.

Connectionless integrity: integrity of a single connectionless data exchange or detection of any modification, insertion, deletion, or duplication of the single message.

Selective field connectionless integrity: integrity of selected fields within the data in a single connectionless data exchange or detection of any modification, insertion, deletion, or duplication of the single message.

Table 2.3
Sub-categories of security proprieties (as defined in X.800).

Non-repudiation with proof of origin: protection against any attempt by the sender to falsely deny sending the data or its contents.

Non-repudiation with proof of delivery protection against any subsequent attempt by the recipient to falsely deny receiving the data or its contents.

to the standard real-life protocols Kerberos [63] (based on symmetric keys, and widely used in the Windows and mac-OS OSs) and SSL/TLS [37, 71] (based on asymmetric keys and certificates, and widely used in the web). A classical survey of such theoretical protocols was produced by Clark and Jacob [28]. The survey presents not only a description of the classical theoretical protocols but also the known attacks that were discovered after their publication.

The standard real-life cryptographic protocols are typically reviewed many times by academy researchers, industry experts, and standards organizations (e.g., IETF, ETSI, ISO, IEEE) before being standardized. New protocols are subjected to security threat modeling and analysis to ensure that they offer protection against commonly known attack patterns. When these protocols are deployed in practice, their robustness is monitored, and over time their security issues are worked out. When one of these standard protocols is evaluated as insecure, a more secure version is made available (e.g., TLS 1.1/1.2/1.3 to replace TLS 1.0), or new protocols are designed to replace the aging ones (e.g., AES replaced the aging DES/3DES).

A cryptographic protocol uses communication and cryptographic primitives to achieve its goals. A cryptographic protocol involves two or more communicating actors, in literature commonly named as principals. Their communications may occur on communication channels of various kinds (e.g., Ethernet, Wi-Fi, TCP/IP communications, public/private networks) by message exchanges. Multiple sessions of a cryptographic protocol can run concurrently.

One or more roles are associated with each principal, depending on the reasons why the principal participates in the protocol. For each role, the protocol defines a set of rules that all the principals playing that role have to stick to. When a principal executes a role during the effective communication, it is commonly named agent in the literature.

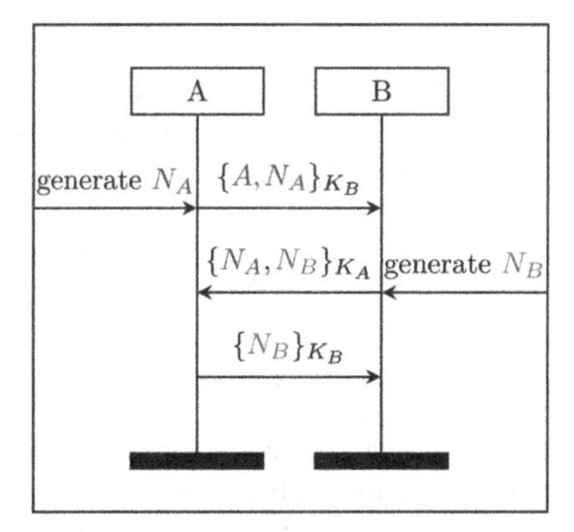

Figure 2.1 Needham-Schroeder Public-key Authentication Protocol.

One significant yet simple example of a theoretical cryptographic protocol is one of the two authentication protocols proposed by Roger Needham and Michael Schroeder in 1978 [62] and known as the Needham-Schroeder Public Key (NSPK) protocol. Many existing protocols are derived from the NSPK protocol, including the widely used Kerberos authentication protocol suite.

The aim of the protocol is to establish mutual authentication between principals *A* (the initiator) and *B* (the receiver). In the original formulation of the protocol, *A* and *B* obtain each other's public key by interacting with a public key server. Here, we present a simplified version, in which it is assumed that *A* and *B* already know each other's public key. As shown in Figure 2.1, the protocol is composed of three messages:

1. in the first step, *A* creates a random value (i.e., a nonce or a challenge) N_A and encrypts this value along with its name *A* with the public key K_B of *B*;
2. when *B* receives such a message, *B* decrypts it, generates another random value N_B and responds to the challenge by encrypting both nonces, with the public key of A (i.e., K_A), in order to show to *A* that *B* was really able to decrypt the original message;
3. in the last step, *A* sends back N_B to *B*, encrypted by K_B, in order to show that it was able to decrypt the second message.

At the end of the session, *A* should have the guarantee that the other party is really *B*, and *B* have the guarantee that the other party is *A*. Moreover, the two nonces N_A and N_B should be shared secrets, only known by *A* and *B*, and can be used to establish a session key to encrypt further messages.

Simple cryptographic protocols are usually described by means of the so-called Alice and Bob notation, which specifies the sequence of messages exchanged, with the indication of who is the sender and who is the intended recipient of each message. For the NSPK protocol, this description reads as follows.

$$\begin{aligned} A \rightarrow B &: \{A, N_A\}_{K_B} \\ B \rightarrow A &: \{N_A, N_B\}_{K_A} \\ A \rightarrow B &: \{N_B\}_{K_B} \end{aligned}$$

A second simple example of a theoretical cryptographic protocol is the one-way authentication based on X.509 certificates, which is described in the X.509 standard [47, 46]. This protocol can be used when a principle A sends some data to another principle B, in order to provide the following security guarantees to the recipient B: (a) the received data were generated by A, were not modified, and are timely (i.e. they are not a reply of already sent data) (b) the received data were intended for B (c) a certain identified part of the received data is confidential (i.e., only known to A and B). The protocol starts from the assumption that A and B know each other's public key (which can be obtained securely by means of X.509 certificates) and it includes a single message sent from A to B:

$$A \rightarrow B : A, T_A, N_A, B, X, \{Y\}_{K_B}, \{h(T_A, N_A, B, X, \{Y\}_{K_B})\}_{K_A^{-1}}$$

where T_A and N_A are, respectively, a time stamp and a nonce, generated by A, X represents the non-confidential data to be sent, Y the confidential data to be sent, K_B is B's public key, K_A^{-1} is A's private key, and $h()$ is a cryptographic hash function. Here, the encrypted hash is a signature that will be checked by B before accepting the message as valid.

In the execution of a cryptographic protocol, an actor is legitimate if allowed to execute it, illegitimate if not allowed to. Besides, an actor is honest if he/she does not deviate from the protocol specification, dishonest he/she deviates, possibly preventing the protocol from reaching the expected security goals. Dishonest actors may be legitimate: dishonest because deviating from the protocol specification, legitimate because allowed to execute the protocol.

It is possible to divide attackers into two classes: passive and active. Passive attackers just intercept, copy and inspect protocol messages, while active ones can also interfere with the protocol, by deleting, altering, reordering and redirecting protocol messages, as well as by forging new protocol messages and inserting them into the conversation.

It is also possible to categorize the attacks on security protocols depending on the weaknesses they exploit [69, 43, 25]. Table 2.4 reports a non-exhaustive list of the most common classes of attacks. Moreover, it is important to underline that the above classification is not a partition in a mathematical sense, i.e., that some attacks may fall into several classes.

Table 2.4
Most common classes of attacks.

Cryptographic flaw attacks: attacks that try to break idealistic properties of cryptographic operations by exploiting weaknesses of the cryptographic algorithms. Analyzing an encrypted message to extrapolate the secret key is an example of a cryptographic flaw attack. These attacks can be refined to take collateral information into account, leaked by a cryptographic primitive or by its particular usage within a security protocol [49].
Internal action flaw attacks: attacks that exploit the absence of some protocol operations that are crucial to guarantee a security property. This kind of weakness falls within the so-called internal action flaw class and may be due both to protocol design errors or to implementation mistakes [22].
Type flaw attacks: attacks based on the absence of proper message type checking. In this kind of attack, the attacker sends a protocol actor a message of different type than what expected, and the actor fails to detect the type mismatch, so misinterpreting the message contents or behaving in an unexpected way [38, 45].
Replay attacks: attacks that are based on re-sending an intercepted message, exploiting the weakness that some actor is not able to distinguish between a fresh message and an old message that is being re-used [48].
Man-in-the-middle attacks: attacks where the attacker stands in the middle between two honest actors and breaks the protocol while relaying messages from one actor to the other [30].
Oracle attacks: attacks where the attacker uses a protocol actor as an oracle to get some information that the attacker could not generate on its own. Then, this information can be used by the attacker to forge new messages that get injected into another parallel protocol session [79].

As an example of an attack on a cryptographic protocol, in Figure 2.2 we show the man-in-the-middle attack to the NSPK Authentication Protocol, which was discovered by Lowe in 1995 [54], 17 years after the protocol

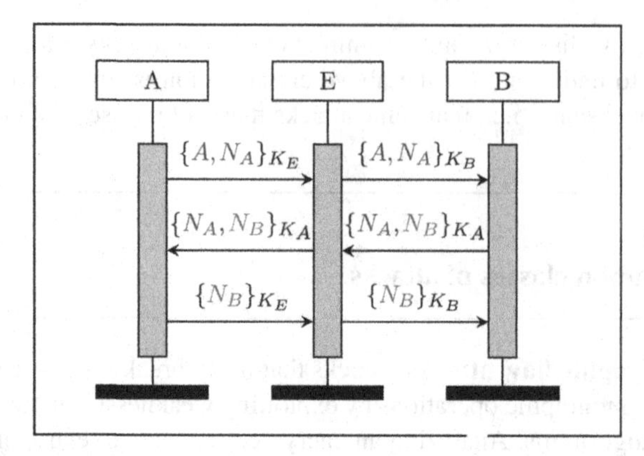

Figure 2.2 Man-in-the-middle attack on the Needham-Shroeder Public Key Authentication protocol.

was initially defined! In the mean time, the flawed version had been used as the basis of commercial products, which, of course, were vulnerable to this attack. In order to break the protocol, an attacker E persuades A to initiate a session with it. Then, it relays the messages to B and convinces B that B is communicating with A, by executing the following steps (as illustrated in Figure 2.2:

1. A sends N_A to E, who decrypts the message with K_E;
2. E relays the message to B, pretending that A wants to communicate with B;
3. B sends N_B
4. E relays it to A
5. A decrypts N_B and confirms it to E, who learns it;
6. E re-encrypts N_B, and convinces B that she's decrypted it.

At the end of the attack, B falsely believes that A started a session with B, and that N_A and N_B are known only to A and B, while in fact A started a session with E and E knows N_A.

Focusing now on the standard, real-life cryptographic protocols, a significant example widely spread in applications such as web, email, and instant messaging, is the SSL/TLS protocol, which aims primarily at providing authentication, privacy and data integrity between two principals. This protocol will be referenced throughout the chapter.

The Secure Sockets Layer (SSL) protocol was firstly defined by Netscape in 1994. However, the first official release of SSL, version 2.0, was out in 1995, because version 1.0 of SSL was never released because it had serious security flaws. While, the updated and final version of the SSL protocol, SSL 3.0, was released in November 1996.

In 1999, the Internet Engineering Task Force (IETF) introduced the Transport Layer Security (TLS) protocol as an upgrade to SSL v3. The TLS 1.0 RFC document (RFC 2246) states the differences between TLS 1.0 and SSL 3.0, while the TLS 1.1 RFC document (RFC 4346) describes minor updates to TLS 1.0 released in April 2006. In August 2008, the TLS 1.2 (RFC 5246) was released with the following main changes: (i) adding cipher-suite-specified pseudorandom functions (PRFs); (ii) adding AES cipher suites; (iii) removing IDEA and DES cipher suites. The current version of TLS, TLS 1.3, was released in August 2018 (RFC 8446), where several unsafe technologies were removed, and the protocol was streamlined for better performance.

The SSL/TLS protocol runs in the application layer. It is composed of two parts: the TLS Record and the TLS Handshake protocol. Specifically, the cryptographic parameters of the session state are produced by the TLS Handshake Protocol, which operates on top of the TLS Record Protocol. When a TLS client and server first start communicating, they agree on a protocol version, select cryptographic algorithms, optionally authenticate each other, and use public-key encryption techniques to generate shared secrets.

In this chapter, for simplicity, we focus only on the TLS Handshake protocol, with reference to TLS 1.0. This Protocol involves the following phases:

- Exchange hello messages to agree on algorithms, exchange random values, and check for session resumption.
- Exchange the necessary cryptographic parameters to allow the client and server to agree on a premaster secret.
- Exchange certificates and cryptographic information to allow the client and server to authenticate themselves.
- Generate a master secret from the premaster secret and exchanged random values.
- Provide security parameters to the record layer.
- Allow the client and server to verify that their peer has calculated the same security parameters and that the handshake occurred without tampering by an attacker.

Specifically, the full handshake protocol messages of TLS 1.0 are presented below and in Figure 2.3, in the order they must be sent. The only message which is not bound by these ordering rules is the Hello Request message, which can be sent at any time, but it should be ignored by the

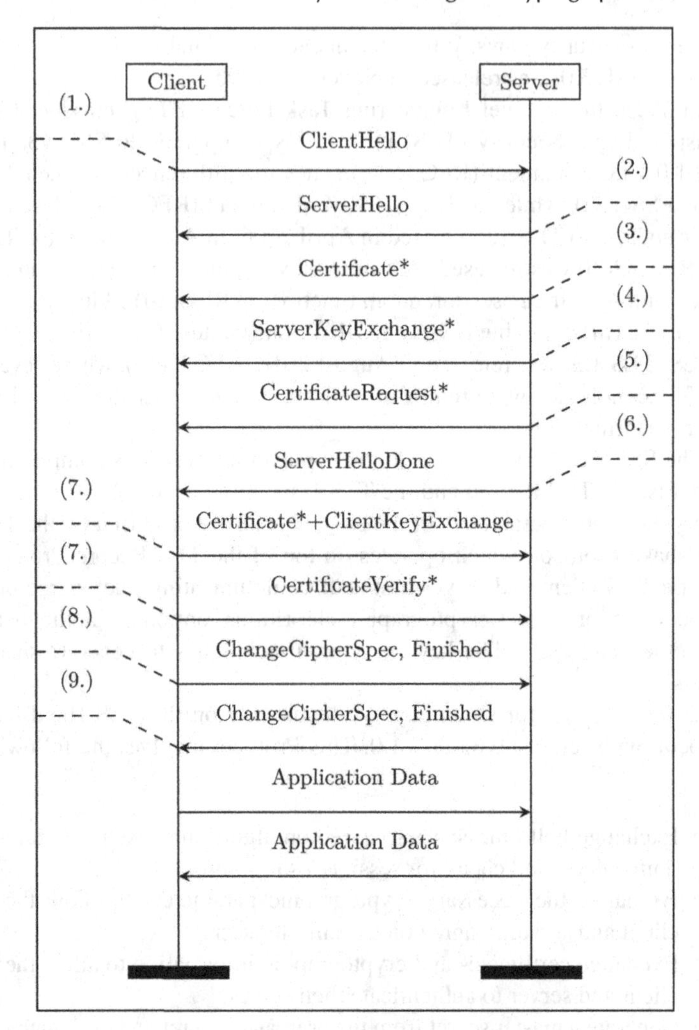

Figure 2.3 Message flow for a full TLS 1.0 handshake.

client if it arrives in the middle of a handshake. The character * is used to flag optional or situation-dependent messages that are not always sent.

A full specification of each message and of the protocol state machine would be too complex to be reported here. They can be found in the protocol standard (RFC 2246).

1. When a client first connects to a server it is required to send the **client hello** as its first message;

2. The server will the **server hello** in response to a client hello message when it was able to find an acceptable set of algorithms;
3. The Server sends **certificate**, whenever the agreed-upon key exchange method is not an anonymous one. This message will always immediately follow the server hello message;
4. The **server key exchange message** is sent by the server only when the server certificate message (if sent) does not contain enough data to allow the client to exchange a premaster secret;
5. A non-anonymous server can optionally request a **certificate** from the client, if appropriate for the selected cipher suite;
6. The **server hello done message** is sent by the server to indicate the end of the server hello and associated messages. After sending this message the server will wait for a client response;
7. The **Client certificate** is only sent if the server requests a certificate. **Client key exchange message** will immediately follow the client certificate message, if it is sent. Otherwise it will be the first message sent by the client after it receives the server hello done message;
8. The **Certificate verify** message is used to provide explicit verification of a client certificate;
9. A **finished message** is always sent immediately after a change cipher spec message to verify that the key exchange and authentication processes were successful. It is essential that a change cipher spec message be received between the other handshake messages and the Finished message.

2.2 FORMAL METHODS

Formal methods [80] are rigorous, mathematically-based techniques used to analyze, design and implement computer-based systems (hardware, software, complex ICT systems). Their mathematical foundation stands in the mathematical models used to represent the systems to analyze, design, or implement. Such models use mathematical concepts, such as sets, functions, relations, etc., to represent system structure or behavior, finally providing unambiguous and precise representations. Formal methods are generally contrasted with informal methods, i.e., the ones that do not employ mathematical models at all, relying on empirical approaches for design, implementation, and verification. Examples of informal methods are the development techniques based on expressing security assumptions, requirements, and algorithms as a mix of English text and time diagrams showing the most typical behavioral scenarios, and that include, as verification techniques, tests manually derived from the requirements, and manual reviews. There are also intermediate methods, classified as semi-formal, that combine rigorous mathematical models with informal ones. For example, UML [64] is a semi-formal modeling language because it includes parts that are formal (e.g., state charts), and parts that are not (e.g., use case diagrams).

One of the main reasons for using formal methods is that they enable mathematical proofs of the correctness of the employed system models, thus finally providing high correctness assurance for the systems that are designed and implemented based on them. Another reason is to provide unambiguous specifications of systems or their requirements in the various stages of development. In fact, an informal model, not having a mathematical model behind, may be subject to slightly different interpretations. Having an unambiguous reference point is particularly important when different parts of the system are developed by different persons or when different stages of the development are performed by different developers. Accordingly, formal methods include formal specification techniques, i.e., mathematically-based languages or notations that can be used to describe systems or their properties unambiguously through formal models, and formal verification techniques, i.e., algorithms that can be used to formally prove the correctness of such models and of the development process.

Developing a system by means of formal methods usually consists of developing several formal specifications. For example, the starting point can be a high-level formal requirements specification, i.e., an unambiguous specification of the requirements of the system to be implemented. From this specification, a high-level system specification can be derived, which provides a high-level description of how the system is organized and works. Then, this high-level description can be refined into other, more detailed, formal descriptions of the system, its components, and their properties. The refinement process continues until the formal models finally obtained are so detailed that an implementation can be directly derived from them.

In such a development process, formal verification can be used to check the correctness of the models that are developed and of the process itself. This is done by analysing each formal specification, in order to check its internal consistency, i.e., absence of contradictions in the model, as well as by comparing different formal specifications in order to check that they are correctly related. For example, the high-level system specification must satisfy the high-level formal requirements expressed by the formal requirements specification. Checking that this is true is an example of formal verification. If both specifications are formal, i.e., formulated in unambiguous mathematical notation, it is even possible to obtain a formal proof of this fact. Another example is when we have two different formal models of the system, at different detail levels. In this case, formal verification can be used to check that the more detailed model is coherent with the less detailed one, i.e., it does not diverge from the corresponding higher-level description.

Formal models are key references not only for formal verification, but also for other development activities. For example, a formal requirements specification can be used as a reference for building a set of tests that aim to check that the system implementation fulfils the system requirements (requirements-based testing). Or, a detailed formal specification of a system module can be used as a reference for developing the implementation of that model. Usually, all these activities are performed automatically or semi-automatically, in order to avoid human errors and to further increase the confidence that the development process is correct, i.e., that the final implementation satisfies the system requirements formulated initially.

2.3 FORMAL METHODS FOR CRYPTOGRAPHIC PROTOCOLS

Formal methods are particularly important for protocols, for example as a way to enable protocol interoperability, i.e., the possibility for different implementations of the same protocol, developed by different software providers, to work together and to achieve the protocol goals. When designing a protocol, a set of formal requirements is formulated, expressing the goals the protocol is required to achieve. Then, a formal protocol specification is formulated, describing the protocol features (messages and procedures) unambiguously. The detail level of this description is chosen so that only the constraints that are strictly necessary for achieving interoperability and the desired goals are expressed, while leaving all the other details free to be chosen by the developers of implementations. Formal protocol specifications can be used as the reference protocol descriptions, e.g., when issuing the protocol standard (e.g., the EGP standard [1] uses a state machine-based formal specification). At this stage, protocol designers can also perform a formal verification in order to prove that the protocol specification is enough to guarantee interoperability of implementations and the fulfilment of the desired protocol goals. Of course, a protocol implementation must be coherent with the protocol specification, otherwise, if it deviates from what is prescribed by the protocol, interoperability with other implementations may no longer hold, and the protocol goals may not be reached. In order to limit this kind of problem, an implementer can use automatic code generation tools that perform the refinement from formal specification to implementation automatically or semi-automatically [74, 76].

Formal methods represent an important opportunity especially for the design and implementation of cryptographic protocols. In this section, the main motivations why the application of formal methods can improve cryptographic protocol engineering are first illustrated. Then, the effective ways how formal methods can be leveraged for cryptographic protocols are discussed.

2.3.1 MOTIVATION

The specification, design and implementation of a cryptographic protocol are complex tasks, difficult to manage by human beings without the support of automated or semi-automated tools. Two main reasons motivate why cryptographic protocol engineering is characterized by such a high complexity. The first reason is common to all distributed systems: it is the intrinsic parallel and asynchronous nature of cryptographic protocols. The second one is more specific to cryptographic protocols: it is the large set of possible different behaviors of the attackers.

Concerning the first motivation, cryptographic protocols have been defined as communication protocols that ensure security properties among multiple parties. The concurrent nature of cryptographic protocols is explicitly implied from this definition. Multiple sessions of the same protocol might be active in parallel, one/more for each participant to the communication, and these sessions may share data such as cryptographic keys and other material. For this reason, attackers may exploit weaknesses by playing with two or more parallel sessions, by injecting data obtained from one session into another session. An example of attack that exploits this kind of bad interaction among parallel processes of the same cryptographic protocol is the so-called oracle attack. In this attack, the attacker would use one of the participants to the communications to generate some information that the attacker would not be able to get by itself. The attacker could then create a new message, based on this information, and send it to a different parallel process of the same protocol.

Therefore, the security of a protocol must be analyzed by considering what happens when an arbitrary number of sessions are executed in parallel. It is well known that this kind of analysis is difficult because the number of possible interleavings of actions from multiple sessions blows up rapidly with the number of parallel processes involved. If the implementation of a protocol is tested in a traditional way without the support of formal verification, some vulnerabilities that might manifest only when some messages are received in a specific order could be overlooked.

Instead, formal methods guarantee that all the possible cases are analyzed. Such an exhaustive analysis has thus become a fundamental component for communication protocol engineering, and it is even more important for cryptographic protocols, since security is directly involved.

The unconstrained behavior of an intruder is the second motivation for leveraging formal methods in this field. Traditional testing requires to predict and manually simulate scenarios where an attacker might try to exploit some possible vulnerabilities of a cryptographic protocol. However, a similar approach would never be able to consider a large number of cases, as each one should be specifically planned and tested. Instead, in principle, only the computational complexity of what an attacker may do can reasonably be constrained. All behaviors that are computationally feasible should be considered, including those that do not follow the protocol rules, and the different number of such behaviors is huge compared to what a test could exercise. Formal methods can consider all possible cases, or a large number of them, rather than just some of them, so having more possibility to find the corner cases that an attacker could exploit.

2.3.2 ROLE OF FORMAL METHODS IN CRYPTOGRAPHIC PROTOCOL ENGINEERING

Formal methods can contribute to cryptographic protocol engineering in four main ways: protocol specification, protocol verification, protocol implementation verification, and protocol monitoring. All these contributions apply to both standard protocols (i.e, protocols that are developed when a new standard has to be established) and to custom protocols.

2.3.2.1 Protocol Specification

As already mentioned, a formal protocol specification can provide the necessary unambiguous, complete, and consistent definition of the protocol rules, which is essential in order to avoid possible misunderstandings of the specifications by different implementers and consequent failure of guaranteeing the expected security properties. A formal protocol specification can be expressed by means of any language that has a formal, i.e. mathematically-based, syntax and semantics. Typical examples are state machines, coupled with language-independent data representations, such as for example the language accepted by the NRL protocol analyzer [58], or other formalisms based on process algebras, such as for example the spi calculus [3] and the applied pi calculus [20]. Protocol specifications can be expressed at different abstraction levels. For example, one in which messages are expressed as abstract data types and one in which messages are expressed as bit strings. More information about such different types of models is provided in Section 2.4.

In addition to specifying the protocol rules, including the behavioral rules of each protocol role and the rules for building the exchanged data representations, formal protocol specifications provide an unambiguous and precise description of the security properties that the protocol is expected to enforce, i.e. the protocol requirements. While, as discussed, most of the properties belong to some standard classes, in order to leverage formal methods it is not enough to just define the classes of properties enforced by the protocol, but it is necessary to give all the details of such properties, so that they cannot be misinterpreted. For example, a single authentication property generally involves not only ensuring that when an actor *A* successfully concludes a session of the protocol with another actor, *A* knows the identity of the other actor, but also that the two actors agree on some data, which usually must remain secret. All this has to be expressed by means of mathematical notation. For example, the Proverif protocol verification tool [20] takes, as its input, a script that includes a protocol description written in the applied pi calculus formal language, and security properties expressed in a tool-specific formal language. Moreover, each security property may hold or not under certain environmental conditions and assumptions about what the attacker is supposed to be able to do, which also has to be specified in order to have a fully formal specification. The Proverif script, for example, also includes statements specifying what the intruder is initially assumed to know.

The starting point for developing formal protocol (and requirements) specifications is generally an informal idea, which itself is not yet suitable to the application of formal methods. This idea gets formalized and encoded in the selected formal language. Of course, formalization is a human activity, which is subject to errors. For this reason, formal protocol specification is tightly coupled with formal protocol verification, which is necessary to provide adequate correctness assurance.

2.3.2.2 Protocol Verification

Having formally defined the protocol security requirements and the protocol rules, possibly at different detail levels, formal protocol verification checks that these specifications are internally consistent and coherent with one another. For example, one can verify that each specification follows the syntax and semantic rules of its specification language (internal consistency), and that if each honest actor of the protocol follows the specified rules of the protocol, the specified security properties hold under certain assumptions about what the attacker can do (coherence between requirements specification and protocol specification). Another form of coherence verification is the verification that a more abstract protocol specification is consistent with the corresponding low-level specification.

A noteworthy observation must be made for providing a correct interpretation of what verification represents exactly. In formal methods, the terms "validation" and "verification" have different meanings. On one side, validation is the operation through which it is checked if a model fulfills user requirements and it is never formal, since user requirements are originally informal and subjective. On the other side, verification checks the internal consistency of the model and compares the model with others at higher level; it can be a formal process, if the involved models are formal as well. To this end, in protocol verification the models are always compared to other models or the formal specification of the requirements, while the subjective and informal requirements are not considered anymore.

Under this important assumption, protocol verification at this stage (i.e., design stage, when abstract models are defined early before the actual real-world implementation) can be performed with different techniques. Some of the most common techniques used for formal verification of cryptographic protocols are model checking [72], theorem proving [67], type systems [50, 42] and game theory [14]. Further information about these techniques will be provided in Section 2.5. The advantage provided by performing formal verification at this stage in cryptographic protocol engineering is that the internal model consistency and the fulfillment of the security requirements are formally verified before the development of the real-world implementation, thus saving time and human resources.

2.3.2.3 Protocol Implementation Verification

As it has been explained for protocol verification, the existence of intermediate levels of abstraction between the formal representation of the requirements and the concrete implementation allows to notice errors in the protocol formulation before developers spend time on writing the source code. However, protocol verification at the model level is not sufficient to have the guarantee of protocol correctness. The reason is that, when the developers write the implementation, they might introduce programming errors and bugs (e.g., errors due to the differences between high-level modeling languages and the typical low-level programming languages), and as a result the expected protocol behavior is altered, leading to protocol vulnerabilities.

Protocol implementation verification aims to fill the gap between protocol design and implementation by avoiding inconsistencies between them. To this end, automated or semi-automated tools are exploited. According to the categorization proposed by Avalle et al. in [8], the two main techniques investigated in the literature are automatic code generation and automatic model extraction. In the former technique, which follows the well-known software engineering paradigm of model-based development, an abstract formal model is developed first, by hand, and the real-world implementation is automatically derived from it on the basis of some user-specified implementation choices. In the latter, the protocol implementation code is written first, by hand, even without necessarily having a formal model yet, and a formal model is derived automatically from the implementation code, enriched with some annotations that guide the model extractor in understanding the protocol rules that are implemented. In both cases, there is a formal model which can be used to perform protocol verification, and to detect protocol flaws. In both cases, the typical human mistakes made in the implementation phase are avoided, because the mapping between abstract model (which is formally verified) and implementation is done automatically by a tool. Additional information about protocol implementation verification is provided in Section 2.6.

2.3.2.4 Protocol Monitoring

Formal methods can be used to perform *runtime verification*, i.e. to monitor the execution of a process implementing a cryptographic protocol. With respect to protocol model or implementation verification, runtime verification has access to practical and more detailed information on the runtime behavior of the process.

As for traditional protocol formal verification, the security properties and the protocol rules must be formally expressed. The specification languages that are used for runtime verification in protocol monitoring generally allow for the possibility to express a temporal evolution of events (i.e., a security property might be guaranteed after a certain sequence of events happened in the communication among the participants). The most common family of specification languages used for this purpose is temporal logic: examples are linear temporal logic , interval temporal logic, and signal temporal logic. The survey by Cassar et al. [26] details how these three types of temporal logic can be used in the context of protocol monitoring.

The typical protocol monitoring setup is made by three components: the process running the protocol, the monitor and the instrumentation mechanism. The monitor is the process that observes the protocol execution and can establish if the security properties formally expressed are satisfied, after its behavior has been observed for a sufficient interval of time. The instrumentation mechanism, instead, records the information related to the behavior of the process running the protocol, decides which pieces of information should be made visible for the monitoring activity, and sends them to the monitor in the form of an ordered stream called "execution trace". The monitor retrieves this ordered stream of events and applies formal verification techniques.

Another possibility is to derive automatically the monitoring specifications from the same kind of protocol specification used as a basis for protocol implementation.

2.4 TYPES OF FORMAL MODELS USED FOR CRYPTOGRAPHIC PROTOCOLS

This section focuses on how cryptographic protocols are modeled formally, and on the different classes of models that are used.

The formal models used for cryptographic protocol engineering are characterized by different abstraction levels. The more the model is abstract, the more it is simple to analyze and easy to manage. However, at the same time, the more a model is abstract the less number of possible attacks on the protocol it can take into account. For this reason, different models at different abstraction levels are generally used in different phases of a cryptographic protocol design and implementation lifecycle.

There exist mainly three kinds of models that have been successfully applied for rigorously analysing and implementing cryptographic protocols: belief models, symbolic models, and computational models. Belief models are the most abstract ones. They are suited especially for analyzing the logic of a protocol in terms of the beliefs that the principals can legitimately achieve through the protocol. Symbolic models represent the actual behavior of a cryptographic protocol in terms of states and actions of the actors, using abstractions that preserve the most important logical flaws that may affect the protocol. Finally, computational models are even more precise, being able to represent a larger class of attacks on cryptographic protocols, but at the expense of higher verification complexity. In the remainder of this section, the different flavors of such classes of models will be detailed.

2.4.1 BELIEF MODELS

Belief logics are logic systems that have been useful in the design and analysis of cryptographic protocols, starting with the pioneering work done by Burrows, Abadi, and Needham on BAN logic [23], and continuing with several related logics: GNY [41], AT [5], SVO [78]. More details about these different logic systems can be found in the survey by Syverson and Cervesato [77]. A belief logic is used to determine how the legitimate beliefs of agents in a protocol run evolve when the protocol events occur. For example, in BAN-like logics, statements like the following ones can be used:

- A said M (protocol agent A sent a message containing M in a protocol run);
- A sees M (protocol agent A received a message containing M in a protocol run and A could get M, e.g., by decrypting the message with a key known by A, and use M);
- A believes P (protocol agent A has right to believe that statement P is true);
- N is fresh (N has never been sent in any previous message of the protocol);
- $\stackrel{K}{\rightarrow} A$ (K is A's public key);
- $A \stackrel{K}{\leftrightarrow} B$ (K is a shared key known only by A and B).

A logic of beliefs includes axioms and inference rules that describe how the beliefs of agents may evolve, i.e. how each agent can deduce new legitimate beliefs. For example, if $\{M\}_{K^{-1}}$ represents the encryption of M with the private key corresponding to public key K, the following inference rule

$$\frac{A \text{ believes } \stackrel{K}{\rightarrow} B,\ A \text{ sees } \{M\}_{K^{-1}}}{A \text{ believes } B \text{ said } M} \tag{2.1}$$

specifies that if agent A has right to believe that K is B's public key and A sees $\{M\}_{K^{-1}}$, then A is also entitled to believe that B said M (because, as K is B's public key, B is the only agent that could encrypt M with the private key corresponding to K).

The logic can be used to determine the set of legitimate beliefs that each agent can deduce after a run of the protocol, starting from an initial set of assumed beliefs. For example, if we assume that agent A knows B's public key K, we can assume the initial belief "A believes $\stackrel{K}{\rightarrow} B$". Then, if the protocol prescribes that B sends $\{M\}_{K^{-1}}$ to A, after A actually receives this message, we can conclude that "A believes B said M", because of inference rule (2.1).

Indeed, belief logics usually do not model protocol messages but rather the *meaning* conveyed by such messages, considering only the encryoted ones (cleartext messages can be forged or modified by the attacker hence they do not contribute to making legitimate beliefs). More precisely, a protocol is modeled by what is called the idealized view of the protocol, which includes the meaning conveyed by each protocol message, and the encryption key, rather than the message itself. For example, this is the idealized view of the NSPK protocol presented in [23]:

$$\begin{aligned} A \rightarrow B &: \{N_A\}_{K_B} \\ B \rightarrow A &: \{\langle A \overset{N_B}{\rightleftharpoons} B \rangle_{N_A}\}_{K_A} \\ A \rightarrow B &: \{\langle A \overset{N_A}{\rightleftharpoons} B \rangle_{N_B}\}_{K_B} \end{aligned}$$

where $\langle X \rangle_{N_A}$ is a statement that combines another statement X with N_A, which is a secret that is used to prove the identity of whoever utters $\langle X \rangle_{N_A}$, while $A \overset{Y}{\rightleftharpoons} B$ means that Y is a secret known only to A and B. In this view, the meaning of the first message is that A communicates N_A to B, while the meaning of the second protocol message, which is $\{N_A, N_B\}_{K_A}$, is that B communicates to A that N_B is a secret known only to them and N_A proves B's identity to A. The last message has a similar meaning, but with inverted roles.

The final beliefs can be compared with the expected ones in order to check whether security properties are satisfied or not. For example, if the aim of the protocol is to agree about a shared key, at the end we would like to establish the beliefs "A believes $A \overset{K}{\leftrightarrow} B$" and "$B$ believes $A \overset{K}{\leftrightarrow} B$", but also "$A$ believes B believes $A \overset{K}{\leftrightarrow} B$" and "$B$ believes A believes $A \overset{K}{\leftrightarrow} B$".

One of the advantages of this model is that belief logics are fairly simple and decidable. Hence, when a protocol has been properly formalized, if its logic is correct, it is possible to automatically prove the expected beliefs, and the resulting proofs are usually short and simple. However, when a proof for a protocol cannot be found, there is no evidence that an attack on the protocol is possible. Moreover, aspects such as how messages are built and how they are processed when received, and which real cryptosystems are used are not modeled. Hence, a protocol proved correct may still have weaknesses in the unmodeled aspects. For this reason, belief logics can be used just to get an initial assessment of a protocol, in order to validate its logic before continuing with more sophisticated analyzes.

Another issue with belief logics is that subtle errors may be introduced in the formalization of the initial beliefs and of the protocol itself, which is difficult to detect and may lead to wrong conclusions about the protocol correctness. For example, the NSPK protocol was proved correct using the BAN logic in 1990 [23]. However, later on, after Lowe found a man-in-the-middle attack on the same protocol [55], Alves-foss showed in [7] that this attack was not detected when developing the above cited BAN logic proof because of a wrong formalization of the protocol used in that proof. In the same paper, he also showed that the protocol properties could be proved by the BAN logic under the fixed formalization. Similar errors when applying BAN logic are documented in [21].

2.4.2 SYMBOLIC MODELS

The foundation for the symbolic analysis of security protocols was laid by Dolev and Yao [34]. They introduced a model in which cryptographic primitives and other operations on data are represented as algebraic function symbols that are assumed to satisfy certain properties, and the adversary has full control over the network but is constrained to compute only through the defined function symbols.

In this abstraction, the protocol messages are algebraic terms built from the defined function symbols. This model assumes perfect cryptography, which means making a number of assumptions about the properties that cyptographic operations have. For example, the only way to decrypt encrypted data is to know the corresponding decryption key; encrypted data do not reveal the key that has been used to encrypt them; encrypted data have enough redundancy so that decryption can detect whether a wrong key has been used to attempt decryption; there is no way to compute the inverse of a cryptographic hash function and different messages are hashed to different values.

For instance, by following this approach, shared-key encryption is basically modeled by two function symbols, *enc* and *dec*, where $enc(x,y)$ represents the encryption of x under key y, $dec(x,y)$ represents the decryption of x with key y, and the following equality holds:

$$dec(enc(x,y),y) = x \tag{2.2}$$

As this is the only equality assumed about these functions, one can decrypt $enc(x,y)$ only when one has the key y, according to the perfect cryptography assumption.

The attacker's capabilities include the possibility to eavesdrop messages transmitted on public channels, and use any cryptographic function on the eavesdropped data. In addition, an attacker can forge new messages and send them (i.e. insert) on public channels. The attacker can also participate in the protocol as any other legitimate actor and it can have its own keys.

The model can represent the behavior of an arbitrary number of parallel protocol sessions involving legitimate actors and the attacker. Generally, the attacker is considered as the communication medium, to emphasise that it has complete control on the communications (see Figure 2.4). Each protocol role is modeled by specifying its operational behavior in the form of a state transition model where not only message send and receive operations are represented, but also how received messages are manipulated and checked.

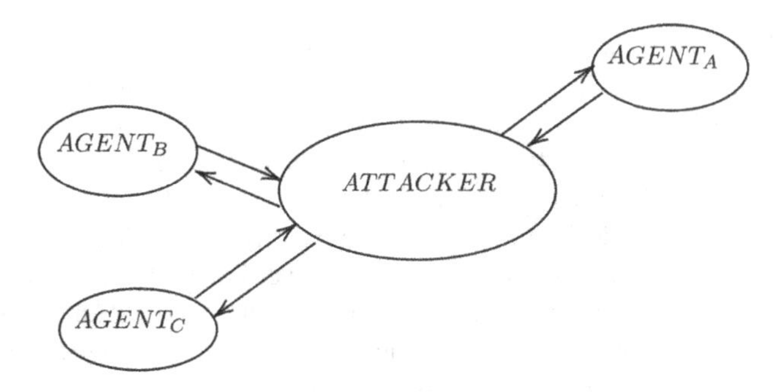

Figure 2.4 Representation of a Dolev-Yao model with the attacker as the medium.

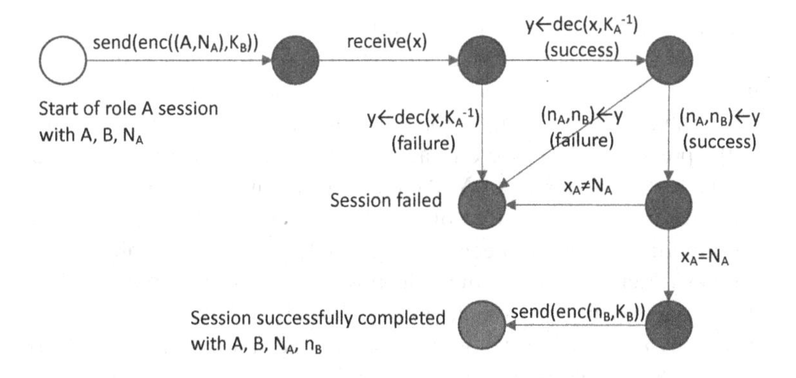

Figure 2.5 Symbolic model of A's role in the NSPK model.

For example, the A role of the NSPK protocol in one protocol session can be modeled by the state-transition model represented in Figure 2.5. Then, the state-transition models of the protocol roles in multiple sessions and of a possible attacker are combined together to form the overall state transition model. When building this model, some assumptions are made about what the attacker initially knows. Then, during the execution of the protocol, the attacker may get additional knowledge from the eavesdropped messages. The overall state transition model is unbounded, because of the unbounded number of parallel sessions that may be started and the unbounded number of ways the attacker has to manipulate messages (e.g., an attacker who knows m and k may decide to send m encrypted with k any number of times, i.e., $enc(m,k)$, or $enc(enc(m,k),k)$, etc., and all these messages are considered as different according to the symbolic model).

In this kind of model, there are two different main ways of formalizing security properties. The first way is to express security properties as trace properties. A trace is the record of a specific run of the protocol, i.e., a path in the overall state-transition model graph. For each trace, a given security property is true or false, and the property is considered to hold for the protocol if it holds for all its traces. For example, a way to formalize the secrecy of a given N is to require that N is never known by the attacker. If there is a trace which arrives at a state in which the attacker knows N, the property is false for that trace, and, hence, for the protocol, and the trace shows how an attacker can attack the protocol.

A second way to formalize security properties in symbolic models is to leverage some concept of equivalence, i.e. state the unobservability of some differences between different versions of the protocol. For example, the secrecy of N could be formalized by requiring that an attacker cannot distinguish a version of the protocol in which N is transmitted from a second version that only differs from the previous one because, instead of N, another value N' is sent. This second way of expressing secrecy is stronger than the previous one, based on trace properties: if a protocol satisfies the equivalence-based secrecy of N, then, under the same assumptions, it also satisfies the trace-based secrecy of N, but the reverse is not necessarily true.

Symbolic models are precise enough to detect a large number of protocol flaws, including for example the possibility of man-in-the-middle attacks, replay attacks, and oracle attacks. For example, a symbolic model was used by Lowe to find the man-in-the-middle attack on the NSPK protocol [55]. Symbolic models are suitable for automated verification using different techniques, as it will be detailed in section 2.5. However, their complexity (they are infinite-state) makes the usual security properties undecidable on them, as it was shown by Durgin et al. in [35]. Although this theoretical result implies the impossibility of building algorithms that can automatically decide whether an arbitrary input protocol model satisfies a given security property, it still admits the construction of automated procedures that can prove or disprove security properties on symbolic models in some cases. Indeed, as it will be discussed in section 2.5, the state-of-the-art verification techniques for symbolic cryptopgraphic protocol models can prove or disprove the usual security properties in most cases, making this approach the most used today.

Like belief models, symbolic models use abstractions that disregard some aspects of a real protocol, such as the features of the real cryptographic operations, and for this reason they do not consider possible weaknesses related to such aspects.

2.4.3 COMPUTATIONAL MODELS

Computational models were developed at the beginning of the 1980s by Goldwasser, Micali, Rivest [40, 39] and Yao [81], laying their foundations on the computability and computational complexity theories. Messages are modeled as bit strings, and cryptographic primitives as functions from bit strings to bit strings. An attacker is modeled as any probabilistic Turing machine having polynomially bounded resources to perform an attack. This is the same model generally used by cryptographers.

In this kind of models, the impossibility of some operations (e.g., the computation of the inverse of a cryptographic hash function), which characterizes perfect cryptography of symbolic models, is replaced by a probabilistic model, according to which the probability that such operations are completed successfully by a probabilistic polynomial-time Turing machine is negligible. This property usually is motivated by a widely accepted computational complexity assumption, e.g., that a certain mathematical function cannot be computed in polynomial time. Consequently, security proofs on computational models do not aim to prove the impossibility for an attacker to achieve a certain goal, rather to prove that the probability that an attacker achieves the goal is negligible. Such proofs generally exploit complexity-theoretic reductions, i.e., implications of the form: "if there exists an attacker that runs in polynomial time and has non-negligible success probability, then there exists an algorithm that solves a hard computational problem in polynomial time with non-negligible probability, which is assumed to be false".

Computational models are less abstract and, hence, more realistic, than symbolic ones. For example, they can model cryptosystem-related weaknesses, in addition to the weaknesses also captured by symbolic models. However, proofs turn out to be more complex and difficult to automate. For this reason, proofs on computational models have been traditionally developed manually, until Laud made a first attempt to mechanize them in 2003 [52], followed by more general techniques developed in the next years [51, 17]. Another approach that has been developed to mechanize such proofs is to infer security properties in the computational model from corresponding properties automatically proved in the symbolic model. This kind of inference is possible under certain assumptions. The study of these assumptions started from the seminal work done by Abadi and Rogaway in [4], and continued with other studies that led to a number of approaches and tools to obtain security proofs in the computational model indirectly from corresponding proofs automatically built for the symbolic model. A survey by Cortier et al. [32] presents these techniques thouroghly. Another detailed account of the literature about direct and indirect techniques to automate security proofs for computational models has been provided by Blanchet in [19].

Although today there are automated theorem provers that can build such proofs automatically, as computational models are more complex than symbolic ones, undecidability affects them as well, and such provers may not reach a result in some cases.

2.5 FORMAL VERIFICATION OF CRYPTOGRAPHIC PROTOCOLS AT DESIGN LEVEL

This section explains the main techniques used for formal verification of cryptographic protocols.

Formal verification is an imperative step in the design of cryptographic protocols and provides a rigid and thorough means of evaluating the correctness of cryptographic protocols. There exists a number of different formal verification techniques to verify cryptographic protocols. In this section, the main different approaches are presented: model checking, theorem proving, type systems, and game theory. The next chapter of this book will provide a more detailed analysis of formal verification techniques for cryptographic protocols.

2.5.1 MODEL CHECKING

Model checking approaches to the analysis of cryptographic protocols have proved remarkably successful [66]. Model checking [29] uses a state exploration technique, in which the state space of a state transition model is defined and then explored by the tool to determine if there are any paths through the space corresponding to violations of the intended properties of the modeled system. For instance, Fiterău-Broştean et al. [36] formalized several security and functional properties drawn from the SSH RFC specifications [53], and then they verified these properties by performing model checking on state-transition models extracted from different SSH implementations. In the context of cryptographic protocol verification, model checking is especially suitable for symbolic models, as they are based on state-transition semantic models. By using this technique, it is possible to look for the possible attacks on the modeled protocol [10, 72]. However, since the search space to be explored is infinite, because of the infinite different ways an attacker can combine eavesdropped data and because of the unbounded number of protocol sessions, a naive approach to model checking is not exhaustive and this kind of search alone cannot guarantee security when no attack is found. This problem has been addressed by research in the context of Dolev-Yao models, and a number of more sophisticated state exploration techniques have been developed that at least partially remove this limitation. Such techniques exploit the fact that an infinite number of different data or execution sequences are in fact equivalent. Exploiting this fact, under certain circumstances, by using these techniques, it is possible to achieve even a proof of security, but this is not possible in all cases, because, as already mentioned, the formal verification problem in the Dolev-Yao model has been shown to be undecidable.

The application of model checking to security protocols has been surveyed extensively by Basin et al. in [10]. For the formal verification of symbolic models, different model checking techniques have been used. Initially, explicit state space exploration by means of general purpose model checkers, such as FDR [55], murphi [60], and spin [56], was proposed. These first attempts were limited to models with a bounded number of sessions and length of messages. Then, more sophisticated tools have been developed, capable of analyzing models without these restrictions. For example, OFMC [11] uses lazy, demand-driven search techniques to analyze even infinite-state symbolic models, while Tamarin [80] analyzes the state space implicitly, leveraging constraint solving based on satisfiability checkers. If the checker fails, Tamarin gives the user the possibility to try to complete the proof interactively.

2.5.2 THEOREM PROVING

A totally different approach to formal verification is theorem proving, which consists of developing a security proof in a deductive system.

This approach is very obvious when using belief logics to model and verify protocols, because in that case the logic system is already defined by the belief logic and by the protocol specification rules. As belief logics are decidable, theorem proving can be automated and it always leads to a positive or negative result. For example, using BAN logic, and the formalization of the NSPK protocol in that logic, as presented by Burrows et al. in [23], a proof that the NSPK protocol satisfies some security properties can be constructed.

For what concerns symbolic models, formal verification by theorem proving is possible by building a formal system that represents the protocol rules and the attacker's capabilities in a way that is equivalent to (i.e., sound and complete w.r.t.) the symbolic model of the protocol. However in this case, apart from the difficulty of building the logic system, which can be solved by doing it automatically staring from other state-transition-oriented representations, the undecidability of the verification problem prevents theorem provers to always find a proof or disproof of a security property. Different solutions have been proposed to overcome this problem. One is to resort to manual or semi-automatic theorem provers. For example, Paulson shows in [67] how Isabelle, a general purpose higher-order logic (HOL) theorem prover can be used to verify security properties in cryptographic protocols. Another solution, adopted for example by Blanchet for Proverif [18], is to develop a proof search algorithm that is fully automatic but only in some cases it provides a result (proof or disproof), while in other cases it fails to return a result. In some cases (e.g., in the work by Havelund [44]), theorem proving and state exploration techniques have been combined together.

One valuable advantage of using theorem proving for symbolic models is that, if successful, it provides a formal proof of correctness. However, when the prover is not automatic, an extremely high expertise is necessary in order to find proofs, while fully automatic solutions exist but they sometimes fail to provide a result. Among the success stories of theorem proving, we can mention that Paulson succeeded in 1999 to prove some security properties of the TLS protocol using a symbolic model and his interactive proving method [68], while more recently, in 2017, some security properties have been proved on TLS 1.3 by Bhargavan et al. [13].

2.5.3 TYPE SYSTEMS

A type system is a lightweight formal method originally developed for reasoning about programs. A type checker which exploits a type system is typically used within a compiler in order to prove the absence of some errors related to data types in the program.

In the early days of programming, type systems were used only to ensure certain basic correctness properties of programs such as the arguments to primitive arithmetic operations are always numbers and differentiating between a string and integer value in the memory. Later on, type systems have been exploited to perform also other types of verification, and not only for programming languages.

In the field of cryptographic protocol verification, the use of type systems has been explored, for example, by Gordon and Jeffrey [50, 42], who developed a series of type systems for verifying authenticity in security protocols, and by Cortier et al. [31], who addressed privacy properties (e.g. anonymity) by means of a similar technique. These approaches reduce the problem of verifying properties in security protocols to the type checking problem. Intuitively, a type system can be used to over-approximate the behavior of a protocol, in such a way that a positive check of its types can be used to prove some security properties.

A notable example of protocol verification performed by means of this technique is the verification of an interoperable implementation of TLS, written in F#, by means of the F7 type system [15]. In this work, TLS is implemented by a set of modules. Initially, these modules are just typed interfaces, with types that capture the security properties. Then, these modules are refined by adding their concrete implementations.

2.5.4 GAME THEORY

Another approach, which has been used to find proofs automatically for computational protocol models, is the automation of game-based cryptographic proofs. A game-based proof is organized as a sequence (or tree) of games, sometimes also named experiments. Each game is characterized by a certain probability an attacker has to win the game. On one end of a chain of games we have a game that describes the competition between a protocol to be verified and a computationally bounded attacker which tries to break it. On the other end of the chain we have another game, for which the probability the attacker has to win is known (e.g. it is known it is negligible). The games in the chain are such that a transition from one game to the next one is characterized by a particular relation between the win probabilities of the two games. For example, the probability an attacker has to win the second game is less than the probability it has to win the first one. Or, the difference of win probability between consecutive games is negligible. If one such chain is found, it is a proof, based on reductions.

Even though cryptographic proofs based on game reductions are powerful, they are complex, and can easily become involved and difficult to manage manually. For this reason, their automation has been addressed by research, and tools that implement it are available. Among them, we can mention cryptoverif [17, 19] and easycrypt [9].

Among the notable results obtained with this approach, we can mention the verification of the TLS Handshake protocol [14, 13].

2.6 TECHNIQUES FOR LINKING DESIGN MODELS TO IMPLEMENTATIONS

Belief, symbolic, and computational models are characterized by a higher level of abstraction than the corresponding real-world protocol implementations. Therefore, this existing gap between abstract formal models and their implementations might have a negative impact on the security level that the usage of formal methods should guarantee for cryptographic protocols.

Among the multiple potential problems derived from this gap, a first issue is that the complexity of the control flow and data types for a real-world protocol implementation is typically much higher than the complexity of the techniques that are used for the definition of abstract models at design level. As such, logical and coding errors might be introduced in the implementation of the protocol; if those errors are not identified in the testing phase before release, they might create a discrepancy between the model and the effective behavior of the protocol. Another problem is that programming languages that can be used for the real implementation deeply differ from each other, e.g., in terms of data structures and operations to manage them. These low-level mechanisms do not find a direct correspondence to the high-level abstract models defined with formal methods. This missing link might be another cause of errors that would not be easily identified.

OpenSSL and OpenSSH are real-life examples where the implementation step introduced errors on a protocol that has been proven secure when the abstract model has been previous analyzed. They are open-source implementations of two widespread cryptographic protocols, respectively SSL (i.e., the name that TLS originally had when developed by Netscape Communications) and SSH. The original protocols have been widely studied and formally verified in literature (e.g., Morissey et al. succeeded in providing a modular and generic proof of security for the application keys established through the TLS handshake protocol [61]). However, their implementations receive several security patches per year, due to low-level implementation bugs that could not be envisioned when designing the models. For example, a notorious OpenSSL bug that has been exploited by man-in-the-middle attacks is the one described in [65]. Several functions inside OpenSSL incorrectly checked the result after calling the EVP˙VerifyFinal function, allowing a malformed signature to be treated as a good one rather than as an error. This fault cannot be found by applying verification techniques at design level, because in the semantics of the model the way the code handles return values is evidently overlooked. Instead, a user enumeration vulnerability flaw was found in OpenSSH, through version 7.7 [33]. The vulnerability occurs because the bailout for an invalid authenticated user is not delayed until after the packet containing the request has been fully parsed.

These considerations clarify that the gap between language-agnostic abstract models and real-world protocol implementations with a specific programming language cannot be easily filled through a manual translation of the model into the software program. For this reason, automated techniques have been developed, so that the gap between them is reduced, and the possibility to have flaws introduced in the implementation phase is also reduced. According to the categorization proposed by Avalle et al. in [8], these techniques are mainly classified into two families (i.e., automatic code generation and model extraction), depending on the starting point for the automated process: abstract model or code implementation.

2.6.1 AUTOMATIC CODE GENERATION

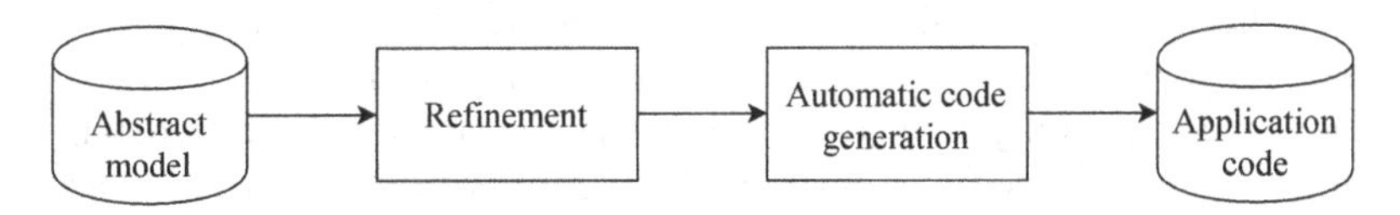

Figure 2.6 Workflow of automatic code generation.

Figure 2.6 depicts the general workflow that is followed by automatic code generation to link the abstraction of formal models and the specificity of a real software implementation. As the workflow suggests, automatic code generation requires the definition of an abstract model as first step, and later the code of the corresponding implementation is produced by an automated tool.

In more detail, starting from an informal description of the cryptographic protocol (i.e., a description of the expected behavior and the security properties that the protocol must fulfill), the user designs a formal, abstract model (e.g., belief, symbolic or computational model) which respects the description, and the model is formally verified. Nevertheless, before a software implementation is generated by employing an automatic tool, the abstract model must go though a refinement operation. More specifically, it is necessary to enrich the model with the implementation choices that will drive the generation of the code. The reason is that a design choice might correspond to multiple alternative implementation choices, among which the programmer typically chooses the most suitable one according to the protocol requirements. These implementation choices must be thus provided to the tool, so that they can be followed in the code generation. At that point, the abstract model enriched by these indications is enough for the employed tool to automatically generate the protocol implementation. An example that follows this procedure is the technique described by Cadé et Blanchet in [24], where the source code is automatically derived from computational models.

Approaches based on automatic code generation must be characterized by two essential features. The first one is that the function that is employed in the automatic code generation for the mapping of the abstract model to the concrete implementation must preserve the abstract security properties of the model. In fact, joining the formal proof of soundness for the code generation function together with the proof of correctness for the security properties of the abstract model (e.g., through other formal verification techniques), then the correctness of the security properties is consequently proved on the concrete implementation. The second feature, instead, is the variety and nature of implementation choices that the user of the automated tool can express. The optimal scenario would be the possibility to express a high range of implementation choices, to guarantee the interoperability of the implementation generated by the tool with the other third party existing implementations. However, there might be cases where some restrictions are placed so that certain aspects of the output software are bounded to choices taken by the developer of the tool itself.

The automation provided by code generation brings over many advantages with respect to naive approaches based on a manual programming of the concrete implementation. Firstly, a central problem related to the gap between abstract model and real-world implementation, i.e., the introduction of programming errors and bugs in the cryptographic protocol, is solved by construction. With the formal proof of the soundness for the code generation function, in fact, the behavior that is shown by the corresponding implementation is proved to be the same as expected, and no security vulnerabilities due to the typical manual coding errors are introduced. Secondly, it is possible to avoid information leakage through side channels by construction as well. Finally, another benefit is that the human user does not have to know the low-level artifacts of a programming language, since they are automatically managed by the tool.

At the same time, some drawbacks are present in this approach. The most evident drawback is that the user is required to know the abstract modeling language that is used for the formalization of the model. If on one side the user must deal with a language at a higher level, on the other side reaching a good level of expertise on a modeling language might be not immediate. Moreover, since the code is automatically generated, all the optimizations an expert developer might introduce are absent, and the efficiency of the resulting software might be not the most suitable one for the target application.

2.6.2 MODEL EXTRACTION

Figure 2.7 depicts the general workflow that is followed by model extraction. The starting and ending points of this process are the opposite of those that are present in the approaches based on automatic code generation.

First of all, software developers write the application code, i.e., they directly write the real-world implementation of the cryptographic protocol. The source code is later enriched with annotations, whose purpose

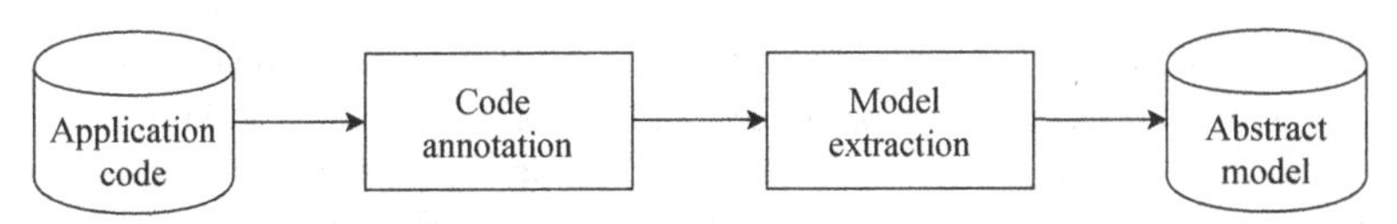

Figure 2.7 Workflow of model extraction.

is to provide additional specifications about formalisms that cannot be conveyed through the constructs of traditional programming languages. Then, an automated tool performs the extraction of an abstract model from this annotated source code, discarding the elements of the source code that are not relevant for the security properties that must be guaranteed by means of the cryptographic protocol. The extracted model is finally verified with a formal prover, so that the formal correctness of the model is guaranteed. As an example, a technique that follows this workflow is described by Aizatulin et al. in [6], where the abstract model is extracted from a source code written in C language.

A characteristic feature of model extraction is the degree of coverage for the programming language that is used to write the source code of the application. The degree of coverage represents the percentage of programming language features that can be used to write the application and that are handled by the automated tool for the extraction of the abstract model. Concerning this aspect, traditional programming languages such as C, C++, C#, Java, or F# are characterized by a syntax with a high complexity, so it is impossible that all the elements of their syntax are covered by techniques based on model extraction. To this end, for each programming language a subset of features has been identified as the features that are properly handled by the majority of this kind of tools.

The most evident benefit of model extraction is that the starting point of the process is the application code written by a human being. As such, an expert developer can introduce manifold optimizations which can improve the efficiency of the protocol implementation when employed in the real world; this was evidently not feasible for the automatic code generation, where at most a limited number of choices could be specified to drive the generation of the code. Moreover, this technique hides the difficulty of learning and managing abstraction languages to the users. Even though programming languages such as C++ and Java themselves have a rich and complex syntax, the developers can limit themselves to learn these languages, because they can consider the formalization of the abstract model as a black box represented by the tool employed for model extraction. Additionally, the tools which are available for finding errors in the definition of a high-level model are much less advanced than the debuggers for common programming languages. In light of this gap, the possibility to use state-of-the-art analyzers of the source code of applications grants the developer a more user-friendly and complete control on the product they are developing.

Nevertheless, model extraction is not exempted from some disadvantages. From a theoretical point of view, this technique might be applied on legacy protocol implementations to get the corresponding abstract model. However, this possibility usually collides with the language features that are used for the legacy application and the restricted subset that is allowed for the effective employment of techniques based on model extraction. In fact, it cannot be expected that legacy code was originally thought for this purpose. Therefore, a manual intervention (e.g., elimination of constructs that are not accepted, or introduction of annotations) is nearly always required in this context. Furthermore, the principles on which model extraction is based are the opposite of the traditional precepts of model-driven software engineering. In that field, the waterfall model, where the code is derived from high-level descriptive requirements through multiple stages of refinement, is the most common pattern that has been used for years in software development. A reverse approach thus puts a bigger emphasis and complexity on the coding activity, which is not guided by a recognized software engineering standard.

2.7 FRAMEWORKS FOR CRYPTOGRAPHIC PROTOCOL ENGINEERING

The development of a robust and strong cryptographic protocol is a burdensome challenge if it is completely managed in a manual way. The support provided by specialized frameworks for cryptographic protocol engineering is the most suitable resource to reduce the presence of flaws and vulnerabilities in the protocol implementations, which might be exploited by attackers to penetrate into a networked system.

In the following, two of the most important frameworks for cryptographic protocol engineering are presented. The first one is SPEAR II, proposed by Saul and Hutchison in [74] to enable a user-friendly interaction with multiple tools for formal analysis according to a so-called multidimensional approach. The second is instead JavaSPI, an environment proposed by Sisto et al. in [76] to allow the user to create a symbolic model of a cryptographic protocol by using a subset of Java features that correspond to the modeling language applied π-calculus [2].

2.7.1 SPEAR II

Multidimensional cryptographic protocol engineering is a unified approach where multiple dimensions (i.e., classes of techniques and tools, characterized by a common goal) are effectively combined in a single application, which provides a comprehensive way to specify, develop and formally analyze a protocol. The main reason that led to the birth of the multidimensional approach is that most of the available techniques for formal analysis cannot detect all the flaws and vulnerabilities that might be present in the design and implementation of a cryptographic protocol, if they are singularly applied. Their combination could simply overcome this problem, because each technique would compensate the drawbacks of the others, and at the end a protocol engineer would get a higher assurance of the protocol correctness.

An example of framework that has been developed according to the principles of multidimensional cryptographic protocol engineering is SPEAR II [74], acronym for "Security Protocol Engineering and Analysis Resources II", and evolution of the original SPEAR I proposed by Bekmann et al. in [12], which was just a proof-of-concept lacking the user-friendliness and completeness that characterize its successor. A human user can interact with the graphical elements of this framework, which offers a great variety of techniques based on formal methods for each stage of the development of a cryptographic protocol, i.e., from the specification of the security properties that it must fulfill until the development of the real-world implementation. In the following, the main dimensions, corresponding to the most important modules of SPEAR II, will be described, so as to show how all the different stages of protocol engineering are managed throughout a multidimensional approach with respect to the application of traditional formal techniques.

Cryptographic protocol design. A dimension that is traversal to all the other components of SPEAR II is the protocol design. The modeling phase is eased thanks to the GYPSIE environment, better described by Saul and Hutchison in [73], which offers a graphical user interface through which the protocol specification is performed at three different levels of abstraction, so that the user can perform operations of varied complexity according to their expertise. More specifically, the high-level view describes the complete flow of messages among the participants of the protocol communication, and the user-friendliness is provided by graphical and textual languages such as SDL (Specification and Description Language) and MSC (Message Sequence Chart). The navigator view offers another high-level prospective of the message flow of the cryptographic protocol, throughout a tree view which offers a full integration with the high-level view, so that drag-and-drop operations between graphical elements of the two views are allowed. Then, the component view works at a lower level of abstraction, because it allows the formal specification of each message of the flow, with another representation based on a hierarchical tree. Finally, the output of the GYPSIE environment is a protocol specification as simple text, LaTeX or a Prolog-like language.

Code generation. The code generation module can automatically generate the Java source code of a real protocol implementation, starting from the output of the GYPSIE environment, i.e., the abstract model describing the message flow of the protocol. Even though such a technique does not present any novelty by itself, the integration of automatic code generation in a unified environment improves the quality of all the other stages of protocol engineering. For example, the additional specifications driving the generation of the source code might be specified with interactive menus in GYPSIE, and if they do not result to be the most suitable ones according to performance and bench-marking measurements, they can be easily modified.

Performance analysis. The performance evaluation of the cryptographic protocol is based on a message rounds calculator. The input of this module is the protocol specification, i.e., the message flow of the communication, which is produced by the GYPSIE environment. Starting from this flow, the message rounds calculator establishes which messages could be sent in parallel for improving the protocol efficiency. In SPEAR II, the performance analysis can be carried out both on synchronous message rounds, if a participant to the communication can send a message only if it has received all the previous ones that were targeting it, or asynchronous (also called optimal) message rounds, if the communication among the participants may be asynchronous. Overall, the results of this performance evaluation can become one of the specifications driving the development of the real-world implementation, because the source code would be directly written considering not only formal aspects of the protocol, but also its future performance. This feature represents an additional way to bridge the gap between the model and its implementation.

Security logics. In SPEAR II, the formal analysis of the protocol model specified through GYPSIE is performed with GYNGER, a Prolog-based analyzer proposed by Mathuria et al. in [57], based on the GNY belief logic, whose access is eased by the presence of a visual GNY environment. GYNGER exploits a forward-chaining approach for the application of GNY inference rules; as such, the GNY inference rules that cannot be adapted for forward-chaining are not implemented in this framework. In particular, 16 of the 88 GNY inference rules are not implemented because of this restriction. However, their absence does not impact the outcome of the formal analysis, because those rules are not needed to perform this type of analysis. Then, GYNGER receives as input the message flow of the protocol (i.e., the model), alongside with the GNY rule set, the initial assumptions and the target goals. If a target goal is successfully reached, a formal proof is generated by the tool, with the indication of which statements were used to derive that goal. This proof is expressed with an English-style language, to further increase the user-friendliness of GYNGER.

Other dimensions. Other dimensions are related to the simulation of use cases where the protocol might be effectively used. For example, the *Meta-Execution* dimension aids the user in correctly modeling the protocol by leveraging it in simulated scenarios, so that the user can directly understand possible problems in the model before the generation of the code. Instead, the *Attack Analysis* dimension is composed by an attack construction engine, whose purpose is to simulate cyber-attacks and help the user understand if the security properties that the protocol should guarantee are effectively guaranteed also under attack.

2.7.2 JAVASPI

JavaSPI [76] is a framework for cryptographic protocol engineering, based on a model-driven approach for the automatic generation of the protocol source code. Inheriting and extending the main characteristics of its predecessor Spi2Java developed by Pironti and Sisto [70], the purpose of JavaSPI is to simplify the modeling of a protocol, allowing the users to use the Java language for both the protocol design and implementation. This also enables the possibility to simulate the execution of the protocol by running a Java debugger on the model, before the development of the corresponding implementation. In the remainder of this section, the workflow of JavaSPI will be detailed, with the aim to underline the improvements it brings over for the traditional protocol engineering.

First, the specification of a protocol symbolic model is made by using a subset of the Java programming language, corresponding to the features that are necessary to reach the same expressiveness as applied π-calculus, a modeling language derived from the Dolev-Yao abstraction. In greater details, two subsets of the Java language are identified and incorporated in this framework. On one side, the "core language" includes the minimum number of Java constructs that are really needed for the mapping to π-calculus, and it is the language that is used by the internal modules of JavaSPI for the derivation of formal proofs or the automatic management of the model, i.e., it is not directly employed by a human user. On the other side, the "extended language" extends, as the name suggests, the core language by introducing additional Java features and it is the subset that is effectively exploited by the users for the specification of the protocol model. All the additional features that are included in the extended language can be reconduced to the subset present in the core language, though. Therefore, this redundancy (i.e., the existence of two separate subsets of the Java language) might be theoretically avoided, since the core language would be enough for the creation of any model with a correspondence in π-calculus. However, in that case the users would have a more limited language to use, and this would impact the user-friendliness of the framework, since experienced Java developers should excessively restrain their knowledge and limit their capabilities.

The formal verification of security properties for the model specified in the core or extended language of JavaSPI is performed by means of the automated and efficient theorem prover ProVerif [16]. Since this theorem prover works on an abstract model formulated with the applied π-calculus language, a translation from the Java-based specification is needed, and it is performed by the Java-ProVerif module of JavaSPI. Nonetheless, the level of abstraction is the same for both the core language of JavaSPI and the applied π-calculus required by ProVerif. Therefore, the translation is straightforward and consists in a simple conversion of the language syntax. However, the input that is received by ProVerif is not exclusively made by the model expressed in π-calculus. In fact, the user should introduce annotations to the Java-based model, representing the security properties that the cryptographic protocol must enforce. In particular, the user can specify two different types of security properties by means of annotations: the first type is represented by secrecy properties for the data exchanged by the parties involved in the protocol communications, whereas the second type is called correspondence of events and includes properties such as party authentication. These annotations are then translated into queries for ProVerif, so that ProVerif can provide a formal proof of the corresponding properties if this proof can be reached, or it can disprove them by providing counterexamples.

JavaSPI also offers a semi-automatic generator for deriving the protocol source code from the abstract model. This module, called Java-Java generator because both the model and the real implementation are written in the Java language, is not fully automated, because a manual refinement of the abstract model must be performed by the user. More specifically, some annotations must be added to the code. These annotations have a different purpose with respect to those needed for the formal verification of security properties: in particular, they consist in the implementation choices that the Java-Java generator must take in the development of the application code.

Within this approach, information specifically focused on the implementation and information exclusively related to the abstract model are separated. As such, during the specification of the model, the user can avoid to care about the future implementation, since additional details will be provided later on by means of annotations, and he can thus only focus on the definition of a correct protocol behavior. The separation of the two main tasks of protocol engineering, i.e., model design and protocol implementation, is made even more evident thanks to the fact that the developers of JavaSPI have proved a set of soundness theorems such that, if a security property is verified for the abstract model by employing ProVerif, then the same property is guaranteed to be fulfilled in the implementation derived with the Java-Java generator.

In light of all the analyzed features of JavaSPI, the conclusion is that this framework optimally counterbalances the typical drawbacks of approaches based on automatic code generation. Users are not required to know π-calculus, and all the implementation choices are separated from the model design because they are conveyed through annotations. Finally, in the whole workflow, the users have the possibility to find problems in the protocol specification in multiple ways before the generation of the source code (e.g., symbolically executing the Java model with a debugger, or running the ProVerif tool), so that protocol engineering becomes less troublesome than how it traditionally was.

2.8 CONCLUSIONS

Cryptographic protocols are widely used in many applications today. However, their design and implementation are notoriously very difficult to get right. In this chapter we have first outlined the main features of cryptographic protocols and analyzed the main reasons for this difficulty. Starting from these considerations, we have showed how formal methods can help cryptographic protocol designers and implementers to avoid errors as much as possible, by enabling accurate verifications based on mathematical models, and even proofs of correctness.

The typical engineering activities related to design and implementation of a cryptographic protocol have been dissected, together with the support tools based on formal methods that are available. Specifically, we have analyzed the typical lifecycle that starts from the elicitation and formalization of security requirements, followed by protocol design and finally implementation. We have also shown how each phase can be supported by verification tools that operate at different levels of abstraction, including the capability of linking descriptions at different levels. Finally, we have illustrated examples of frameworks that combine together different types of tools, operating at different abstraction levels.

REFERENCES

1. Exterior Gateway Protocol formal specification. RFC 904, April 1984.
2. Martín Abadi and Cédric Fournet. Mobile values, new names, and secure communication. *SIGPLAN Not.*, 36(3):104–115, January 2001.
3. Martin Abadi and Andrew D. Gordon. A calculus for cryptographic protocols: The spi calculus. *Information and Computation*, 148(1):1–70, 1999.
4. Martin Abadi and Phillip Rogaway. Reconciling two views of cryptography (the computational soundness of formal encryption). *Journal of Cryptology*, 20(3):395–395, 2007.
5. Martín Abadi and Mark R. Tuttle. A semantics for a logic of authentication (extended abstract). In *Proceedings of the Tenth Annual ACM Symposium on Principles of Distributed Computing*, PODC '91, page 201–216, New York, NY, USA, 1991. Association for Computing Machinery.
6. Mihhail Aizatulin, Andrew D. Gordon, and Jan Jürjens. Extracting and verifying cryptographic models from c protocol code by symbolic execution. In *Proceedings of the 18th ACM Conference on Computer and Communications Security*, CCS '11, page 331–340, New York, NY, USA, 2011. Association for Computing Machinery.
7. Jim Alves-Foss. The use of belief logics in the presence of causal consistency attacks. In *Proceedings of the Nineteenth National Computer Security Conference*, pages 406–417, 1997.
8. Matteo Avalle, Alfredo Pironti, and Riccardo Sisto. Formal verification of security protocol implementations: a survey. *Formal Aspects of Computing*, 26(1):99–123, 2014.
9. Gilles Barthe, François Dupressoir, Benjamin Grégoire, César Kunz, Benedikt Schmidt, and Pierre-Yves Strub. *EasyCrypt: A Tutorial*, pages 146–166. Springer International Publishing, Cham, 2014.
10. David Basin, Cas Cremers, and Catherine Meadows. *Model Checking Security Protocols*, pages 727–762. Springer International Publishing, Cham, 2018.
11. David Basin, Sebastian Mödersheim, and Luca Viganò. Ofmc: A symbolic model checker for security protocols. *International Journal of Information Security*, 4(3):181–208, 2005.

12. JP Bekmann, P De Goede, and ACM Hutchison. Spear: Security protocol engineering and analysis resources. In *DIMACS Workshop on Design and Formal Verification of Security Protocols*, pages 3–5, 1997.

13. Karthikeyan Bhargavan, Bruno Blanchet, and Nadim Kobeissi. Verified models and reference implementations for the TLS 1.3 standard candidate. In *IEEE Symposium on Security and Privacy (S&P'17)*, pages 483–503, San Jose, CA, May 2017. IEEE. Distinguished paper award.

14. Karthikeyan Bhargavan, Cédric Fournet, Markulf Kohlweiss, Alfredo Pironti, Pierre-Yves Strub, and Santiago Zanella-Béguelin. Proving the tls handshake secure (as it is). In Juan A. Garay and Rosario Gennaro, editors, *Advances in Cryptology - CRYPTO 2014*, pages 235–255, Berlin, Heidelberg, 2014. Springer Berlin Heidelberg.

15. Karthikeyan Bhargavan, Cédric Fournet, Markulf Kohlweiss, Alfredo Pironti, and Pierre-Yves Strub. Implementing tls with verified cryptographic security. In *2013 IEEE Symposium on Security and Privacy*, pages 445–459, 2013.

16. Bruno Blanchet. An efficient cryptographic protocol verifier based on prolog rules. In *Proceedings of the 14th IEEE Workshop on Computer Security Foundations*, CSFW '01, page 82, USA, 2001. IEEE Computer Society.

17. Bruno Blanchet. A computationally sound mechanized prover for security protocols. *IEEE Transactions on Dependable and Secure Computing*, 5(4):193–207, October–December 2008. Special issue IEEE Symposium on Security and Privacy 2006. Electronic version available at http://doi.ieeecomputersociety.org/10.1109/TDSC. 2007.1005.

18. Bruno Blanchet. Automatic verification of correspondences for security protocols. *Journal of Computer Security*, 17(4):363–434, December 2009.

19. Bruno Blanchet. Mechanizing game-based proofs of security protocols. In Tobias Nipkow, Olga Grumberg, and Benedikt Hauptmann, editors, *Software Safety and Security - Tools for Analysis and Verification*, volume 33 of *NATO Science for Peace and Security Series – D: Information and Communication Security*, pages 1–25. IOS Press, May 2012.

20. Bruno Blanchet. Modeling and verifying security protocols with the applied pi calculus and ProVerif. *Foundations and Trends in Privacy and Security*, 1(1–2):1–135, October 2016.

21. Colin Boyd and Wenbo Mao. On a limitation of ban logic. In Tor Helleseth, editor, *Advances in Cryptology — EUROCRYPT '93*, pages 240–247, Berlin, Heidelberg, 1994. Springer Berlin Heidelberg.

22. Jakub Breier, Dirmanto Jap, Xiaolu Hou, and Shivam Bhasin. On side channel vulnerabilities of bit permutations in cryptographic algorithms. *IEEE Transactions on Information Forensics and Security*, 15:1072–1085, 2020.

23. Michael Burrows, Martin Abadi, and Roger Needham. A logic of authentication. *ACM Transactions on Computer Systems*, 8(1):18–36, Feb 1990.

24. David Cadé and Bruno Blanchet. Proved generation of implementations from computationally secure protocol specifications. *Journal of Computer Security*, 23(3):331–402, 2015.

25. U. Carlsen. Cryptographic protocol flaws: know your enemy. In *Proceedings of the Computer Security Foundations Workshop VII*, pages 192–200, Los Alamitos, CA, USA, Jun 1994. IEEE Computer Society.

26. Ian Cassar, Adrian Francalanza, Luca Aceto, and Anna Ingólfsdóttir. A survey of runtime monitoring instrumentation techniques. *Electronic Proceedings in Theoretical Computer Science*, 254:15–28, Aug 2017.

27. CCITT (Consultative Committee on International Telegraphy and Telephony). *Recommendation X.435: Message Handling Systems: EDI Messaging System*, 1991.

28. John Andrew Clark and Jeremy Lawrence Jacob. *A Survey of Authentication Protocol Literature: Version 1.0*. Citeseer, 1997. Query date: 14/01/2011.

29. Edmund M Clarke Jr, Orna Grumberg, Daniel Kroening, Doron Peled, and Helmut Veith. *Model Checking*. MIT Press, 2018.

30. Mauro Conti, Nicola Dragoni, and Viktor Lesyk. A survey of man in the middle attacks. *IEEE Communications Surveys Tutorials*, 18(3):2027–2051, 2016.

31. Véronique Cortier, Niklas Grimm, Joseph Lallemand, and Matteo Maffei. A type system for privacy properties. In *Proceedings of the 2017 ACM SIGSAC Conference on Computer and Communications Security*, CCS '17, page 409–423, New York, NY, USA, 2017.

32. Véronique Cortier, Steve Kremer, and Bogdan Warinschi. A survey of symbolic methods in computational analysis of cryptographic systems. *J. Autom. Reason.*, 46(3–4):225–259, Apr 2011.

33. CVE. CVE-2018-15473. https://cve.mitre.org/cgi-bin/cvename.cgi?name=CVE-2018-15473, 2018. [Online; accessed 03-December-2021].

34. D. Dolev and A. C. Yao. On the security of public key protocols. In *Proceedings of the 22nd Annual Symposium on Foundations of Computer Science*, SFCS '81, page 350–357, USA, 1981. IEEE Computer Society.

35. N. Durgin, P. Lincoln, J. Mitchell, and A. Scedrov. Undecidability of bounded security protocols. In *Workshop on Formal Methods and Security Protocols (FMSP'99)*, July 1999.

36. Paul Fiterau-Broştean, Toon Lenaerts, Erik Poll, Joeri de Ruiter, Frits Vaandrager, and Patrick Verleg. Model learning and model checking of ssh implementations. In *Proceedings of the 24th ACM SIGSOFT International SPIN Symposium on Model Checking of Software*, SPIN 2017, page 142–151, New York, NY, USA, 2017.

37. Alan O. Freier, Philip Karlton, and Paul C. Kocher. The Secure Sockets Layer (SSL) Protocol Version 3.0. RFC 6101, August 2011.

38. Han Gao, Chiara Bodei, and Pierpaolo Degano. A formal analysis of complex type flaw attacks on security protocols. In José Meseguer and Grigore Roşu, editors, *Algebraic Methodology and Software Technology*, pages 167–183, Berlin, Heidelberg, 2008. Springer Berlin Heidelberg.

39. Shafi Goldwasser and Silvio Micali. Probabilistic encryption & how to play mental poker keeping secret all partial information. In *Proceedings of the Fourteenth Annual ACM Symposium on Theory of Computing*, STOC '82, page 365–377, New York, NY, USA, 1982. for Computing Machinery.

40. Shafi Goldwasser, Silvio Micali, and Ronald L. Rivest. A digital signature scheme secure against adaptive chosen-message attacks. *SIAM Journal on Computing*, 17(2):281–308, 1988.

41. L. Gong, R. Needham, and R. Yahalom. Reasoning about belief in cryptographic protocols. In *Proceedings. 1990 IEEE Computer Society Symposium on Research in Security and Privacy*, pages 234–248, 1990.

42. Andrew D Gordon and Alan Jeffrey. Typing correspondence assertions for communication protocols. *Theoretical Computer Science*, 300(1-3):379–409, 2003.

43. S. Gritzalis and D. Spinellis. Cryptographic protocols over open distributed systems: A taxonomy of flaws and related protocol analysis tools. In Peter Daniel, editor, *Safe Comp 97*, pages 123–137, London, 1997. Springer London.

44. Klaus Havelund and Natarajan Shankar. Experiments in theorem proving and model checking for protocol verification. In Marie-Claude Gaudel and James Woodcock, editors, *FME'96: Industrial Benefit and Advances in Formal Methods*, pages 662–681, Berlin, Heidelberg, 1996. Springer Berlin Heidelberg.

45. J. Heather, G. Lowe, and S. Schneider. How to prevent type flaw attacks on security protocols. In *Proceedings 13th IEEE Computer Security Foundations Workshop. CSFW-13*, pages 255–268, 2000.

46. International Standard Organization (ISO). International standard iso/iec 9594-8:2020, information technology - open systems interconnection - part8: The directory: Public-key and attribute certificate frameworks, 2020.

47. ITU. Recommendation x.509 (10/19), information technology - open systems interconnection - the directory: Public-key and attribute certificate framework, 2019.

48. Anca D. Jurcut, Tom Coffey, and Reiner Dojen. Design guidelines for security protocols to prevent replay and parallel session attacks. *Computers and Security*, 45:255–273, 2014.

49. Geir M Køien. Why cryptosystems fail revisited. *Wireless Personal Communications*, 106:85–117, 2019.

50. David E Langworthy, Gavin Bierman, Andrew D Gordon, Donald F Box, Bradford H Lovering, Jeffrey C Schlimmer, and John D Doty. Type system for declarative data scripting language, February 3, 2015. US Patent 8, 949, 784.

51. P. Laud. Symmetric encryption in automatic analyses for confidentiality against active adversaries. In *IEEE Symposium on Security and Privacy, 2004. Proceedings. 2004*, pages 71–85, 2004.

52. Peeter Laud. Handling encryption in an analysis for secure information flow. In *Proceedings of the 12th European Conference on Programming*, ESOP'03, page 159–173, Berlin, Heidelberg, 2003. Springer-Verlag.

53. Chris M. Lonvick and Tatu Ylonen. The Secure Shell (SSH) Authentication Protocol. RFC 4252, January 2006.

54. Gavin Lowe. An attack on the needham-schroeder public-key authentication protocol. *Information Processing Letters*, 56(3):131–133, 1995.

55. Gavin Lowe. Breaking and fixing the needham-schroeder public-key protocol using fdr. In Tiziana Margaria and Bernhard Steffen, editors, *Tools and Algorithms for the Construction and Analysis of Systems*, pages 147–166, Berlin, Heidelberg, 1996. Springer Berlin Heidelberg.

56. Paolo Maggi and Riccardo Sisto. Using spin to verify security properties of cryptographic protocols. In *International SPIN Workshop on Model Checking of Software*, pages 187–204. Springer, 2002.

57. AM Mathuria, Reihaneh Safavi-Naini, and PR Nickolas. On the automation of gny logic. In *Proceedings of the 18th Australian Computer Science Conf.*, volume 17, pages 370–379, Glenelg, South Australia, 1995.

58. Catherine Meadows. The nrl protocol analyzer: An overview. *The Journal of Logic Programming*, 26(2):113–131, 1996.

59. Simon Meier, Benedikt Schmidt, Cas Cremers, and David Basin. The tamarin prover for the symbolic analysis of security protocols. In Natasha Sharygina and Helmut Veith, editors, *Computer Aided Verification*, pages 696–701, Berlin, Heidelberg, 2013. Springer Berlin Heidelberg.

60. J.C. Mitchell, M. Mitchell, and U. Stern. Automated analysis of cryptographic protocols using mur/spl phi/. In *Proceedings. 1997 IEEE Symposium on Security and Privacy (Cat. No.97CB36097)*, pages 141–151, 1997.

61. Paul Morrissey, Nigel P. Smart, and Bogdan Warinschi. A modular security analysis of the TLS handshake protocol. In *Advances in Cryptology - ASIACRYPT 2008, 14th International Conference on the Theory and Application of Cryptology and Information Security, Melbourne, Australia, December 7-11, 2008. Proceedings*, volume 5350 of *Lecture Notes in Computer Science*, pages 55–73. Springer, 2008.

62. Roger M. Needham and Michael D. Schroeder. Using encryption for authentication in large networks of computers. *Commun. ACM*, 21(12):993–999, December 1978.

63. Dr. Clifford Neuman, Sam Hartman, Kenneth Raeburn, and Taylor Yu. The Kerberos Network Authentication Service (V5). RFC 4120, July 2005.

64. OMG. OMG Unified Modeling Language (OMG UML), Version 2.5.1, 2017.

65. OpenSSL. Incorrect checks for malformed signatures. https://www.openssl.org/news/secadv/20090107.txt, 2008. [Online; accessed 03-December-2021].

66. Reema Patel, Bhavesh Borisaniya, Avi Patel, Dhiren Patel, Muttukrishnan Rajarajan, and Andrea Zisman. Comparative analysis of formal model checking tools for security protocol verification. In Natarajan Meghanathan, Selma Boumerdassi, Nabendu Chaki, and Dhinaharan Nagamalai, editors, *Recent Trends in Network Security and Applications*, pages 152–163, Berlin, Heidelberg, 2010. Springer Berlin Heidelberg.

67. Lawrence C. Paulson. The inductive approach to verifying cryptographic protocols. *Journal of Computer Security*, 6(1–2):85–128, January 1998.

68. Lawrence C. Paulson. Inductive analysis of the internet protocol tls. *ACM Transactions on Information and System Security*, 2(3):332–351, aug 1999.

69. Alfredo Pironti, Davide Pozza, and Riccardo Sisto. Automated formal methods for security protocol engineering. *Cyber Security Standards, Practices and Industrial Applications: Systems and Methodologies*, page 138, 01 2011.

70. Alfredo Pironti and Riccardo Sisto. An experiment in interoperable cryptographic protocol implementation using automatic code generation. In *Proceedings of the 12th IEEE Symposium on Computers and Communications (ISCC 2007), July 1-4, Aveiro, Portugal*, pages 839–844.

71. Eric Rescorla. The Transport Layer Security (TLS) Protocol Version 1.3. RFC 8446, August 2018.

72. Peter YA Ryan. The design and verification of security protocols. *Technical Report DRA/CIS3/SISG/CR/96/1.0, Defense Research Agency*, 1996.

73. Elton Saul and Andrew Hutchison. In *A Generic Graphical Specification Environment for Security Protocol Modelling"*, pages 311–320, Boston, MA. Springer US.

74. Elton Saul and Andrew Hutchison. Enhanced security protocol engineering through a unified multidimensional framework. *IEEE Journal on Selected Areas in Communications*, 21(1):62–76, 2003.

75. R. Shirey. Rfc2828: Internet security glossary. Technical Report, USA, 2000.

76. Riccardo Sisto, Piergiuseppe Bettassa Copet, Matteo Avalle, and Alfredo Pironti. Formally sound implementations of security protocols with javaspi. *Formal Aspects Comput.*, 30(2):279–317, 2018.

77. Paul Syverson and Iliano Cervesato. The logic of authentication protocols. In Riccardo Focardi and Roberto Gorrieri, editors, *Foundations of Security Analysis and Design*, pages 63–137, Berlin, Heidelberg, 2001. Springer Berlin Heidelberg.

78. P.F. Syverson and P.C. van Oorschot. On unifying some cryptographic protocol logics. In *Proceedings of 1994 IEEE Computer Society Symposium on Research in Security and Privacy*, pages 14–28, 1994.

79. Ilaria Venturini. Oracle attacks and covert channels. In Mauro Barni, Ingemar Cox, Ton Kalker, and Hyoung-Joong Kim, editors, *Digital Watermarking*, pages 171–185, Berlin, Heidelberg, 2005. Springer Berlin Heidelberg.

80. Jim Woodcock, Peter Gorm Larsen, Juan Bicarregui, and John Fitzgerald. Formal methods: Practice and experience. *ACM Computing Surveys*, 41(4), oct 2009.

81. Andrew C. Yao. Theory and application of trapdoor functions. In *Proceedings of the 23rd Annual Symposium on Foundations of Computer Science*, SFCS '82, page 80–91, USA, 1982. IEEE Computer Society.

3 An Introduction to Tools for Formal Analysis of Cryptographic Protocols

Murat Moran
Giresun University, Giresun, Turkey

Pascal Lafourcade
University Clermont Auvergne, Clermont-Ferrand, France

Maxime Puys
University Clermont Auvergne, Clermont-Ferrand, France

David Williams
University of Portsmouth, Portsmouth, UK

CONTENTS

DOI: 10.1201/9781003090052-3

It is necessary for practitioners of formal methods to be aware of the array of tools available that support the formal verification of cryptographic protocols. This chapter presents several such automated tools, namely: Maude-NPA, Tamarin, ProVerif, CasperFDR, Scyther, CL-AtSe, OFMC and SATMC. A summary of their capabilities is presented to aid in directly comparing the tools, alongside an overview of the real world protocols to which the tools have been successfully applied, including: TLS, IKE, EMV and Kerberos. We use the NSPK protocol to illustrate the similarities and differences between the tools, presenting the way in which the protocol, the intruder, and the protocol properties are modeled and analyzed. We intend to support practitioners in determining the correct tool for their context by consistently presenting the input and outputs of the tools and highlighting their similarities, differences and merits.

3.1 INTRODUCTION

The goal of this chapter is to present automated tool support for the formal verification of cryptographic protocols. For this, we use the Needham-Schroeder Public Key (NSPK) protocol as a means of comparison. In its shortest form, the NSPK protocol exchanges only 3 messages. While it was published in 1978 [54] and apparently proven secure in the BAN logic, G. Lowe found a flaw 17 years later [47] using an automated formal verification tool. The NSPK protocol will aid us in illustrating the similarities and differences between such tools in the way that the protocol, the intruder, and the protocol properties are modeled and analyzed. We begin by giving an overview of the NSPK protocol, the automated tools, and their success stories before continuing to give a more detailed description of each tool and how they may be used to model and analyze the NSPK protocol as a running example.

3.1.1 THE NSPK PROTOCOL

The Needham-Schroeder Public Key protocol (NSPK) [54] is a well-known cryptographic protocol, in which the protocol participants (Alice and Bob are denoted respectively by A and B) wish to establish mutual authentication using public key cryptography. The following three messages version of the protocol assumes that the initiator A and responder B know each other's public keys ($pk(A)$ and $pk(B)$, respectively). We denote by N_A a nonce generated by Alice and by N_B a nonce generated by Bob. The protocol works as follows:

1. $A \rightarrow B : \{N_A, A\}_{pk(B)}$
2. $B \rightarrow A : \{N_A, N_B\}_{pk(A)}$
3. $A \rightarrow B : \{N_B\}_{pk(B)}$

The initiator A sends to the responder B their randomly chosen nonce N_A and identity A encrypted with the responder's public key pk_B (line 1). Having received this message, B decrypts the message to obtain the nonce

N_A. B, then, selects a new nonce N_B, encrypts it along with N_A using A's public key, and sends it to A (line 2). A then encrypts N_B with B's public key and sends it over to B (line 3). As they are the only ones who can decrypt the messages encrypted under their own public keys, A should (supposedly) believe that they have indeed been talking to B, and similarly B should (supposedly) believe they have indeed been talking to A. However, this is not the case as identified by Lowe in [47].

3.1.1.1 The NSPK Protocol Flaw by G. Lowe

Despite an apparent proof of correctness, G. Lowe discovered a flaw in the NSPK protocol [47]. The earlier proof assumed only a single session, whereas the flaw exploits the fact that two concurrent sessions of the protocol can be executed by the participants. The result of the attack is that a responder (Bob) may finish the protocol execution believing they were in communication with some party (Alice) that had not initiated a protocol run with them (i.e., Alice had not initiated a protocol run Bob). The attack works as follows.

$1a.\quad A \rightarrow E : \{N_A, A\}_{pk(E)}$
$1b.\quad E(A) \rightarrow B : \{N_A, A\}_{pk(B)}$
$2b.\quad B \rightarrow E(A) : \{N_A, N_B\}_{pk(B)}$
$2a.\quad E \rightarrow A : \{N_A, N_B\}_{pk(B)}$
$3a.\quad A \rightarrow E : \{N_B\}_{pk(E)}$
$3b.\quad E(A) \rightarrow B : \{N_B\}_{pk(B)}$

$A \rightarrow E$ denotes Alice sending a message to Eve, whereas $E(A) \rightarrow B$ denotes Eve impersonating Alice when sending a message to Bob.

In Lowe's attack Alice talks to Eve; however, Bob thinks that he is talking with Alice and has established a shared secret N_B with Alice only, but Eve also learns it.

3.1.1.2 Intruder Model

Lowe's attack on the NSPK protocol demonstrates that a protocol can only be proven secure against a certain intruder. Should assumptions made of the capabilities of the intruder change, then an attack might be possible and then a new proof should be constructed. Automated tools have proved useful in identifying flaws and/or proving their absence, which is otherwise a notoriously error-prone task. Attacks involving multiple interleaved protocol runs, such as Lowe's attack on the NSPK protocol, can be particularly difficult to conceive without automated tool support.

The intruder model often assumed in the analysis of cryptographic protocols was proposed by Dolev and Yao in 1983 [34]. The Dolev-Yao (DY) model presents a powerful intruder with full control of the network. Such an intruder can listen, intercept, block, modify, play, and replay any messages sent by the other protocol participants. The intruder's initial knowledge T_0 is extended with all messages sent over the network as well as further deductions that can be made from such knowledge by:

- Replaying any message learned;
- Splitting a pair of messages;
- Constructing a pair from two known messages;
- Decrypting an encrypted message with knowledge of the corresponding decryption key;
- Encrypting a message with knowledge of the encryption key.

The intruder is however unable to break cryptographic primitives (e.g., decrypt a message without knowledge of the decryption key). These simple actions are modeled in the following deduction system.

$$(A)\quad \frac{u \in T_0}{T_0 \vdash u} \qquad\qquad (UL)\quad \frac{T_0 \vdash \langle u, v\rangle}{T_0 \vdash u}$$

$$(P)\quad \frac{T_0 \vdash u \qquad T_0 \vdash v}{T_0 \vdash \langle u, v\rangle} \qquad\qquad (UR)\quad \frac{T_0 \vdash \langle u, v\rangle}{T_0 \vdash v}$$

$$(C)\quad \frac{T_0 \vdash u \qquad T_0 \vdash v}{T_0 \vdash \{u\}_v} \qquad\qquad (D)\quad \frac{T_0 \vdash \{u\}_v \qquad T_0 \vdash v}{T_0 \vdash u}$$

The DY intruder model is the model used by formal verification tools for cryptographic protocols.

3.1.1.3 Security Properties

The flaw in NSPK protocol is a flaw in the secrecy of N_B, because Eve learns it, but also on the authentication because Bob thinks to talk to Alice while it is not the case. More generally, once the protocol is modeled and the intruder is defined, the remaining point to model is the security properties. The first property to model is the **secrecy**. There is an attack against the secrecy property for a secret message s if the intruder can learn the value of s. Then a protocol is secure if for any execution where the intruder cannot learn the secret. This property is a reachability property: is it violated if the execution "reaches" a state where the intruder obtains this information with its knowledge and its capabilities.

Another security property is the **authentication** of the communication. This property ensures that one participant is talking to another one. There exist several definitions of this property as shown by G. Lowe [49]. This property is often modeled with the implication of events, where an event is raised once some stages are reached in a protocol. The property is satisfied for all possible traces. Therefore an attack is a trace where the implication of the events is not true.

Privacy is an additional security property, outside of the intended properties of the NSPK protocol, that is supported by some of the tools. This is useful for voting schemes where a voter does not want that someone can deduce her vote. For this, we model that if the voters swap their votes then the execution of the protocol can not distinguish the two situations. This property is called **observational equivalence**. Only a few tools like FDR, ProVerif and Tamarin can deal with this more complex property. One of the challenges in formal verification is to define techniques, tools and theories that can handle more complex properties.

3.1.2 FORMAL VERIFICATIONS APPROACHES

Several theoretical frameworks have been developed to perform formal verification of cryptographic protocols. Some tools are using tree automata theory like T4ASP, some are using process algebra like CASPER-FDR, some are using logic like SATMC or ProVerif, some are using ad-hoc constraints solvers like CL-ATSE, some are using rewriting techniques like Maude or Tamarin, and some others are developing specific techniques like OFMC or Scyther. Nevertheless, tools can be classified into two categories:

- **Prover:** Some tools like ProVerif, Tamarin, T4ASP, or Scyther can prove that a protocol ensures a certain security property. They often need to relax one of the hypotheses of the impossibility result [17] that proves that formal verification is in general undecidable. They are often proving the security of the protocol for an unbounded number of sessions. For this, they are doing over approximations or have strong hypotheses. However, attacks they found might be due to their over approximations.
- **Attack finder:** Some tools are dedicated to finding attacks. Usually, these tools consider only a bounded number of sessions like CL-ATSE, OFMC, or Scyther. However, not finding any attack with them does not mean that the protocol is secure.

We select only free, maintained, and trace-based symbolic tools. Our aim in the rest of the chapter is to describe how the NSPK is modeled by each tool. It allows us to concretely show the difference between them. We summarize the differences among the chosen tools in Table 3.1. We do not consider tools that are analysing cryptographic protocols in the computational model like CryptoVerif, Easycrypt. For a comparison of all these tools, the reader can refer to [1].

Table 3.1

Summary of symbolic tools capabilities, where "Trace" means that the tool can dead with trace properties (secrecy and authentication) "Unbounded" means that the tool can prove unbounded number of sessions, "Equivalence" means that the tool can verify equivalence properties (like anonymity, privacy), "XOR" and "DH" mean that the tool can analyze protocols that use equational theories of XOR or Diffie-Hellmann.

Tools	Trace	Unbounded	Equivalence	XOR	DH
Maude-NPA [37]	✓	✓	✓	✓	✓
Tamarin [50]	✓	✓	✓	✓	✓
ProVerif [18]	✓	✓	✓	✓	✓
CasperFDR [48]	✓		✓	✓	✓
Scyther [31]	✓	✓			
CL-AtSe [59]	✓				
OFMC [14]	✓				
SATMC [15]	✓				

3.1.3 SUCCESS STORIES

The following are the success stories of each tool that we investigate in this chapter.

3.1.3.1 CasperFDR

In 1999, Donovan et. al [35] published a list of 52 protocols analyzed using Casper and FDR. Since then, new protocols were added, notably:

- TLS 1.0 [42] also known as Transport Layer Security, is the successor of the now-deprecated Secure Sockets Layer (SSL). This protocol and its derivatives are widely used in applications such as email, instant messaging, and voice over IP, but its use in securing HTTPS remains the most publicly visible. The TLS protocol aims primarily to provide privacy and data integrity between two or more communicating computer applications. It runs in the application layer and is composed of two layers: the TLS record and thc TLS handshake protocols.
- IKEv2 [57] also known as Internet Key Exchange is the protocol used to set up a security association in the IPsec protocol suite. IKE builds upon the Oakley protocol and ISAKMP. It uses X.509

certificates for authentication (either pre-shared or distributed using DNS, preferably with DNSSEC) and a Diffie–Hellman key exchange to set up a shared session secret from which cryptographic keys are derived.

- For more complex protocols, such as cryptographic voting systems, where Casper is not suitable for checking specific system properties such as anonymity, CSP and FDR have been shown to be quite flexible and suitable in defining security properties and verifying such properties automatically [51, 52].

3.1.3.2 ProVerif

ProVerif's website features 40 publications at the time of writing and 148 using the tool

(https://bblanche.gitlabpages.inria.fr/proverif/proverif-users.html). Although the tool is still under active development, many recent protocols or recent versions have been verified so far:

- TLS 1.2 and TLS 1.3 [16]
- 5G EAP-TLS [60] also known as Extensible Authentication Protocol (EAP) is an authentication framework frequently used in network and internet connections (e.g., IEEE 802.1X for WPA and WPA2). This specific version of EAP features a TLS handshake and is performed as part of 5G networks.
- Kerberos [2, 44] is a computer network authentication protocol that works based on tickets to allow nodes communicating over a non-secure network to prove their identity to one another in a secure manner. It provides mutual authentication to both the user and the server and protection against eavesdropping and replay attacks. Kerberos builds on symmetric-key cryptography and requires a trusted third party, and optionally may use public-key cryptography during certain phases of authentication.
- Signal [43] is a cryptographic protocol that can be used to provide end-to-end encryption for voice calls and instant messaging conversations. The protocol combines the Double Ratchet algorithm, prekeys, and a triple Elliptic-curve Diffie–Hellman (3-DH) handshake, and uses Curve25519, AES-256, and HMAC-SHA256 as primitives.
- Automated analysis of cryptographic voting systems is another challenging problem domain in terms of defining desired properties and mechanized analysis of such complex systems. However,

ProVerif and applied pi calculus have been proved to be very useful in the analysis of such systems [6, 56, 53].

3.1.3.3 Scyther

On their GitHub page, Scyther features no less than 243 protocol models. Among the most emblematic, protocols that have been analyzed with Scyther are:

- IKEv2 [26]
- The ISO/IEC standards 9798-2, 9798-3, and 9798-4 [28, 10] define a suite of 17 authentication protocols based on symmetric-key cryptography, on cryptographic signatures, and on cryptographic check functions.
- SSH [45], also known as Secure Shell is a cryptographic network protocol often used for remote command-line, login, and remote command execution, but any network service can be secured with SSH. SSH provides a secure channel over an unsecured network by using a client-server architecture, connecting an SSH client application with an SSH server. The encryption used by SSH is intended to provide confidentiality and integrity of data over an unsecured network, such as the Internet.
- The tool has also been used to find automatically new multi-protocol attacks on many existing protocols [25].
- The compromising adversaries version of Scyther has been applied to a large amount of Authenticated Key Exchange (AKE) protocols, such as HMQV, UM, NAXOS, and many other protocols[9, 13, 8].

3.1.3.4 Tamarin

On their GitHub page, Tamarin features an impressive repository of 887 protocol models. Tamarin is a widely used tool, and its authors are often implied in the verification of world-class cryptographic protocols:

- EMV is a payment method based upon a technical standard for smart payment cards and for payment terminals and automated teller machines which can accept them. EMV originally stood for "Europay, Mastercard, and Visa", the three companies that created the standard. In [12, 15], Tamarin is used to find flaws this international protocol standard for smart-card payment.
- Wi-Fi Protected Access (WPA2) is the standard for the security of Wi-Fi communications. It replaces WPA and WEP's previous

standard, with the main introduction of a four-way handshake protocol and AES-CCMP mode. In [30], the authors used Tamarin to perform a security analysis and were able to recover the existing attacks. It allows them to fix the protocol.

- Authentication and Key Agreement (AKA) is a security protocol used in GSM networks. In particular, Tamarin has also been used in [27] for verifying 5G authentication and key agreement protocol. The discovered issues were submitted to 3GPP to improve the standard.
- Selene is a novel voting protocol that supports individual verifiability, Vote-Privacy, and Receipt-Freeness. In [20, 11], Tamarin has been used for the formal design and proof of vote privacy and receipt-freeness properties.
- TLS 1.3 [29]

3.1.3.5 Avispa

The AVISPA website features an impressive library of 71 protocol models[1]. However, at the time of writing, the project's website has been taken down. The library can still be consulted on *Web Archive*[2]. This includes various well-known and used protocols such as:

- EAP (many variants available)[3], also known as Extensible Authentication Protocol (EAP) is an authentication framework frequently used in network and internet connections (e.g., IEEE 802.1X for WPA and WPA2). EAP is an authentication framework for providing the transport and usage of material and parameters generated by EAP methods. There are many methods defined by RFCs, and a number of vendor-specific methods and new proposals exist.
- IKEv2[4]
- Kerberos[5]
- SSH:[6]
- TLS 1.0[7] The protocol runs in the application layer and is composed of two layers: the TLS record and the TLS handshake protocols, only the handshake was studied with AVISPA.

[1] http://www.avispa-project.org/library/avispa-library.html
[2] https://web.archive.org/web/20211005213356/http://www.avispa-project.org/
[3] http://www.avispa-project.org/library/EAP˙TLS.html
[4] http://www.avispa-project.org/library/IKEv2-DS.html
[5] http://www.avispa-project.org/library/Kerb-preauth.html
[6] http://www.avispa-project.org/library/ssh-transport.html
[7] http://www.avispa-project.org/library/TLS.html

3.1.3.6 Maude-NPA

Maude-NPA's website features 19 publications at the time of writing and multiple industry-mature protocols verified. As the tool is still under active development, the followings are two recent protocols or their recent versions that have been verified with Maude-NPA:

- PKCS#11: [40] is a standard defining a platform-independent API to cryptographic tokens, such as hardware security modules (HSM) and smart cards. It includes the most commonly used cryptographic object types (RSA keys, X.509 certificates, DES/Triple DES keys, etc.) and all the functions needed to use, create/generate, modify and delete those objects.
- IBM CCA Security API [39] also known as IBM 4758 Common Cryptographic Architecture (CCA) protocol is an API for accessing IMB's hardware cryptographic co-processor. It provides a variety of cryptographic processes such as encryption, hashes, signatures, etc implemented as part of the hardware. Protocols from this API appear in several studies as it relies heavily on the XOR operator.

3.2 CASPERFDR

Communicating Sequential Processes (CSP) [41] is a calculus that can be used to produce models of security protocols, which can be analyzed using FDR [38]. While CSP and FDR have been used successfully to identify flaws in security protocols, Casper [48] is a compiler that eases these procedures using pre-defined agent behaviors. It has a more abstract and domain specific notation than CSP; short protocol descriptions can be verified using FDR model checker.

The input file given to Casper must specify the following: the messages sent between agents, tests performed by the agents, data types, the ultimate aim of the protocol/specification, any algebraic equivalences, and the intruder's abilities [55].

In the following sections, we aim to explain how Casper compiler works in the analysis of security protocol using the NSPK security protocol as an example.

3.2.1 PROTOCOL DESCRIPTION

The first part of the input file defines each step of the protocol, where `A` and `B` represent agents, and `{m}{PK(A)}` models asymmetric encryption of the message m using the agent *A*'s public key. In the NSPK example, the protocol description is given as follows:

```
0.     -> A : B
1. A -> B : {na, A}{PK(B)}
2. B -> A : {na, nb}{PK(A)}
3. A -> B : {nb}{PK(B)}
```

When there is no particular sender, in step 0 for instance, it is assumed that the agent *A* receives an identity *B* as a message from the environment.

3.2.2 ENVIRONMENT

The following defines the variables e.g. agent names, nonces, etc., and functions used in the protocol, such as; public-secret key functions, which take agent names as arguments and return the corresponding key for that agent. These are called free variables in Casper and will be initiated with actual values when running an actual protocol run. For instance, in step 1 of the protocol, agent *A* sends the fresh nonce *na* and her identity *A* encrypted under the agent *B*'s public key to agent *B*.

```
#Free variables
A, B : Agent
na, nb : Nonce
PK : Agent -> PublicKey
SK : Agent -> SecretKey
```

To define valid public and secret key pairs that are inverses of each other, the following assignment needs to be included
`InverseKeys = (PK, SK).`

3.2.3 PROTOCOL PARTICIPANTS

For each protocol participant, her role in the protocol (initiator, responder, or server), and initial knowledge need to be described in the `#Processes` section of the input file.

```
#Processes
INITIATOR(A, na) knows PK, SK(A)
RESPONDER(B, nb) knows PK, SK(B)
```

Above are the roles played by each agent in the NSPK protocol. Agent *A* in the protocol is represented by the CSP process `INITIATOR(A, na)`, and similarly agent *B* as a responder by `RESPONDER(B, nb)`. Here, `knows` defines the initial knowledge of each agent. For example, agent *A* is expected to know her identity *A*, the nonce *na*. When initialized with `INITIATOR()` process, *A* is expected to know her public key *PK* and the corresponding secret key *SK(A)*.

It is assumed that an agent can send a message if she knows the recipient's identity and all the required components of the message. Additionally, she can learn information pieces from the messages she has received in previous steps of the protocol runs. If an agent is required to decrypt an encrypted message, she has to possess the corresponding secret key. In case an agent is expected to run the protocol several times, each instance process is initialized with the value of the agent's name, however with a different nonce from the previous ones. Lastly, an agent accepts a message only if all fields represented by variables already in the agent's knowledge contain the expected values [55].

3.2.4 INTRUDER MODEL

Casper compiles an intruder that is capable of performing various actions in the system, such as; intercepting, sending, signaling, faking, and leaking messages or facts over the built-in communication channels. It can also compose and decompose facts, encrypt and decrypt messages, and derive new facts according to a set of derivation rules, which are already specified in Casper. In the input file, the intruder's name and its initial knowledge need to be defined. Once a fact that is supposed to be secret is captured by the intruder, it will signal the fact over the channel `leak`.

In the NSPK, for instance, the intruder's identity, which is *Mallory*, and the set of his initial knowledge are defined as follows:

```
#Intruder Information
Intruder = Mallory
IntruderKnowledge = {Alice, Bob, Mallory, Nm, PK, SK(Mallory)}
```

3.2.5 SYSTEM COMPOSITION

For the actual system to be checked, the agents and their role in the actual system, data types, and intruder's ability, which is defined previously need to be specified. The following defines actual variables just like free variables. In the NSPK example, the actual system consists of three agents, namely, *Alice*, *Bob* and *Mallory* (the intruder), and three unique nonces: *Na*, *Nb* and *Nm*.

```
#Actual variables
Alice, Bob, Mallory : Agent
Na, Nb, Nm : Nonce
```

The public and secret key functions, *PK* and *SK* respectively, are defined using the keyword `symbolic` as follows.

```
#Functions
symbolic PK, SK
```

Next, we initialize *INITIATOR*, *RESPONDER* and *SERVER* (if needed) processes with the agents' name, taking part in the system, and their private fresh nonces. For the NSPK example, Alice is considered to be an initiator with nonce *Na*, and Bob to be the only responder with nonce *Nb*. These are actual variables and they take place of free variables in the protocol. For instance, *Na* takes place of *na* in the protocol description.

```
#System
INITIONATOR(Alice, Na)
RESPONDER(Bob, Nb)
```

3.2.6 SPECIFYING PROTOCOL PROPERTIES

A variety of security protocol specifications is supported by Casper: `Secret`, `StrongSecret`, `Agreement`, `NonInjectiveAgreement`, `WeakAgreement`, `Alive-ness`, `TimedAgreement`, `TimedNonInjectiveAgreement`, `TimedWeakAgreement`, and `TimedAliveness`) (see [55, 48] for more details about these specifications). Below, we cover secrecy and agreement specifications in detail.

In terms of secrecy, having a specification line such as `Secret(A, s, [B_1,..., B_n])`, where A is an agent, s is a variable for a secret, and $B_1,\dots,B_n$ are the variables representing other agents' role in the protocol with whom the secret is shared, means that agent A expects the value of s to be kept secret throughout the protocol run. The secrecy of the value given to s by A is violated, if A completes a protocol run and the intruder obtains the value of s without taking none of the roles represented by $B_1,\dots,B_n$ legitimately. In more detail, if an event like $signal.Claim_Secret.s$ is performed, then the event $leak.s$ should not occur during the protocol run, which would indicate that the value assigned to s is not secret anymore.

Regarding the authentication specification, `Agreement(A, B, [v_1, ..., v_n])` indicates that A is authenticated to B, and these agents agree on the values given to the variables $v_1,\dots,v_n$ and on the roles they took in the protocol run. In more detail, if *B* thinks he has completed a protocol run with *A*, performing a *signal.Commit* event, then *A* should have been running the protocol previously with *B*, having performed a *signal.Running* event.

Running the Casper compiler with the input file creates an output file written in CSP, which, then, can be loaded into FDR. For each authentication and secrecy specification in the input file, Casper produces refinement assertions, which can be verified by FDR.

In the NSPK example, this compilation creates the following refinement assertions: the first one is to check whether the system provides secrecy, and the other two are to check whether the system satisfies the authentication specification.

```
#Assertions
SECRET_SPEC [T= SYSTEM_S
AuthenticateINITIATORToRESPONDERAgreement_na_nb [T= SYSTEM
AuthenticateRESPONDERToINITIATORAgreement_na_nb [T= SYSTEM_2
```

For the secrecy specification for the NSPK, if *A* thinks that *na* is a secret, then it can only be known by herself and *B*. Hence, if *B* is not the intruder, then the intruder should never learn the secret value *na*. Below, the first two lines of the `#Specification` define the secrecy of *na* and *nb* in CSP. The following two lines starting with `#Agreement` define the authentication specification for the NSPK. The first `Agreement` checks whether *A* is correctly authenticated to *B*, and *A* and *B* agree on the values of *na* and *nb* (one-way authentication). Passing the second agreement specification ensures mutual authentication between *A* and *B*.

```
#Specification
Secret(A, na, [B])
Secret(B, nb, [A])
Agreement(A, B, [na, nb])
Agreement(B, A, [na, nb])
```

3.2.7 RESULTS

Casper indicates the specifications that are satisfied with `PASS` and the specifications that are not satisfied with `FAIL`. Running Casper to check whether the privacy and mutual authentication specifications are satisfied by the NSPK protocol produces a short informative output.

```
-- #Specification
--
```

```
-- -- FAIL
-- Secret(b, nb, [a])
-- -- FAIL
-- Secret(b, na, [a])
--
-- -- FAIL
-- Agreement(a, b, [na, nb])
-- -- PASS
-- Agreement(b, a, [na, nb])
```

As stated before, the output file produced by Casper can be exported to FDR for further analysis with counter-example traces if the specification fails. For the NSPK's secrecy property (the first assertion), FDR produces the following trace.

```
#Counter Example
signal.Claim_Secret.Alice.Na.{Mallory}
signal.Claim_Secret.Bob.Nb.{Alice}
leak.Nb
```

The counter-example above with the internal messages hidden shows that, as the system process `SYSTEM_S`, the right-hand side of the refinement, performs a sequence of events that it is not supposed to perform, the refinement check fails. With the help of Casper's `interpret` function, the trace above is restated as:

```
Alice believes Na is a secret shared with Mallory.
Bob believes Nb is a secret shared with Alice.
The intruder knows Nb.
```

CSP hides some of the messages, and if we debug further down, the following trace can be produced.

```
env.Alice.(Env0,Mallory,<>)
intercept.Alice.Mallory.(Msg1,Encrypt.(PK_.Mallory,<Na,Alice>),<>)
fake.Alice.Bob.(Msg1,Encrypt.(PK_.Bob,<Na,Alice>),<>)
intercept.Bob.Alice.(Msg2,Encrypt.(PK_.Alice,<Na,Nb>),<>)
fake.Mallory.Alice.(Msg2,Encrypt.(PK_.Alice,<Na,Nb>),<>)
intercept.Alice.Mallory.(Msg3,Encrypt.(PK_.Mallory,<Nb>),<Na>)
fake.Alice.Bob.(Msg3,Encrypt.(PK_.Bob,<Nb>),<Na>)
leak.Nb
```

Again using Casper's `interpret` function, it can be interpreted as the following:

```
0.              ->     Alice     : Mallory
1.    Alice     -> I_Mallory     : {Na, Alice}{PK_(Mallory)}
1. I_Alice      ->     Bob       : {Na, Alice}{PK_(Bob)}
2.    Bob       -> I_Alice       : {Na, Nb}{PK_(Alice)}
2. I_Mallory    ->     Alice     : {Na, Nb}{PK_(Alice)}
3.    Alice     -> I_Mallory     : {Nb}{PK_(Mallory)}
3. I_Alice      ->     Bob       : {Nb}{PK_(Bob)}
The intruder knows Nb
```

Here, `I_Alice` represents the intruder's impersonating Alice by sending a fake message to Bob as in message 1, and she can receive the messages originally intended for Alice as in message 2. The attack above is Lowe's attack on the NSPK [47], in which the intruder retrieves a nonce in one protocol run and uses it to impersonate an agent to set up a second protocol run, a false session, with another agent (a.k.a. man-in-the-middle-attack). Hence, the messages sent between Alice and Bob, encrypted under the agreed key based on `Na` and `Nb`, are not secret anymore.

3.3 PROVERIF

ProVerif is a command-line protocol verifier for automatic analysis of security protocols. ProVerif has been developed using Objective Caml (OCaml) and it can handle input files encoded in many languages, such as; typed, which is the focus of this chapter, and untyped pi calculus. It supports several cryptographic primitives such as; symmetric and asymmetric encryption, digital signatures, hash functions, bit-commitment, and non-interactive zero-knowledge proofs [18]. ProVerif can handle reachability properties (e.g. secrecy), correspondence assertions (e.g. authentication), and observational equivalence (e.g. the secrecy properties that require indistinguishability between instances of a protocol). It can also produce an execution trace if the desired property does not hold.

ProVerif can verify the NSPK protocol in various complexities of the protocol. Here, we present an analysis of a simplified version of the NSPK. Assuming the participants, Alice and Bob, know each other's public key (omitted trusted server) before the execution of the protocol, the following sections describe the formalization of the simplified (3-message) NSPK protocol.

3.3.1 ENVIRONMENT

In the typed pi calculus, the declarations below describe the rules for public and private channels, cryptographic primitives (symmetric and asymmetric encryptions, and digital signatures), and their appropriate properties in the model. Public channels are declared using the keyword `free`, and private channels that are not known by the intruder with `private`. User-defined types are declared with `type`. `fun` is a constructor, which is used to model cryptographic primitives/functions. Similarly `reduc forall` is a destructor that breaks down terms built by constructors. For instance, for a ciphertext, such as $aenc(m, pk)$, message m encrypted under a public key pk, only the agents with the appropriate or corresponding secret key, sk, can decrypt it. Similarly, one can recover a message that is signed under a private signing key, say *sskey*, using the function `getmess`. Additionally, an agent who possesses the correct public signing key can verify whether the signature is valid or the message is authentic by extracting the message with the function `checksign`. Lastly, symmetric key encryption or shared key encryption is defined similarly: only the agent possessing the shared/secret key that is used to encrypt the message can decrypt the message.

```
(* Public key encryption *)
type pkey.
type skey.

fun pk(skey): pkey.
fun aenc(bitstring, pkey): bitstring.
reduc forall x: bitstring, y: skey; adec(aenc(x, pk(y)),y) = x.

(* Signatures *)
type spkey.
type sskey.

fun spk(sskey): spkey.
fun sign(bitstring, sskey): bitstring.
reduc forall x: bitstring, y: sskey;
getmess(sign(x,y)) = x.
reduc forall x: bitstring, y: sskey;
checksign(sign(x,y), spk(y)) = x.

(* Shared key encryption *)
fun senc(bitstring, bitstring): bitstring.
reduc forall x: bitstring, y: bitstring; sdec(senc(x,y),y) = x.
```

3.3.2 PROTOCOL PARTICIPANTS

Instead of a single process, the main protocol is divided into sub-processes (macros). They start with `let`, and describe individual processes defining protocol participant's behavior in the model. `in` denotes incoming channel, `out` is outgoing channel, `new` creates new instances of terms, and `let` models an assignment in process descriptions.

The following processes, `processA` and `processB` below, Alice and Bob, in this case, behave according to the NSPK protocol description. Accordingly, as an initiator Alice obtains her public key, creates a fresh random number `Na`, encrypts them with Bob's public key, and sends the ciphertext to Bob over the channel `c`. She then receives a message `m`, encoding `{Na, Nb}pk(A)`, decrypts it with her unique private key, and checks whether the nonce `Na` is the one she has generated previously. Finally, she encrypts it with Bob's public key, `pk(B)`, and sends the ciphertext to Bob. Similarly, `processB` below models Bob's behavior according to its responder role in the NSPK protocol.

```
let processA(pkB: pkey, skA: skey) =
in(c, pkX: pkey);
new Na: bitstring;
out(c, aenc((Na, pk(skA)) , pkX));
in(c, m: bitstring);
let (=Na, NX: bitstring) = adec(m, skA) in
out(c, aenc(NX, pkX)).

let processB(pkA: pkey, skB: skey) =
in(c, m: bitstring);
let (NY: bitstring, pkY: pkey) = adec(m, skB) in
new Nb: bitstring;
out(c, aenc ((NY, Nb), pkY));
in(c, m3: bitstring);
if Nb = adec(m3, skB) then 0.
```

3.3.3 INTRUDER MODEL

Like Casper/FDR, ProVerif avoids the need to manually model how the intruder behaves. Its environment captures all such malicious intruder behavior. ProVerif provides a standard powerful Dolev-Yao intruder that may access all public channels, and perform malicious actions on the messages flowing over the public channels. It can construct and decompose messages seen as well as read, modify, delete and inject them. The intruder can also perform cryptographic operations when required cryptographic keys are in its possession. Additionally, we can determine what information to be made available to the intruder. For instance, dishonest participants can share their private data with the intruder by giving the intruder access to their secret keys or by making their private communication channels open to the public — the intruder can access all channels unless they are declared private.

3.3.4 SYSTEM COMPOSITION

The main process brings all participants together along with their freshly generated nonces and cryptographic key pairs. Hence, it is a composition of the sub-processes. The following defines the main process of the NSPK protocol, in which public and secret key pairs are constructed for each participant with `new`, and public keys are provided to the participants on a public channel, also available to the intruder. The last line of the main process with `!` in front of the processes enables the processes representing Alice and Bob to participate in an unbounded numbers of sessions.

```
process
new skA: skey; let pkA = pk(skA) in out(c, pkA);
new skB: skey; let pkB = pk(skB) in out(c, pkB);
( (!processA(pkB, skA)) | (!processB(pkA, skB)) )
```

3.3.5 SPECIFYING PROTOCOL PROPERTIES

ProVerif allows several security protocol properties to be modeled and verified automatically using reachability, correspondence assertions, and observational equivalence. Among these, the secrecy property is defined as a reachability property in the Applied Pi, i.e., verifying whether the intruder can reconstruct a message that is supposed to be secret[8]. To analyze such secrecy of a term or message, a query, such as `query attacker(Na)`, is asked, where `Na` is a presumably secret nonce in the NSPK case.

To verify authentication properties, the events for each participant (such as `beginAparam(pkey)`, `endAparam(pkey)`) need to be declared, modeling the beginning and end of a protocol run respectively. Then, we query the sequences of events to check whether when Alice reaches the end of the protocol believing that she has completed the protocol run with Bob by comparing Bob's public keys, and sends her public key by the event `endAparam(pkey)`, then Bob has indeed started a session with Alice, denoted by the occurrence of the event `beginAparam(pkey)`. The same query is asked for Bob for the other part of the mutual authentication property.

[8] There are many variations of privacy definitions in ProVerif, and this is the general definition of privacy we capture in this chapter.

```
query x: pkey; inj-event(endAparam(x)) ==>
inj-event(beginAparam(x)).
query x: pkey; inj-event(endBparam(x)) ==>
inj-event(beginBparam(x)).
```

3.3.6 RESULTS

Executing input files with ProVerif produces query results showing whether the queries, such as; secrecy of Na and mutual authentication above are satisfied (the query is true or false). When ProVerif finds no attack on a security protocol, then its model guarantees that property. That is, ProVerif is sound or correct. However, it may not be possible to prove the desired property that is supposed to be held by the protocol with ProVerif. Hence, ProVerif is not complete. Additionally, it may produce false-positive attacks due to over-approximation, which is an essential part of it, and also it may not terminate, yielding any result about the verification.

```
...
RESULT not attacker(secretBNa[ ]) is false.
RESULT inj-event(endBparam(x_21032,y_21033)) ==>
inj-event(beginBparam(x_21032,y_21033)) is false.
...
The attacker has the message secretBNa.
A trace has been found.
RESULT not attacker_bitstring(secretBNa[]) is false.
```

Above, shown as RESULT, are the results of analysing the NSPK indicating that the secrecy of B regarding the nonce Na is violated, and the injective authentication of A to B is false. The second result shows that Bob may think that he has completed a protocol run with Alice, while Alice has never stated a session with Bob earlier implying the violation of the injective authentication of Alice and Bob, and hence Lowe's attack on NSPK. Investigating the details of these attacks further is possible with attack traces produced by ProVerif using `set traceDisplay = long.` parameter and describes the attack step-by-step. As it is too long to include, we omit the whole trace here.

3.4 SCYTHER

Upon analysis of a given protocol, Scyther will either generate counterexample traces illustrating (classes) of protocol attacks (*i.e.*, violations of protocol security properties with a graphical representation of the attack) or else verify the absence of attacks [31]. Verifying a protocol with an unbounded number of sessions with guaranteed termination is achieved through Scyther's pattern refinement algorithm, which concisely represents (infinite) sets of traces [32].

3.4.1 ENVIRONMENT

The messages exchanged during a protocol session are represented by a set of terms comprising basic and compound terms. Basic terms include global constants, freshly generated values specified using the `fresh` declaration (*e.g.*, `fresh ni: Nonce` generates a random number `ni` of type `Nonce`) and variables to which agents can store received terms (e.g., `var nr: Nonce;` specifies the variable `nr` as type `Nonce`).

```
protocol ns3(I,R)
{
role I
{
fresh ni: Nonce;
var nr: Nonce;
...
```

Compound terms comprise the basic terms and terms constructed by tupling or encrypting a term with some key. A public-key infrastructure (PKI) is predefined with corresponding public and private key pairs. For example, `{nr}pk(R)` (see Section 3.4.2) represents a nonce generated by the responder `nr` encrypted under their public key, `pk(R)`. The term `{nr}pk(R)` can only be decrypted by an agent who knows the corresponding secret key `sk(R)`.

3.4.2 PROTOCOL PARTICIPANTS

Each protocol participant role is specified in terms of a finite list of protocol events, including `send` and `recv` events, control flow events, and claims (see Section 3.4.5). The role of the protocol responder in the NSPK protocol, `role R`, is specified as three protocol events `recv_1`, `send_2` and `recv_3` in which the responder receives a nonce challenge from the initiator, sends the appropriate response alongside a nonce challenge of their own and expects to receive the appropriate response in the third message. The suffixes (e.g., `_1`) of the `send` and `recv` keywords are helpful labeling that aid in associating the corresponding send and receive events between roles. For example, the receive event `recv_1(I,R,{I,ni}pk(R))` modeling the first protocol message in role `R` has a corresponding send event in role `I`, namely `send_1(I,R,{I,ni}pk(R))`.

```
...
role R
{
var ni: Nonce;
fresh nr: Nonce;
recv_1(I,R, {I,ni}pk(R) );
claim(R,Running,I,ni,nr);
send_2(R,I, {ni,nr}pk(I) );
recv_3(I,R, {nr}pk(R) );
...
```

3.4.3 INTRUDER MODEL

Scyther assumes a Dolev-Yao style intruder, in which every sent message is immediately observed by the intruder, and all messages received by the honest participants are supplied by the intruder. The intruder may generate arbitrary constants of the type of each variable and possesses a finite set of terms for the agents under their control including their private keys. Scyther handles tupling implicitly and encryption explicitly (through the introduction of encryption and decryption events **encr** and **decr**). Two other intruder events **init** and **know** correspond to terms stemming from the intruder's knowledge whether known initially or built-up throughout the message exchange.

3.4.4 SYSTEM COMPOSITION

The complete system $Sys(Q)$ is composed of the honest participants and intruder model within the context of a given protocol Q. If there exists a violation of a security property (that can be expressed in terms of traces) of a protocol Q, it will be witnessed by exploring the set of traces of $Sys(Q)$; if no security property violation is witnessed in the traces of $Sys(Q)$ then no attack exists for protocol Q.

3.4.5 SPECIFYING PROTOCOL PROPERTIES

The specification of the protocol participant roles may be decorated with claim events to aid in specifying intended security properties. Claims are necessary for Scyther to carry out checks against them. For example, `claim_i1(I,Secret,ni)` models that `ni` is meant to remain secret.

Scyther's predefined claim types include:

- `Secret` : This claim requires a parameter term. Secrecy of this term is claimed.
- `Alive` : Aliveness (of all roles).
- `Weakagree` : Weak agreement (of all roles).
- `Niagree` : Non-injective agreement on messages.
- `Nisynch` : Non-injective synchronization.

Two additional claim types, `Commit` and `Running` are signal claims that support the specification of non-injective agreement with a role on a set of data items. Commit marks the effective `claim`, whose correctness requires the existence of a corresponding `Running` signal in the trace. For example, to specify that `I` and `R` agree on the nonces `ni` and `nr`, `claim(I,Commit,R,ni,nr);` is inserted at the end of role `I` and `claim(R,Running,I,ni,nr);` is inserted immediately before the last send that causally precedes the claim in the `I` role.

```
...
claim_r1(R,Secret,ni);
claim_r2(R,Secret,nr);
claim_r3(R,Alive);
claim_r4(R,Weakagree);
claim_r5(R,Commit,I,ni,nr);
claim_r6(R,Niagree);
claim_r7(R,Nisynch);
}
}
```

In Scyther, the authentication and secrecy mechanisms are automatically managed by the tool and the user only needs to use the above keywords.

3.4.6 RESULTS

The graph produced by Scyther (see Figure 3.1) illustrates two parallel runs of the protocols. The protocol run `Run #2` casts `Agent2` in the role of the initiator `I` and `Eve` (an identity controlled by the intruder) in the role of the responder. The first message sent to `Eve` by `Agent2` is `{Agent2,ni#2}pk(Eve)`. As `sk(Eve)` is in the initial intruder knowledge it can be used to decrypt the message sent to `Eve`

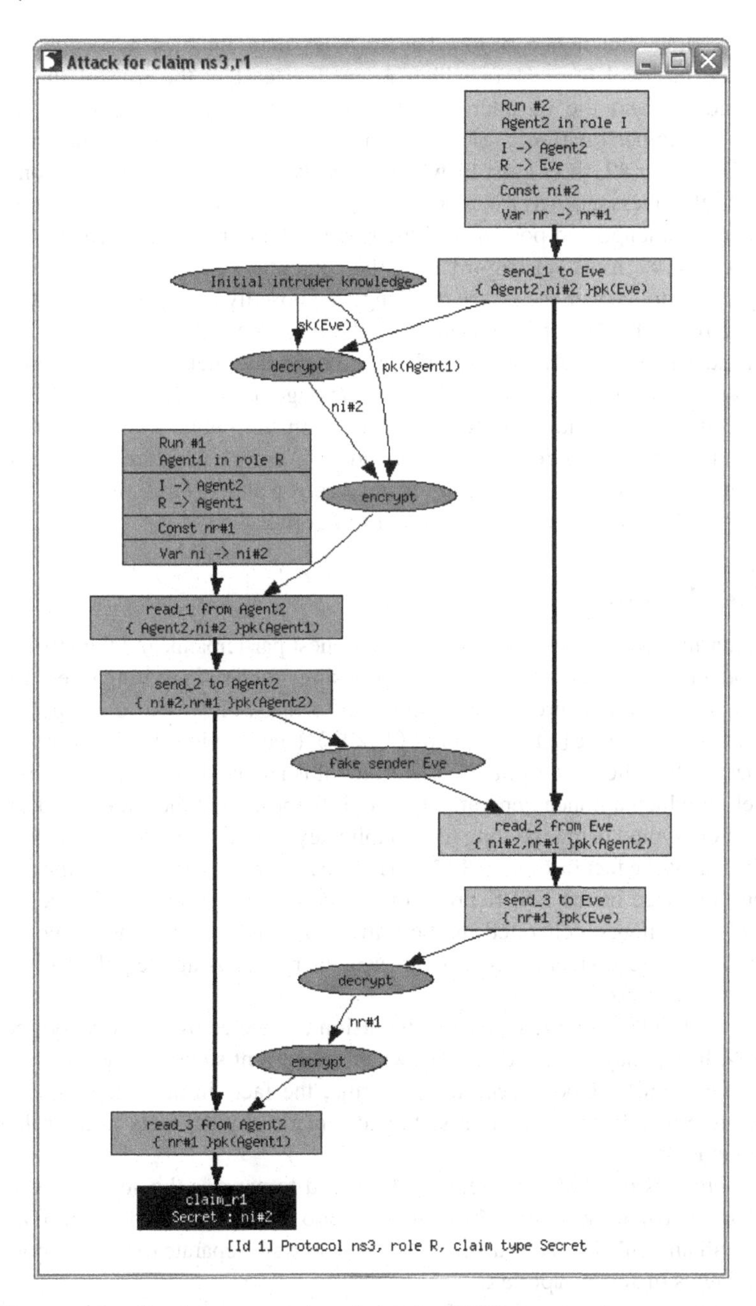

Figure 3.1 Scyther output: Attack on the NSPK protocol.

and extract the nonce ni#2. The intruder then initiates a second run of the protocol Run #1 in which Agent1 is cast in the role of the responder R and the intruder uses the nonce received in Run #2 to begin the impersonation of Agent2 in Run #1. The intruder constructs {Agent2,ni#2}pk(Agent1) which is received by Agent1, who responds with the message {ni#2,nr#1}pk(Agent2) to answer the initiator's nonce challenge and pose one of their own. The intruder subsequently replays {ni#2,nr#1}pk(Agent2) to the authentic Agent2 in Run #2 who proceeds to answer the nonce challenge supposedly set by Eve (but in reality posed by Agent1) by returning the message {nr#1}pk(Eve). The intruder now has sufficient knowledge to conclude their impersonation of Agent2; the intruder decrypts the final message of Run #2 using sk(Eve) to extract nr#1 and encrypts this value with the public key of Agent1. When {nr#1}pk(Agent1) is received by Agent1 they are satisfied to claim that ni#2 is now a secret shared Agent1 and Agent2, when in fact ni#2 is known to the intruder and not to Agent2.

3.5 TAMARIN

Tamarin models a system composed of honest participants and the intruder model in the context of a given protocol using multi-set rewriting rules. The system's state is expressed as a multi-set of facts, on which rules operate. The fact In(aenc{'1',ni,I}pk(ltkR)) (see Section 3.5.2) expresses that the first message of the NSPK protocol is received on the public channel, in which a nonce generated by the initiator ni and their identity I are sent encrypted using the long term public key of the responder, pk(ltkR). Similarly, the fact Out(aenc{'2',ni,~nr,R}pkI) expresses that the second message of the NSPK protocol is received on the public channel, in which the nonce generated by the initiator ni and the fresh nonce generated by the responder identity nr are sent encrypted using the public key of the initiator, pkI.

Each rule has a premise and conclusion connected by an arrow symbol -->. Rules may be executed only when the current state includes all facts in the premise. Upon execution of a rule, the facts in the rule's premise are removed from the current state and replaced by the facts in the rule's conclusion.

Prefixes provide a convenient shorthand to express the sort of a variable: ~x denotes x:fresh, $x denotes x:pub and #i denotes i:temporal. fresh and pub are each subsorts of a top sort msg, separate to the timepoint variables of sort temporal.

Properties are specified in Tamarin using a guarded fragment of a many-sorted first-order logic with a sort for timepoints (prefixed with #). This logic supports quantification over both messages and timepoints using:

universal and existential quantification over both messages and timepoints, implication, negation, disjunction, conjunction, temporal ordering, and equality.

3.5.1 ENVIRONMENT

The `Register_pk` rule models the registration of a public key. `Fr` is a built-in fact denoting a freshly generated random number, which may be used as a nonce or key. Here the prefix ~ denotes that the random number `ltkA` is fresh, its label suggests that it is intended to be a long-term key belonging to A.

```
rule Register_pk:
[ Fr(~ltkA) ]
-->
[ !Ltk($A, ~ltkA), !Pk($A, pk(~ltkA)), Out(pk(~ltkA)) ]
...
```

3.5.2 PROTOCOL PARTICIPANTS

Protocol participants may act as either an initiator or a responder in a protocol run. The rule `R_1` specifies a responder's receipt of the first message of the protocol and their subsequent sending of the second message of the protocol.

```
...
rule R_1:
let m1 = aenc{'1', ni, I}pk(ltkR)
m2 = aenc{'2', ni, ~nr, $R}pkI
in
[ !Ltk($R, ltkR), In( m1 ), !Pk(I, pkI), Fr(~nr)]
--[ IN_R_1_ni( ni, m1 ), OUT_R_1( m2 ),
Running(I, $R, <'init',ni,~nr>)]->
[ Out( m2 ), St_R_1($R, I, ni, ~nr) ]
...
```

If `aenc{'1',ni,I}pk(ltkR)` is received then `aenc{'2',ni,~nr,$R}pkI` is sent in response. Rule `R_2` specifies that upon receipt of the final message `aenc{'3', nr}pk(ltkR)` both `nr` and `ni` are claimed to be secret.

```
...
rule R_2:
```

```
[ St_R_1(R, I, ni, nr), !Ltk(R, ltkR),
In( aenc{'3', nr}pk(ltkR) ) ]
--[ Commit(R, I, <'resp',ni,nr>)]->
[ Secret(R,I,nr), Secret(R,I,ni) ]
...
```

When the final message aenc{'3', nr}pk(ltkR) is received by the responder R, supposing they already sent the earlier for this protocol run and they know the secret key corresponding to the public key used to encrypt the message, then they commit to the fact they have concluded a run of the protocol as a responder with I using nonces ni and nr and claims secrecy of both of the nonces.

In a preceding rule I_2, the initiator similarly commits to having concluded a run of the protocol as an initiator using nonces ni and nr and claims secrecy of both of the nonces.

3.5.3 SPECIFYING PROTOCOL PROPERTIES

```
...
rule Secrecy_claim:
[ Secret(A, B, m) ] --[ Secret(A, B, m) ]-> []
...
```

The Secrecy_claim rule announces that some secret m is claimed to be secret Secret(A,B,s), by consuming this fact generated by one of the rules specifying a step of the protocol, e.g., rule R_2. The nonce_secrecy lemma to be proven expresses that it cannot be the case that a message claimed to be kept secret from the intruder is in fact known by the intruder.

```
...
lemma nonce_secrecy:
" /* It cannot be that */
not(
Ex A B s #i.
/* somebody claims to have setup a shared secret, */
Secret(A, B, s) @ i
/* but the adversary knows it */
& (Ex #j. K(s) @ j)
)"
end
```

The injective agreement, as modeled below, specifies that whenever a protocol participant (*i.e.*, actor) commits to running a session then the other party must have already

Running and Commit signal events are introduced into the protocol description to mark the points at which a protocol participant asserts authentication to have been initiated and successfully achieved, respectively. The Running and Commit signal events are parametrized by the actor that is making the assertion, the peer with which actor understands they are running the protocol and certain parameters specific to the protocol runparams upon which actor agrees with peer.

To authenticate to a protocol participant a that some other participant b was involved in a particular protocol run, participant a requires that:

1. b was alive, *i.e.*, participated in some recent communication
2. b participated in a run taking a to be the other participant, and
3. b agrees on certain parameters specific to the protocol run

```
// Injective agreement from the perspective of both
// the initiator and the responder.
lemma injective_agree:
" /* Whenever somebody commits to running a session, then*/
All actor peer params #i.
Commit(actor, peer, params) @ i
==>
/* there is somebody running a session with
the same parameters */
(Ex #j. Running(actor, peer, params) @ j & j < i
/* and there is no other commit on the
same parameters */
& not(Ex actor2 peer2 #i2.
Commit(actor2, peer2, params) @ i2 & not(#i = #i2)
)
)
"
```

3.5.4 RESULTS

Tamarin can run both in console mode and in a graphical web interface. Executing input files with Tamarin will produce proofs or attack traces for properties expressed as lemmas (see Section 3.5.3). In command-line mode, proofs will either appear as *verified*, meaning a secure result, or as *falsified - found trace*, meaning that an attack has been found. In the case of the GUI, the proof for each lemma will appear colored in green or red depending on the verdict and an image representing the attack trace will open if found.

```
summary of summaries:

analyzed: examples/classic/NSPK3.spthy

types (all-traces): verified (33 steps)
nonce_secrecy (all-traces): falsified - found trace (16 steps)
injective_agree (all-traces): falsified - found trace (14 steps)
session_key_setup_possible (exists-trace): verified (5 steps)
```

Above is the results of analysing the NSPK, indicating that the secrecy of B regarding the nonce `Na` is violated (`nonce_secrecy`), and the injective authentication of A to B is false (`injective_agree`). The second result shows that Bob may think that he has completed a protocol run with Alice, while Alice has never stated a session with Bob earlier implying the violation of the injective authentication of Alice and Bob, and hence Lowe's attack on the NSPK.

Attack traces are fairly complicated to understand with Tamarin. However, thanks to the GUI pictures, an easy to follow explanation of the attack is provided (see Figure 3.2 for the attack on secrecy of `Na`).

3.6 AVISPA

AVISPA (Automated Validation of Internet Security Protocols and Applications) is a command-line cryptographic protocol verifying framework proposed in 2005 [3]. It is the successor of the AVISS (Automated Validation of Infinite-State Systems) tool designed in 2002 [4]. It has the originality of being composed of four backend protocol analyzers: CL-Atse [59], OFMC [14], SATMC [15], and TA4SP [19]. These four back-end analyzers share the same specification language so that the same protocol model may be analyzed by either of them. Moreover, each of the tools relies on different formal methods and thus has different capabilities from the others:

- CL-Atse (Constraint-Logic-based Attack Searcher), developed by M. Turuani, analyzes protocols over a finite set of sessions, translating traces into constraints. Constraints are simplified thanks to heuristics and redundancy elimination techniques allowing to decide whether secrecy and authentication properties have been violated or not.
- OFMC (Open-source Fixed point Model-Checker), developed by S. Mödersheim, applies symbolic analysis alongside model-checking to perform protocol falsification and bounded analysis also over a finite set of sessions. The state space is explored in a

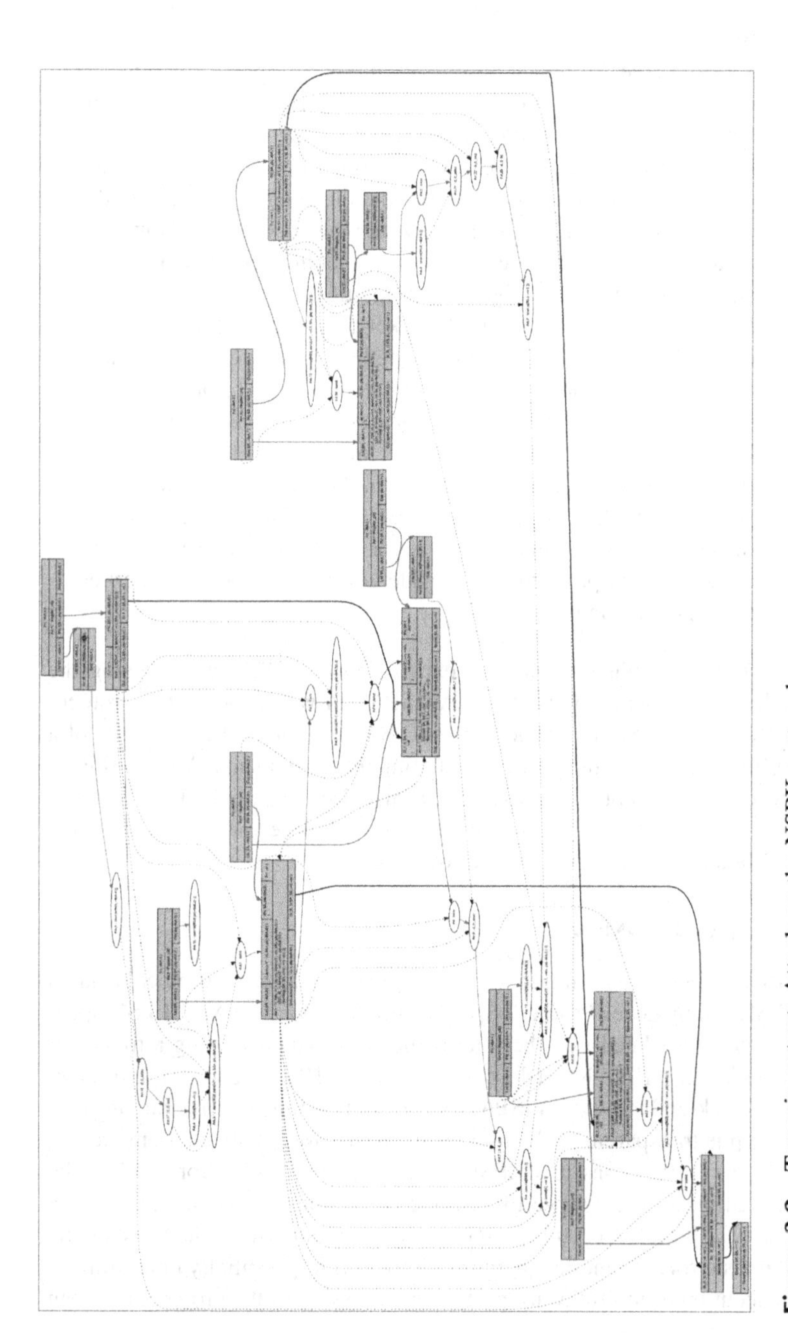

Figure 3.2 Tamarin output: Attack on the NSPK protocol.

demand-driven way to verify secrecy and authentication properties.

- TA4SP (Tree Automata based on Automatic Approximations for the Analysis of Security Protocols), developed by Y. Boichut, approximates the intruder knowledge through the use of regular tree rewriting. For secrecy properties, it supports both over-approximation and under-approximation to show that either the protocol is flawed or safe for any number of sessions. However, no attack trace is provided by the tool and only the secrecy is considered in presence of algebraic properties.
- SATMC (SAT-based Model-Checker for Security Protocols) reduces the problem of checking whether a protocol is vulnerable to attacks of bounded length to the satisfiability of a propositional formula which is then solved by a state-of-the-art SAT solver. This is done by combining a reduction technique of protocol insecurity problems to planning problems and SAT-reduction techniques developed for planning and LTL that allows for leveraging state-of-the-art SAT solvers. It allows for verification of secrecy and authentication properties alongside complex temporal properties (e.g. fair exchange).

In the following sections, we explain how AVISPA works using the NSPK protocol as a toy example. AVISPA can verify the NSPK protocol in various complexities of the protocol. Here, we present an analysis of a simplified version of the NSPK. Assuming the participants, Alice and Bob, know each other's public key (omitted trusted server) before the execution of the protocol, the following sections describe the formalization of the simplified (3-message) the NSPK protocol.

3.6.1 ENVIRONMENT

AVISPA relies on a common ad-hoc language to describe protocols called HLPSL (High Level Protocol Specification Language). An HLPSL specification is translated into the Intermediate Format (IF), using a translator called hlpsl2if. IF is a lower-level language than HLPSL and is read directly by the back-ends to the AVISPA Tool. Note that this intermediate translation step is transparent to the user, as the translator is called automatically. Protocol specifications in HLPSL are divided into roles. Some roles (the so-called basic roles) serve to describe the actions of one single agent in a run of a protocol or sub-protocol. Others (composed roles) instantiate these basic roles to model an entire protocol run (potentially consisting of the execution of multiple sub-protocols), a session of the protocol between multiple agents, or the protocol model itself. This latter role is often called the Environment role. If this Environment role, one will first instantiate all protocol constants such as agents and keys. Then will be defined constants with type protocol_id. These specific constants will act as goals for the intruder (secrets to obtaining or messages to usurpation) and will be bound to actual variables during the protocol. Then intruder's initial knowledge is defined, followed by the protocol as a composition of sessions playing in parallel, one session corresponding to the normal execution of the protocol with fixed participants.

```
role environment() def=
const a, b          : agent,
ka, kb, ki    : public_key,
na, nb,
alice_bob_nb,
bob_alice_na : protocol_id

intruder_knowledge = {a, b, ka, kb, ki, inv(ki)}

composition
session(a,b,ka,kb)
/\ session(a,i,ka,ki)
/\ session(i,b,ki,kb)
end role
```

3.6.2 SYSTEM COMPOSITION

The session role allows defining a single execution of the protocol between some participants. At this stage, participants are already variables as sessions are instantiated in the Environment role shown earlier. In the case of the NSPK, a session will imply the executing once the role of Alice alongside the role of Bob. Channels will be defined as variables local to the session. Within AVISPA, channels are unidirectional and specific for each role. Thus a bidirectional channel between Alice and Bob will be represented by four AVISPA channels: SA for Alice to send, RA for Alice to receive, SA for Bob to send, and finally RB for Bob to receive.

```
role session(A, B: agent, Ka, Kb: public_key) def=
local SA, RA, SB, RB: channel (dy)

composition
alice(A,B,Ka,Kb,SA,RA)
```

```
/\ bob  (A,B,Ka,Kb,SB,RB)
end role
```

3.6.3 PROTOCOL PARTICIPANTS

The protocol is divided into sub-processes (roles) representing the behavior of each participant, interacting with their sending and receiving channels. HLPSL is based on temporal logic. As such, we generally model protocols by describing the system state and then specifying the ways in which that state may change. The description of the state changes is then given by specifying so-called transition predicates which relate the values of variables in one state (intuitively, the current one) and another (the future, or next state). The current state of the protocol in the role is stored as a natural variable usually called State. Also, HLPSL refers to the variables in the next state as *primed variables*: for a variable X, X refers to its value in the previous state and X' to its value in the current.

Then, the role is described as a set of transitions between states. According to the definition of the NSPK protocol, as an initiator, Alice obtains her and Bob's public key. Then, upon starting a call from the tool, she creates a fresh random number `Na'`, encrypts it with Bob's public key, and sends the ciphertext to Bob over the channel `SND`. She then receives a message `{Na, Nb'}_Ka` encrypted with her public key, where specifying Na will require the first term to match her known version of Na and Nb′ being whatever value she receives as the second term. Finally, she encrypts it with Bob's public key, `Kb`, and sends the ciphertext to Bob. Bob's role will be written similarly. The secret keyword allows to bind the actual freshly generated value of Na′ to the protocol_id variable na defined in the Environment role. This will be used later in the security properties section.

```
role alice (A, B: agent,
Ka, Kb: public_key,
SND, RCV: channel (dy))
played_by A def=
local State : nat,
Na, Nb: text

init State := 0

transition

0.  State  = 0 /\ RCV(start) =|>
```

```
State':= 2 /\ Na' := new() /\ SND({Na'.A}_Kb)
/\ secret(Na',na,{A,B})
/\ witness(A,B,bob_alice_na,Na')

2.  State  = 2 /\ RCV({Na.Nb'}_Ka) =|>
State':= 4 /\ SND({Nb'}_Kb)
/\ request(A,B,alice_bob_nb,Nb')
end role
```

3.6.4 SPECIFYING PROTOCOL PROPERTIES

AVISPA has a fixed definition of properties (referenced as goals for the intruder) and does not allow a number of security protocol properties to be modeled and verified automatically using reachability, correspondence assertions, and observational equivalence, like more recent tools. AVISPA supports *secrecy* using the secrecy_of keyword and *authentication* using the authentication_on keyword. Both of these keywords apply on protocol_id terms defined in the protocol. Secrecy will be broken if AVISPA finds a trace where at some point the intruder obtains the variable referenced. As for other tools, to verify authentication properties, the events for each participant need to be declared. Keywords witness and request respectively describe when a participant acknowledges *sending* and *receiving* some term from/to another participant. Then, AVISPA will try to find a trace where a request is not preceded by its matching witness. After specifying the goals, a call to the environment role will launch the whole analysis.

```
goal
secrecy_of na, nb
authentication_on alice_bob_nb
authentication_on bob_alice_na
end goal

environment()
```

3.6.5 INTRUDER MODEL

The way AVISPA works prevents modeling how intruder behaves in the protocol, and its interactions with other protocol participants. AVISPA provides a standard powerful Dolev-Yao intruder model, able to access all public channels, and perform malicious actions on the messages flowing over the public channels. It can construct and decompose messages seen as well as read, modify, delete and inject them. The intruder can also perform cryptographic operations when required cryptographic keys are in its possession. Intruder's knowledge is clearly specified in the environment role, for instance allowing dishonest participants by sharing their private data with the intruder.

3.6.6 RESULTS

Executing input files with AVISPA produces query results showing whether the queries, such as; secrecy of `Na` and mutual authentication above are satisfied (the query is true or false). Unlike ProVerif, AVISPA performs under-approximations, thus, when AVISPA finds an attack on a security protocol, then its model guarantees that the attack is correct, but not all attacks are guaranteed to be found. An exception to this is TA4SP which also performs over-approximations. However, TA4SP only handles secrecy, does not provide attack traces, and can be inconclusive. To the best of our knowledge, we were not able to obtain results with SATMC (it is worth noting that AVISPA was developed in the early 2000s for 32 bits architectures and is not maintained since). Below are shown the results of the NSPK protocol for OFMC (left), CL-Atse (center), and TA4SP (right). Both OFMC and CL-Atse find the usual attack on the NSPK and TA4SP is inconclusive.

```
$ avispa nspk.hspl --ofmc
% OFMC
% Version of 2006/02/13

SUMMARY
UNSAFE

DETAILS
ATTACK_FOUND

PROTOCOL
/path/to/nspk.hspl.if

GOAL
secrecy_of_nb

BACKEND
OFMC

COMMENTS

STATISTICS
parseTime: 0.00s
searchTime: 0.02s
visitedNodes: 10 nodes
depth: 2 plies

ATTACK TRACE
i -> (a,6): start
(a,6) -> i: {Na(1).a}_ki
i -> (b,3): {Na(1).a}_kb
(b,3) -> i: {Na(1).Nb(2)}_ka
i -> (a,6): {Na(1).Nb(2)}_ka
(a,6) -> i: {Nb(2)}_ki
i -> (i,17): Nb(2)
i -> (i,17): Nb(2)
```

```
$ avispa nspk.hspl --cl-atse
SUMMARY
UNSAFE

DETAILS
ATTACK_FOUND
TYPED_MODEL

PROTOCOL
/path/to/nspk.hspl.if

GOAL
Secrecy attack on (n5(Nb))

BACKEND
CL-AtSe

STATISTICS
analyzed    : 9 states
Reachable   : 8 states
Translation: 0.00 seconds
Computation: 0.00 seconds

ATTACK TRACE
i -> (a,6):  start
(a,6) -> i:  {n9(Na).a}_ki
& Secret(n9(Na),set_74)

i -> (a,3):  start
(a,3) -> i:  {n1(Na).a}_kb
& Secret(n1(Na),set_62)
& Witness(a,b,bob_alice_na,n1(Na))

i -> (b,4):  {n9(Na).a}_kb
(b,4) -> i:  {n9(Na).n5(Nb)}_ka
& Secret(n5(Nb),set_70)
& Witness(b,a,alice_bob_nb,n5(Nb));

i -> (a,6):  {n9(Na).n5(Nb)}_ka
(a,6) -> i:  {n5(Nb)}_ki
```

```
$ avispa nspk.hspl --ta4sp
SUMMARY
INCONCLUSIVE

DETAILS
OVER_APPROXIMATION
UNBOUNDED_NUMBER_OF_SESSIONS
TYPED_MODEL

PROTOCOL
/path/to/nspk.hspl.if.ta4sp

GOAL
SECRECY
Property with identitier: nb

BACKEND
TA4SP

COMMENTS
Use an under-approximation
in order to show a potential
attack.
The intruder might know some
critical information.

STATISTICS
Translation: 0.00 seconds
Computation 0.44 seconds

ATTACK TRACE
No Trace can be provided
with the current version.
```

3.7 MAUDE-NPA

Maude-NPA is a tool developed in 2006 [36]. It relies on the rewriting tool Maude to model protocols, intruder capabilities and security properties. This approach allows to model secrecy and authentication properties. Users needs to understand the foundations of this approach in order to design correct models. This offers a flexibility to prove the security. In Maude the basic units of specification and programming are called modules. A module consists of syntax declarations, providing appropriate language to describe the system at hand, and of statements, asserting the properties of such a system. In the context of protocol verification, Maude-NPA requires functional modules. Functional modules define data types and operations on them by means of equational theories.

In the following sections, we explain how Maude-NPA works using the NSPK protocol as a toy example. Detailed explanations can be found in Maude-NPA's manual [9]. Maude-NPA can verify the NSPK protocol in various complexities of the protocol. Here, we present an analysis of a simplified version of the NSPK. Assuming the participants, Alice and Bob, know each other's public key (omitted trusted server) before the execution of the protocol, the following subsections describe the formalization of the simplified (3-message) NSPK protocol.

3.7.1 ENVIRONMENT

The fist part is to define the environment for the equational properties to model public encryption as it is required in NSPK. Then we need to define the type of messages for instances the keys, nonces and pairs of messages. All of these information are defined as functional modules as mentioned before. Equational properties for encryption and decryption are defined using the `eq` keyword. The `[variant]` attribute specifies that these equations should not be used as regular Maude equations, typically used for simplification, but are instead equations to be used for the variant-based unification algorithm available in Maude. On the other hand, functions signatures are defined as operators with the `[frozen]` operator. This operator will mainly prevent Maude to explore irrelevant paths. The `frozen` attribute must be included in all operator declarations in Maude-NPA specifications, excluding constants and macros. User types are defined as `sorts`. Predefined types also exist: `Msg` is the main message type and can be inherited from. `Fresh` is used to describe nonces. Finally, `Public` is used to identify terms that are publicly available, and therefore assumed known by the intruder.

[9]http://maude.cs.illinois.edu/w/images/9/90/Maude-NPA˙manual˙v3˙1.pdf

```
fmod PROTOCOL-EXAMPLE-ALGEBRAIC is
protecting PROTOCOL-EXAMPLE-SYMBOLS .

------------------------------------------------------
--- Overwrite this module with the algebraic
--- properties of your protocol
------------------------------------------------------

var Z : Msg .
var Ke : Key .

*** Encryption/Decryption Cancellation
eq pk(Ke,sk(Ke,Z)) = Z [variant] .
eq sk(Ke,pk(Ke,Z)) = Z [variant] .

endfm

fmod PROTOCOL-EXAMPLE-SYMBOLS is
--- Importing sorts Msg, Fresh, Public, and GhostData
protecting DEFINITION-PROTOCOL-RULES .

------------------------------------------------------
--- Overwrite this module with the syntax of your
--- protocol
--- Notes:
--- * Sort Msg and Fresh are special and imported
--- * Every sort must be a subsort of Msg
--- * No sort can be a supersort of Msg
------------------------------------------------------

--- Sort Information
sorts Name Nonce Key .
subsort Name Nonce Key < Msg .
subsort Name < Key .
subsort Name < Public .

--- Encoding operators for public/private encryption
op pk : Key Msg -> Msg [frozen] .
op sk : Key Msg -> Msg [frozen] .

--- Nonce operator
```

```
op n : Name Fresh -> Nonce [frozen] .

--- Associativity operator
op _;_ : Msg  Msg  -> Msg [gather (e E) frozen] .
```

3.7.2 PROTOCOL PARTICIPANTS AND SYSTEM COMPOSITION

Then we have to define honest participants (Alice and Bob) and the intruder. They are defined in the symbol module with types and other constants and use the `Name` type defined above. Yet, these are not all the possible principal names. Since Maude-NPA is an unbounded session tool, the number of possible principals is unbounded. This is achieved by using variables of type `Name` instead of constants.

```
--- Principals
op a : -> Name . --- Alice
op b : -> Name . --- Bob
op i : -> Name . --- Intruder
endfm
```

The next step is to define the protocol, itself, by defining variables used in the protocol description, relying on formerly defined types. And then describing the role of Alice and the role of Bob. The protocol itself and the intruder capabilities are both specified in the PROTOCOL-SPECIFICATION module. They are specified using strands or processes. Here we give a brief introduction to specifying protocol strands. The symbol + means that the message is outputted on the network (controlled by the intruder in the Dolev-Yao intruder model) and the symbol - means that a term is received by the participant. A vertical bar | is used to distinguish between present and future when the strand appears in a state description. All messages appearing before the bar were sent/received in the past, and all messages appearing after the bar will be sent/received in the future. When a strand is used in a protocol specification as opposed to a state description the bar is irrelevant, and by convention it is assumed to be at the beginning of the strand, right after the initial `nil`.

```
fmod PROTOCOL-SPECIFICATION is
protecting PROTOCOL-EXAMPLE-SYMBOLS .
protecting DEFINITION-PROTOCOL-RULES .
protecting DEFINITION-CONSTRAINTS-INPUT .

-------------------------------------------------------
```

```
--- Overwrite this module with the strands
--- of your protocol
----------------------------------------------------------

var Ke : Key .
vars X Y Z : Msg .
vars r r' : Fresh .
vars A B : Name .
vars N N1 N2 : Nonce .

eq STRANDS-PROTOCOL
= :: r ::
[ nil | +(pk(B,A; n(A,r))), -(pk(A,n(A,r); N)), +(pk(B, N)),
 nil ] &
:: r ::
[ nil | -(pk(B,A; N)), +(pk(A, N; n(B,r))), -(pk(B,n(B,r))),
 nil ]
[nonexec] .
```

3.7.3 INTRUDER MODEL

The intruder is also modeled by rewriting rules. the first ones are modeling the pairing properties, after we have the encryption and decryption rules and finally the fact that the intruder controls the network and learns all the messages sent. As usual, the deduction rules include tuple making, left and right projection, encryption, signature and term generation. Unlike other existing tools, there are no built-in Dolev-Yao model in Maude-NPA. It is possible to fully customize intruder's behavior besides writing additional equational theories.

```
eq STRANDS-DOLEVYAO
= :: nil :: [ nil | -(X), -(Y), +(X ; Y), nil ] &
:: nil :: [ nil | -(X ; Y), +(X), nil ] &
:: nil :: [ nil | -(X ; Y), +(Y), nil ] &
:: nil :: [ nil | -(X), +(sk(i,X)), nil ] &
:: nil :: [ nil | -(X), +(pk(Ke,X)), nil ] &
:: nil :: [ nil | +(A), nil ]
[nonexec] .
```

3.7.4 SPECIFYING PROTOCOL PROPERTIES

Finally, we have to model the secrecy properties and the authentication properties in the Maude language as follows. Each property contains several sections separated by the symbol ||. Only the first two sections can be filled in and correspond, respectively, to the attack state's expected set of strands and expected intruder knowledge. The other sections usually have the symbol nil. The first section containing the attacker's expected final states usually corresponds to a normal execution of the protocol. Intruder's final knowledge is usually used in case of secrecy properties. It is also possible to include never patterns in properties. Such a pattern describes a strand that must not be encountered during protocol execution. It is generally used for authentication properties. For example, suppose that we want to find a state in which Alice has executed an instance of a protocol, apparently with Bob, but Bob has not executed the corresponding instance with Alice. This can be specified using the following attack pattern with a never pattern.

```
eq ATTACK-STATE(0)
= :: r ::
[ nil, -(pk(b,a; N)), +(pk(a,N; n(b,r))), -(pk(b,n(b,r)))
| nil ]
|| n(b,r) inI, empty
|| nil
|| nil
|| nil
[nonexec] .

eq ATTACK-STATE(1)
= :: r ::
[ nil, -(pk(b,a; N)), +(pk(a,N; n(b,r))), -(pk(b,n(b,r)))
| nil ]
|| empty
|| nil
|| nil
|| never *** for authentication
(:: r' ::
[ nil, +(pk(b,a; N)), -(pk(a,N; n(b,r))) | +(pk(b,n(b,r))),
nil ]
& S:StrandSet
|| K:IntruderKnowledge)
[nonexec] .

endfm

--- THIS HAS TO BE THE LAST LOADED MODULE !!!!
select MAUDE-NPA .
```

3.7.5 RESULTS

Executing input files with Maude-NPA produces attack states results showing whether the queries, such as; secrecy of `Nb` and mutual authentication above are satisfied. Either an attack state will be found in the execution tree or all possible executions will end without any violating the properties. In case of an attack, a strand visualization is produced.

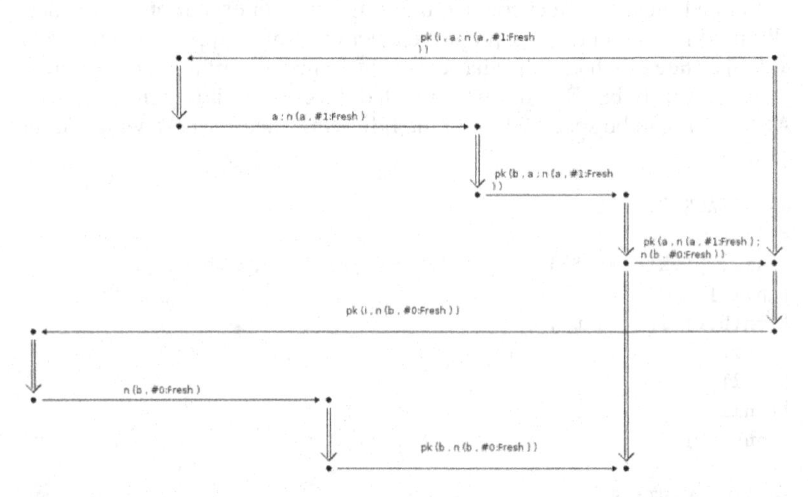

Figure 3.3 Strand visualization of attack given by Maude-NPA.

Figure 3.3 displays the attack found for the NSPK protocol. Following the arrows will show the usual attack introduced in Section 3.1.1.1. Above are the results of analysing the NSPK, indicating that the secrecy of B regarding the nonce `Na` is violated, and the injective authentication of A to B is false. The second result shows that Bob may think that he has completed a protocol run with Alice, while Alice has never stated a session with Bob earlier implying the violation of the injective authentication of Alice and Bob, and hence Lowe's attack on the NSPK.

REFERENCES

1. SAPIC+: protocol verifiers of the world, unite! In *31st USENIX Security Symposium (USENIX Security 22)*, Boston, MA, Aug. 2022. USENIX Association.

2. H. Al Hamadi, C. Yeun, J. Zemerly, M. Al-Qutayri, and A. Gawanmeh. Verifying mutual authentication for the DLK protocol using ProVerif tool. *International Journal for Information Security Research*, 3, 03 2013.

3. A. Armando, D. A. Basin, Y. Boichut, Y. Chevalier, L. Compagna, J. Cuéllar, P. H. Drielsma, P. Héam, O. Kouchnarenko, J. Mantovani, S. Mödersheim, D. von Oheimb, M. Rusinowitch, J. Santiago, M. Turuani, L. Viganò, and L. Vigneron. The AVISPA tool for the automated validation of internet security protocols and applications. In K. Etessami and S. K. Rajamani, editors, *Computer Aided Verification, 17th International Conference, CAV 2005, Edinburgh, Scotland, UK, July 6-10, 2005, Proceedings*, volume 3576 of *Lecture Notes in Computer Science*, pages 281–285. Springer, 2005.

4. A. Armando, D. A. Basin, M. Bouallagui, Y. Chevalier, L. Compagna, S. Mödersheim, M. Rusinowitch, M. Turuani, L. Viganò, and L. Vigneron. The AVISS security protocol analysis tool. In E. Brinksma and K. G. Larsen, editors, *Computer Aided Verification, 14th International Conference, CAV 2002,Copenhagen, Denmark, July 27-31, 2002, Proceedings*, volume 2404 of *Lecture Notes in Computer Science*, pages 349–353. Springer, 2002.

5. A. Armando and L. Compagna. SATMC: a SAT-based model checker for security protocols. In *European Workshop on Logics in Artificial Intelligence*, pages 730–733. Springer, 2004.

6. M. Backes, C. Hritcu, and M. Maffei. Automated verification of remote electronic voting protocols in the applied pi-calculus. In *CSF*, 2008.

7. M. Barbosa, G. Barthe, K. Bhargavan, B. Blanchet, C. Cremers, K. Liao, and B. Parno. Sok: Computer-aided cryptography. In *42nd IEEE Symposium on Security and Privacy, SP 2021, San Francisco, CA, USA, 24-27 May 2021*, pages 777–795. IEEE, 2021.

8. D. Basin and C. Cremers. Modeling and analyzing security in the presence of compromising adversaries. In *Computer Security - ESORICS 2010*, volume 6345 of *Lecture Notes in Computer Science*, pages 340–356. Springer, 2010.

9. D. Basin and C. Cremers. Know your enemy: Compromising adversaries in protocol analysis. *ACM Transactions on Information and System Security*, 17(2):7:1–7:31, Nov. 2014.

10. D. Basin, C. Cremers, and S. Meier. Provably repairing the ISO/IEC 9798 standard for entity authentication. In P. Degano and J. D. Guttman, editors, *Principles of Security and Trust - First International Conference, POST 2012, Held as Part of the European Joint Conferences on Theory and Practice of Software, ETAPS 2012, Tallinn, Estonia, March 24 - April 1, 2012, Proceedings*, volume 7215 of *Lecture Notes in Computer Science*, pages 129–148. Springer, 2012.

11. D. Basin, S. Radomirovic, and L. Schmid. Alethea: A provably secure random sample voting protocol. In *2018 IEEE 31st Computer Security Foundations Symposium (CSF)*, pages 283–297, 2018.

12. D. Basin, R. Sasse, and J. Toro-Pozo. Card brand mixup attack: Bypassing the PIN in non-Visa cards by using them for visa transactions. In *30th USENIX Security Symposium (USENIX Security 21)*, pages 179–194. USENIX Association, Aug. 2021.

13. D. A. Basin and C. J. Cremers. Degrees of security: Protocol guarantees in the face of compromising adversaries. In *Computer Science Logic, 24th International Workshop, CSL 2010, 19th Annual Conference of the EACSL, Brno, Czech Republic, August 23-27, 2010. Proceedings*, volume 6247 of *Lecture Notes in Computer Science*, pages 1–18. Springer, 2010.

14. D. A. Basin, S. Mödersheim, and L. Viganò. OFMC: A symbolic model checker for security protocols. *International Journal of Information Security*, 4(3):181–208, 2005.

15. D. A. Basin, R. Sasse, and J. Toro-Pozo. The EMV standard: Break, fix, verify. In *42nd IEEE Symposium on Security and Privacy, SP 2021, San Francisco, CA, USA, 24-27 May 2021*, pages 1766–1781. IEEE, 2021.

16. K. Bhargavan, B. Blanchet, and N. Kobeissi. Verified models and reference implementations for the tls 1.3 standard candidate. In *2017 IEEE Symposium on Security and Privacy (SP)*, pages 483–502. IEEE, 2017.

17. C. V. Birjoveanu. Secrecy for bounded security protocols: Disequality tests and an intruder with existentials lead to undecidability. In *Proceedings of the 2009 Fourth Balkan Conference in Informatics*, BCI '09, page 22–27, USA, 2009. IEEE Computer Society.

18. B. Blanchet, B. Smyth, V. Cheval, and M. Sylvestre. *ProVerif 2.03:Automatic Cryptographic Protocol Verifier,User Manual and Tutorial.*

19. Y. Boichut and T. Genet. Feasible trace reconstruction for rewriting approximations. In F. Pfenning, editor, *Term Rewriting and Applications, 17th International Conference, RTA 2006, Seattle, WA, USA, August 12-14, 2006, Proceedings*, volume 4098 of *Lecture Notes in Computer Science*, pages 123–135. Springer, 2006.

20. A. M. Bruni, E. Drewsen, and C. Schürmann. Towards a mechanized proof of selene receipt-freeness and vote-privacy. In *E-VOTE-ID*, 2017.

21. R. Chadha, V. Cheval, Ş. Ciobâcă, and S. Kremer. Automated verification of equivalence properties of cryptographic protocols. *ACM Transactions on Computational Logic*, 17(4), sep 2016.

22. V. Cheval. Apte: An algorithm for proving trace equivalence. In E. Ábrahám and K. Havelund, editors, *Tools and Algorithms for the Construction and Analysis of Systems*, pages 587–592, Berlin, Heidelberg, 2014. Springer Berlin Heidelberg.

23. V. Cheval, S. Kremer, and I. Rakotonirina. Deepsec: Deciding equivalence properties in security protocols theory and practice. In *2018 IEEE Symposium on Security and Privacy (SP)*, pages 529–546, 2018.

24. V. Cortier, A. Dallon, and S. Delaune. Sat-equiv: An efficient tool for equivalence properties. In *2017 IEEE 30th Computer Security Foundations Symposium (CSF)*, pages 481–494, 2017.

25. C. Cremers. Feasibility of multi-protocol attacks. In *Proc. of The First International Conference on Availability, Reliability and Security (ARES)*, pages 287–294, Vienna, Austria, April 2006. IEEE Computer Society.

26. C. Cremers. Key exchange in IPsec revisited: formal analysis of IKEv1 and IKEv2. In *Proceedings of the 16th European conference on Research in computer security*, ESORICS, pages 315–334, Berlin, Heidelberg, 2011. Springer-Verlag.

27. C. Cremers and M. Dehnel-Wild. Component-based formal analysis of 5g-aka: Channel assumptions and session confusion. In *26th Annual Network and Distributed System Security Symposium, NDSS 2019, San Diego, California, USA, February 24-27, 2019*. The Internet Society, 2019.

28. C. Cremers and M. Horvat. *Improving the ISO/IEC 11770 Standard for Key Management Techniques*, pages 215–235. Springer International Publishing, Cham, 2014.

29. C. Cremers, M. Horvat, S. Scott, and T. van der Merwe. Automated analysis and verification of tls 1.3: 0-rtt, resumption and delayed authentication. In *2016 IEEE Symposium on Security and Privacy (SP)*, pages 470–485, 2016.

30. C. Cremers, B. Kiesl, and N. Medinger. A formal analysis of IEEE 802.11's WPA2: Countering the kracks caused by cracking the counters. In *29th USENIX Security Symposium (USENIX Security 20)*, pages 1–17. USENIX Association, Aug. 2020.

31. C. J. F. Cremers. The scyther tool: Verification, falsification, and analysis of security protocols. In A. Gupta and S. Malik, editors, *Computer Aided Verification, 20th International Conference, CAV 2008, Princeton, NJ, USA, July 7-14, 2008, Proceedings*, volume 5123 of *Lecture Notes in Computer Science*, pages 414–418. Springer, 2008.

32. C. J. F. Cremers. Unbounded verification, falsification, and characterization of security protocols by pattern refinement. In P. Ning, P. F. Syverson, and S. Jha, editors, *Proceedings of the 2008 ACM Conference on Computer and Communications Security, CCS 2008, Alexandria, Virginia, USA, October 27-31, 2008*, pages 119–128. ACM, 2008.

33. C. J. F. Cremers, P. Lafourcade, and P. Nadeau. Comparing state spaces in automatic security protocol analysis. In V. Cortier, C. Kirchner, M. Okada, and H. Sakurada, editors, *Formal to Practical Security - Papers Issued from the 2005-2008 French-Japanese Collaboration*, volume 5458 of *Lecture Notes in Computer Science*, pages 70–94. Springer, 2009.

34. D. Dolev and A. Yao. On the security of public key protocols. *IEEE Transactions on Information Theory*, 29(2):198–208, 1983.

35. B. Donovan, P. Norris, and G. Lowe. Analyzing a library of security protocols using casper and fdr. In *Proceedings of the Workshop on Formal Methods and Security Protocols*, pages 36–43. Citeseer, 1999.

36. S. Escobar, C. Meadows, and J. Meseguer. A rewriting-based inference system for the nrl protocol analyzer and its meta-logical properties. *Theoretical Computer Science*, 367(1):162–202, 2006. Automated Reasoning for Security Protocol Analysis.

37. S. Escobar, C. Meadows, and J. Meseguer. *Maude-NPA: Cryptographic Protocol Analysis Modulo Equational Properties*, pages 1–50. Springer Berlin Heidelberg, Berlin, Heidelberg, 2009.

38. P. Gardiner, M. Goldsmith, J. Hulance, D. Jackson, B. Roscoe, B. Scattergood, and B. Armstrong. FDR2 user manual. http://www.fsel.com/documentation/fdr2/html/index.html.

39. A. González-Burgueño, S. Santiago, S. Escobar, C. Meadows, and J. Meseguer. Analysis of the ibm cca security api protocols in maude-npa. In L. Chen and C. Mitchell, editors, *Security Standardisation Research*, pages 111–130, Cham, 2014. Springer International Publishing.

40. A. González-Burgueño, S. Santiago, S. Escobar, C. Meadows, and J. Meseguer. Analysis of the pkcs#11 api using the maude-npa tool. In L. Chen and S. Matsuo, editors, *Security Standardisation Research*, pages 86–106, Cham, 2015. Springer International Publishing.

41. C. A. R. Hoare. Communicating sequential processes. *Communications of the ACM*, 21:666–677, August 1978.

42. K. Karaduzovic-Hadziabdic. modeling and analysing the tls protocol using casper and fdr. In *2012 IX International Symposium on Telecommunications (BIHTEL)*, pages 1–6. IEEE, 2012.

43. N. Kobeissi, K. Bhargavan, and B. Blanchet. Automated verification for secure messaging protocols and their implementations: A symbolic and computational approach. In *2017 IEEE European Symposium on Security and Privacy (EuroS&P)*, pages 435–450. IEEE, 2017.

44. R. Küsters and T. Truderung. Using proverif to analyze protocols with diffie-hellman exponentiation. In *2009 22nd IEEE Computer Security Foundations Symposium*, pages 157–171. IEEE, 2009.

45. P. Lafourcade and M. Puys. Performance evaluations of cryptographic protocols verification tools dealing with algebraic properties. In J. García-Alfaro, E. Kranakis, and G. Bonfante, editors, *Foundations and Practice of Security - 8th International Symposium, FPS 2015, Clermont-Ferrand, France, October 26-28, 2015, Revised Selected Papers*, volume 9482 of *Lecture Notes in Computer Science*, pages 137–155. Springer, 2015.

46. P. Lafourcade, V. Terrade, and S. Vigier. Comparison of cryptographic verification tools dealing with algebraic properties. In P. Degano and J. D. Guttman, editors, *Formal Aspects in Security and Trust, 6th International Workshop, FAST 2009, Eindhoven, The Netherlands, November 5-6, 2009, Revised Selected Papers*, volume 5983 of *Lecture Notes in Computer Science*, pages 173–185. Springer, 2009.

47. G. Lowe. An attack on the Needham-Schroeder public-key authentication protocol. *Information Processing Letters*, 56(3):131–133, 1995.

48. G. Lowe. Casper: a compiler for the analysis of security protocols. In *Proceedings 10th Computer Security Foundations Workshop*, pages 18–30, Jun 1997.

49. G. Lowe. A hierarchy of authentication specifications. In *Proceedings of the 10th IEEE Workshop on Computer Security Foundations*, CSFW '97, page 31, USA, 1997. IEEE Computer Society.

50. S. Meier, B. Schmidt, C. Cremers, and D. Basin. The TAMARIN Prover for the Symbolic Analysis of Security Protocols. In N. Sharygina and H. Veith, editors, *Computer Aided Verification, 25th International Conference, CAV 2013, Princeton, USA, Proc.*, volume 8044 of *Lecture Notes in Computer Science*, pages 696–701. Springer, 2013.

51. M. Moran, J. Heather, and S. Schneider. Verifying anonymity in voting systems using CSP. *Formal Aspects of Computing*, pages 1–36, 2012.

52. M. Moran, J. Heather, and S. Schneider. Automated verification of Three-Ballot and VAV voting systems. *Software and Systems Modeling Journal*, 2015.

53. M. Moran and D. S. Wallach. Verification of star-vote and evaluation of FDR and proverif. *CoRR*, abs/1705.00782, 2017.

54. R. M. Needham and M. D. Schroeder. Using encryption for authentication in large networks of computers. *Communications of the ACM*, 21(12), Dec. 1978.

55. P. Y. A. Ryan, S. A. Schneider, M. H. Goldsmith, G. Lowe, and A. W. Roscoe. *The modeling and Analysis of Security Protocols : the CSP Approach*. Addison-Wesley Professional, first edition, 2000.

56. B. Smyth. *Formal Verification of Cryptographic Protocols with Automated Reasoning*. PhD thesis, School of Computer Science, University of Birmingham, 2011.

57. R. A. Soltwisch. *The Inter-domain Key Exchange Protocol: A Cryptographic Protocol for Fast, Secure Session-key Establishment and Re-authentication of Mobile Nodes After Inter-domain Handovers*. PhD thesis, 2006.

58. A. Tiu and J. Dawson. Automating open bisimulation checking for the spi calculus. In *2010 23rd IEEE Computer Security Foundations Symposium*, pages 307–321, 2010.

59. M. Turuani. The CL-Atse Protocol analyzer. In F. Pfenning, editor, *Term Rewriting and Applications*, pages 277–286, Berlin, Heidelberg, 2006. Springer Berlin Heidelberg.

60. J. Zhang, L. Yang, W. Cao, and Q. Wang. Formal Analysis of 5G EAP-TLS Authentication Protocol Using ProVerif. *IEEE Access*, 8:23674–23688, 2020.

4 Formal Verification of Cryptographic Protocols with Isabelle/HOL

Pasquale Noce
HID Global, Italy

CONTENTS

Starting from a historical example, namely the Needham-Schroeder public key protocol, this chapter discusses the modeling of cryptographic

DOI: 10.1201/9781003090052-4

protocols as infinite state machines to achieve the highest possible confidence in their actual enforcement of expected security properties. Accordingly, the verification of such properties demands the construction of mathematical proofs, based on generalized mathematical induction, in a logical calculus. The number and the subtlety of the cases arising in those proofs, along with the required logical calculus expressiveness, render interactive theorem provers, such as Isabelle/HOL, the only viable option. This chapter illustrates two methods allowing for formal protocol verification according to the above approach, both based on Isabelle/HOL, namely Paulson's inductive method and the author's own relational method, by considering their application to the Needham-Schroeder-Lowe protocol and to a sample PAKE protocol, respectively, and detailing how they enable to model relevant real-world entities (agents, messages, attacker's capabilities, protocol rules and properties).

4.1 INTRODUCTION

This section introduces the reasons for verifying cryptographic protocols modeled as infinite state machines, as well as the tools required for achieving this goal, namely interactive theorem provers and generalized mathematical induction.

4.1.1 LESSONS FROM HISTORY: THE NEEDHAM-SCHROEDER PUBLIC KEY PROTOCOL

It is often said, and rightly so, that history is an authoritative teacher (though unfortunately just as often unheard). If possible, this discussion will provide still another confirmation of that aphorism by starting with a very instructive historical example.

In 1978 [15], Needham and Schroeder proposed a protocol, henceforth referred to as the *Needham-Schroeder public key protocol*, making use of public key cryptography with the aim of enabling any two parties communicating over a network:

- to perform a mutual authentication, and
- to share a pair of nonces known to them only, for example usable to generate a session key.

This protocol assumes that each of the two involved parties, say Alice and Bob, has:

- a publicly known identifier, namely A and B, respectively, and

- an asymmetric key pair, namely (SK_A, PK_A) and (SK_B, PK_B), respectively.

In addition to Alice and Bob, the protocol includes a trusted authentication server, whose role is to supply each participant with the other one's public key. Once these preparatory operations are over, Alice and Bob are able to perform their own handshake, which provides for the following steps.

1. Alice $\rightarrow$ Bob: $\{N_A, A\}_{PK_B}$

   Alice sends Bob a message consisting of her nonce N_A and her identifier A, encrypted with Bob's public key PK_B.

2. Bob $\rightarrow$ Alice: $\{N_A, N_B\}_{PK_A}$

   Bob replies with a message consisting of Alice's nonce N_A and his own nonce N_B, encrypted with Alice's public key PK_A.

3. Alice $\rightarrow$ Bob: $\{N_B\}_{PK_B}$

   Finally, Alice replies with Bob's nonce N_B encrypted with Bob's public key PK_B.

In their paper, Needham and Schroeder wisely remarked that such a protocol may be prone to extremely subtle logical errors, as such unlikely to be detected by empirical means, rather demanding suitable correctness verification techniques. However, for a long time, intuition came out in favor of the correctness of their public key protocol: after all, if neither private key is compromised, Alice and Bob are the only ones who can decrypt the messages addressed to them, which seems to leave no room for attacks.

Unfortunately, in 1995 [11] (nearly two decades later!), Lowe found this protocol to be vulnerable to a man-in-the-middle insider attack. Let Chuck be a party enabled to take part in the protocol, as such endowed with an identifier C and a key pair (SK_C, PK_C), but willing to act as an insider. To accomplish his malicious intent, he just has to wait for some other party, say Alice, to contact him, in which case he can misbehave as follows.

1. Alice $\rightarrow$ Chuck: $\{N_A, A\}_{PK_C}$

   Alice wants to talk with Chuck, so she generates nonce N_A and sends the initial message to him, encrypting it with Chuck's public key

PK_C.

2. Chuck → Bob: $\{N_A, A\}_{PK_B}$

   Chuck rather wants to deceive Bob, so he decrypts Alice's message with his own private key SK_C, re-encrypts it with Bob's public key PK_B, and forward it to Bob.

3. Bob → Alice: $\{N_A, N_B\}_{PK_A}$

   Bob sees Alice's identifier A in the received message. So, he generates nonce N_B and sends his reply to Alice, encrypting it with her public key PK_A.

4. Alice → Chuck: $\{N_B\}_{PK_C}$

   Alice sees nonce N_A, which she addressed to Chuck, in the received message. Thus, she sends her reply to Chuck, encrypting it with his public key PK_C.

5. Chuck → Bob: $\{N_B\}_{PK_B}$

   Finally, again, Chuck decrypts Alice's latest message with his own private key SK_C, re-encrypts it with Bob's public key PK_B, and forward it to Bob.

Now Bob, trusting the protocol, mistakenly believes that Alice has authenticated to him and shares secret nonces N_A, N_B with him alone, whereas actually, Alice never authenticated to Bob and the nonces are known to Chuck, too. In other words, Chuck's insider attack has succeeded in subverting both security goals of the Needham-Schroeder public key protocol, at Bob's expense.

To perform his attack, Chuck needs to be able to progressively extend the set of the messages of which he has knowledge by means of the following atomic operations.

1. Decryption of Alice's former message $\{N_A, A\}_{PK_C}$ with Chuck's own private key SK_C, resulting in Chuck's knowledge of plaintext $\{N_A, A\}$.

2. Split of compound message $\{N_A, A\}$ into its components, resulting in Chuck's knowledge of Alice's nonce N_A.
3. Encryption of message $\{N_A, A\}$ with Bob's public key PK_B, resulting in Chuck's knowledge of ciphertext $\{N_A, A\}_{PK_B}$.
4. Decryption of Alice's latter message $\{N_B\}_{PK_C}$ with Chuck's own private key SK_C, resulting in Chuck's knowledge of Bob's nonce N_B.
5. Encryption of nonce N_B with Bob's public key PK_B, resulting in Chuck's knowledge of ciphertext $\{N_B\}_{PK_B}$.

Therefore, Lowe's attack on the Needham-Schroeder public key protocol entails five transitions of the attacker's knowledge, which on the whole require the attacker to be capable of encrypting or decrypting messages with any key he knows, as well as of splitting compound messages.

In [11], Lowe did not stop with describing his attack, but also suggested a correction, namely to include Bob's identifier B in Bob's reply. The resulting fixed protocol, referred to as the *Needham-Schroeder-Lowe (public key) protocol*, looks as follows.

1. Alice $\rightarrow$ Bob: $\{N_A, A\}_{PK_B}$
2. Bob $\rightarrow$ Alice: $\{N_A, N_B, B\}_{PK_A}$
3. Alice $\rightarrow$ Bob: $\{N_B\}_{PK_B}$

If Chuck tries the same attack, Alice will receive message $\{N_A, N_B, B\}_{PK_A}$ from Bob, whereas the expected format for the reply was $\{N_A, N_B, C\}_{PK_A}$, because N_A is the nonce that she had addressed to Chuck. Thus, she will abandon the run, without ever sending her reply $\{N_B\}_{PK_C}$, and eventually so will Bob, as he never receives the final acknowledgment $\{N_B\}_{PK_B}$. Hence, Lowe's attack does not work any longer.

4.1.2 FORMAL PROTOCOL VERIFICATION AND INTERACTIVE THEOREM PROVERS

Even though Lowe's attack is no longer applicable, this is not sufficient for concluding that the Needham-Schroeder-Lowe protocol is secure, namely that unless the participants' private keys are compromised, it achieves both of the above security goals. Actually, the protocol might still be vulnerable to some other attack. The only way to *prove* its security, in the mathematical sense of the term, is building a mathematical model of the protocol allowing for any kind and any number of computationally feasible transitions of the attacker's knowledge, and then verifying that even so, there does not exist any reachable state where some security goal fails to be met.

This comes at the cost of having to deal with an *infinite* state machine, which rules out the possibility to exhaustively check every reachable state, as one does with model checking, rather demanding the construction of mathematical proofs in some logical calculus. What is more, such a powerful attacker might seem a waste, insofar as he could unlimitedly extend his knowledge even in evidently unfruitful ways. For example, the capability to concatenate any known messages without restriction would enable the attacker to do something so silly as building limitless repetitions of an agent's identifier, such as $\{A,A\}$, $\{A,A,A\}$, and so on.

Nonetheless, these drawbacks are made necessary by the unjustifiable arbitrariness of any constraint possibly introduced to prevent them. Given a finite set, however much large and accurately defined, of possible attacker's knowledge evolutions, even if none of them turned out to lead to a successful attack, who on earth could be certain that no other possible evolution, among the infinite ones left out, would do so instead? This concern is made even stronger by the bitter observation that logical flaws in cryptographic protocols may be very subtle and hard to find with an empirical approach, as manifested by the fact that Lowe's attack was published *17 years after* the Needham-Schroeder public key protocol had been proposed – a long-term deadline for real-world attackers to devise and exploit that attack on their own in the meantime!

Still another lesson taught by this story is that an adequate (if demanding, as just seen) degree of confidence in protocol security proofs can be achieved much more effectively using *mechanical theorem provers* – namely, computer programs enabling their users to prove theorems –, rather than traditional, pen-and-paper proofs. Probably, the best confirmation of this statement comes precisely from practicing formal protocol verification with some theorem prover. In fact, the number and the subtlety of the cases, quite often unexpected, springing up in such formal proofs, which indeed are effects of the hardness of logical flaw detection, clearly show how likely it would be to overlook some of those cases if one rather embarked on a pen-and-paper proof.

The ideal mechanical theorem prover would be entirely automatic: one would give it a mathematical formula, and it would return a proof of that formula in case it is a theorem, or otherwise it would report with a shrug that no proof exists. Unfortunately, automatic theorem provers can be built only for basic logical calculi, such as *propositional logic*. In fact, already in *first-order logic*, the problem of deciding whether formulae are theorems turns out to be generally undecidable, namely there exists no algorithm solving it for any given formula (cf. [27], section 6.2). As if that were not enough, even in propositional logic this problem, though being decidable, is computationally intractable for large input formulae, since no algorithm solving it in polynomial time is known. In fact, the problem of deciding the validity of propositional formulae is equivalent to the *Boolean satisfiability (SAT) problem*, which is NP-complete (cf. [27], section 7.4). For the reader willing to know more about classical propositional/first-order logic, a good reference is [9].

In practice, the construction of mathematical models usually requires a logical calculus at least as expressive as first-order logic, and the task of modeling cryptographic protocols, which as seen above is by no means trivial, is certainly no exception. Consequently, a mechanical theorem prover adequate to this task cannot but be an *interactive theorem prover*, or *proof assistant*, which rather requires human-machine collaboration.

In what follows, this discussion will focus on formal protocol verification with *Isabelle*, a generic proof assistant enabling to express mathematical formulae in a formal language and to prove them in a logical calculus. In the interaction with Isabelle, the user's role is to input formal definitions and non-automatic formal proof steps, using the powerful *Isabelle/jEdit* Prover IDE and the human-readable, structured proof language *Isar*. In turn, Isabelle automatically checks the correctness of each step, and offers a wide range of *proof tactics* enabling to accomplish proof steps amenable to automatic processing.

Isabelle has been applied to outstanding mathematical developments, such as that of a formal proof published in 2015 [6] by Hales et al. eventually proving the *Kepler conjecture*, a geometry problem remained unsolved for centuries and included among Hilbert's problems. Another example is the first ever machine-assisted proof, published in 2013 [26] by Paulson, of Gödel's second incompleteness theorem. As regards the computer security domain, the *Common Criteria for Information Technology Security Evaluation* [4] explicitly mention Isabelle among the formal systems usable to generate the developer's formal specifications required for IT product certifications at the highest *Evaluation Assurance Levels (EALs)*.

Several instances of Isabelle are available, differing in the logical calculus used to express and prove mathematical formulae. The most commonly used instance, which is also the one on which the formal protocol verification methods discussed in what follows are based, is *Isabelle/HOL*, whose underlying logical calculus, namely *higher-order logic*, is even more expressive than first-order logic. Further information on Isabelle/HOL can be found in the related documentation, which is updated whenever new Isabelle versions are released. Particularly, for the latest version at the time of writing, namely Isabelle2021-1, [16] and [28] are excellent starting points for beginners.

A good survey of the various mechanical theorem provers currently available, including Isabelle/HOL, can be found in [14]. Isabelle/HOL is based on λ-calculus with simple types and natural deduction, and supports (co)datatypes, (co)inductive definitions, and recursive functions with complex pattern matching. Like Isabelle/HOL, the *HOL* family of interactive theorem provers uses classical higher-order logic. *Coq* rather uses *intuitionistic logic*, which differs from classical logic in not including the laws of excluded middle and double negation elimination and is based on the Calculus of Inductive Constructions. Isabelle/HOL, HOL systems, and Coq are endowed with large mathematical libraries and satisfy the *de Bruijn criterion*, namely proof correctness relies upon a small checker (referred to as the *proof kernel*). An example of an interactive theorem prover not adhering to this criterion is *ACL2*, which is based on classical first-order logic.

4.1.3 GENERALIZED MATHEMATICAL INDUCTION

As observed previously, if a cryptographic protocol is modeled as an infinite state machine, its properties cannot be verified by checking every single reachable state – at least if one wishes to complete sooner or later! Indeed, mathematics does provide a different, powerful tool to reason about infinite sets, namely *mathematical induction*. Its most basic form is the traditional one applying to the set of natural numbers: if a given property holds for m, and for any $n \geq m$ it holds for $n + 1$ if it does for n, then that property holds for every $n \geq m$.

Of course, this form of induction cannot be applied to protocols, since their states are not natural numbers. Luckily, it turns out to be just a (notable) specialization of a far more general form, which can be applied to mathematical objects of any sort and captures the following idea. Imagine a floor wide enough to accommodate infinite dominoes, each of which may either stand or have fallen. You can see the dominoes closest to you, so you know which ones are standing and which ones have fallen, but there are infinite farther ones out of reach, such as one named b. How could you prove that b has fallen? Well, no matter how far b is, it would suffice to know that there exists some finite sequence of dominoes leading from a fallen one a near you up to b, such that any two consecutive dominoes x, y are so close that x's fall causes y to fall, too.

Although this image might seem too picturesque to have ever occurred to a mathematician, it renders well a fairly general form of induction, namely induction over a *well-founded relation* (cf. [17], section 6.4). Actually, it is the basis for several specialized induction rules applying to various kinds of mathematical objects, including traditional mathematical induction over natural numbers, in which case the well-founded relation is the one binding each natural number to its successor.

4.2 THE INDUCTIVE METHOD

This section illustrates the inductive method for formal protocol verification by considering its application to the Needham-Schroeder-Lowe protocol [22] [17].

4.2.1 INDUCTIVELY DEFINED SETS AND THE INDUCTIVE METHOD

As a result of the previous considerations, the way forward for modeling protocols as infinite state machines is to represent their states as mathematical objects allowing for the use of some extended form of induction deriving from induction over a well-founded relation. An example of such objects is the elements of an *inductively defined set*, which is the least set closed under some given rules. For instance, the set of even numbers can be defined inductively as the least set containing 0 and closed under the increment of any one of its elements by 2 (cf. [17], section 7.1). In Isabelle/HOL, this concept can be formalized via the following inductive definition, which consists of two *introduction rules* named *zero* and *step*.

inductive-set *even* :: *nat set* **where**
zero: $0 \in even$ |
step: $n \in even \Longrightarrow Suc\ (Suc\ n) \in even$

In more detail, the header line specifies that a set of natural numbers, named *even*, is being defined inductively. Rule *zero* states that 0 is contained in *even*, whereas rule *step* states that, if any natural number n is contained in *even*, then so is $n + 2$ (function *Suc* takes a natural number as input and returns its successor, so $Suc\ (Suc\ n) = n + 2$). As a result, $0 \in even$ by rule *zero*, then $2 \in even$ by rule *step*, then $4 \in even$ again by rule *step*, and so on for any other even number.

Rule *step* also provides an example of the following important, general features of Isabelle/HOL syntax, which will repeatedly occur throughout the remainder of this chapter.

- Variable n is a *free variable*, which means that every occurrence of n within the rule can be substituted for an arbitrary natural number, and the resulting formula still holds. Hence, by universally quantifying over n, a true formula would be obtained as well.

However, it would not be *operationally* equivalent to the original rule, because a variable bound by a quantifier, unlike a free one, cannot be replaced with an actual value – indeed, to do so, the quantified variable has to be substituted for a free one by applying some valid rule.

- The $\Longrightarrow$ arrow separates the *assumptions* of a rule from its *conclusion*. Rule *step* contains a single assumption, $n \in even$, and its conclusion is $Suc\ (Suc\ n) \in even$. Thus, it expresses that, if the assumption $n \in even$ holds for a given value of n, then the conclusion $Suc\ (Suc\ n) \in even$ also is true for that same value of n.

Interestingly enough, as a result of any such inductive definition, Isabelle/HOL automatically generates a corresponding induction rule, which in the case of inductive set *even* is named *even.induct* and looks as follows.

$$[\![x \in even;\ P\ 0;\ \bigwedge n.\ [\![n \in even;\ P\ n]\!] \Longrightarrow P\ (Suc\ (Suc\ n))]\!] \Longrightarrow P\ x$$

In Isabelle/HOL syntax, a rule containing multiple assumptions can be written by enclosing all its assumptions in a bracket pair $[\![$ and $]\!]$ and separating them from each other with semicolons. Accordingly, rule *even.induct* contains three assumptions, namely $x \in even$, $P\ 0$, and $\bigwedge n.\ [\![n \in even;\ P\ n]\!] \Longrightarrow P\ (Suc\ (Suc\ n))$, and the conclusion $P\ x$. In turn, the third assumption contains two assumptions, $n \in even$ and $P\ n$, and the conclusion $P\ (Suc\ (Suc\ n))$. Variables P and x are free, like n in rule *step*, whereas the variable n occurring in the third assumption is bound: prefix $\bigwedge n.$ means "for an arbitrary but fixed n". Like in first-order logic, except for the lack of parentheses enclosing x (in Isabelle/HOL syntax, atomic function arguments need not be enclosed in parentheses), formula $P\ x$ means "property P holds for term x".

For any natural number x satisfying its *major premise* $x \in even$ (cf. [17], section 5.7), rule *even.induct* enables to prove $P\ x$ by proving its remaining assumptions, namely $P\ 0$ and $\bigwedge n.\ [\![n \in even;\ P\ n]\!] \Longrightarrow P\ (Suc\ (Suc\ n))$. More precisely, this rule can be applied to infer $P\ x$ for any x such that $x \in even$, and as a result of its application, some statements are left to be proven, named *subgoals* in the Isabelle jargon and consisting of the remaining assumptions $P\ 0$ and $\bigwedge n.\ [\![n \in even;\ P\ n]\!] \Longrightarrow P\ (Suc\ (Suc\ n))$, one for each introduction rule in the inductive definition of set *even*. The same holds in general, for any inductive set S: rule *S.induct* can be applied to deduce $P\ x$ from $x \in S$, and as a result of its application, the remaining assumptions in *S.induct*, one for each introduction rule contained in the definition of S, constitute as many subgoals left to be proven.

Rule *even.induct* states that property P holds for any even number x if it holds for 0, as well as for $Suc\ (Suc\ n)$ assuming that it does for even number n. Hence, it can easily be viewed as a specialization of the previous "dominoes" induction rule, where the property of having fallen is replaced by P, a by 0, the arbitrary consecutive dominoes x, y by n, $Suc\ (Suc\ n)$, and b by x. In the Isabelle jargon, the application of such an induction rule, arising from an inductive definition, is referred to as *rule induction* (cf. [17], section 7.1.3).

Thus, modeling the possible states of a cryptographic protocol as the elements of some appropriate inductively defined set, and the possible state transitions as the related introduction rules, would enable to verify the protocol properties by means of rule induction. In the late nineties, Paulson developed this idea to the point of devising a general method for formal protocol verification, the *inductive method*, which he successfully applied to a number of classic protocols, such as the Needham-Schroeder-Lowe and Otway-Rees protocols [22] [17], the Yahalom protocol [24] [25], Kerberos version 4 [1], and TLS version 1.0 [23].

Particularly, the Needham-Schroeder-Lowe protocol is considered in what follows, since it can serve as a gentle introduction to the general features of the inductive method thanks to its simplicity. A sample formalization of this protocol according to the inductive method is also included in the Isabelle2021-1 distribution [21].

4.2.2 AGENTS, MESSAGES, EVENTS, STATES

First of all, how many legitimate participants should be modeled? Perhaps just two, Alice and Bob? On closer inspection, such a model would not agree with the overall aim of the protocol, stated in the very title of [15]: enabling "authentication in large networks of computers". But how many network nodes should exactly be considered for the model to be realistic? Well, since the model is not constrained to be finite, the emergence of a sorites paradox can be prevented by simply considering an infinite population of friendly agents, one for each natural number, plus an adverse agent, briefly referred to as the *spy*.

datatype *agent* = *Friend nat* | *Spy*

This definition introduces a new type, named *agent*, whose values can be built using either of two *constructors*, *Friend* and *Spy*. The former one, used to model friendly agents, is declared as a function taking a natural number as input and returning a value of type *agent*; thus, friendly agents are formalized as terms *Friend 0*, *Friend 1*, and so on. The latter constructor is a nullary function used to model the spy; that is, the spy is simply formalized as term *Spy*. A universal property of

is that they are injective functions with pairwise disjoint ranges; in the case of type *agent*, this entails that $Friend\ i \neq Friend\ j$ for any $i \neq j$ and $Friend\ i \neq Spy$ for any i.

Apparently, such a definition might risk to result in an unrealistic model, as it seems to rule out insider attacks a priori. For example, how could the protocol be proven to resist Lowe's attack if every potential participant, including Chuck, is supposed to be friendly? However, this issue can be solved by taking care to define the spy's capabilities so as to include everything that legitimate agents can do – actually, much more: everything that is computationally feasible. The convenience of modeling the spy as a distinct agent is that in this way, a state breaching the protocol security goals is simply one where the spy comes to know something that should have been secret, or manages to forge something that should have been genuine.

Providing the spy with maximum computational power also makes it sufficient to deal with a single attacker in the model. Of course, the number of the attackers to be faced does matter in the real world: for example, some hacking organization could afford more computational power than a single hacker. Nonetheless, if the spy is modeled as an attacker endowed with maximum computational power, extending the model with further adverse agents would not augment the overall attack capabilities taken into consideration.

Keys and nonces, like agents, are infinite and are identified by natural numbers. Messages comprise agent identifiers, nonces, keys, cryptograms, and compound messages built via message concatenation. Notation $\{\!|X_1, \ldots, X_{n-1}, X_n|\!\}$ stands for $MPair\ X_1\ (\ldots\ (MPair\ X_{n-1}\ X_n))$.

type-synonym *key* = *nat*

datatype *msg* =
Agent agent |
Nonce nat |
Key key |
Crypt key msg |
MPair msg msg

Since *Crypt*, like any other type constructor, is an injective function, any two cryptograms $Crypt\ K\ X$, $Crypt\ K'\ X'$ match just in case $K = K'$ and $X = X'$. Again, this might seem an inappropriate constraint making the model unrealistic, since in real-world cryptosystems there usually exist huge numbers of different encryption key-plaintext pairs resulting in identical cryptograms. For example, in the Elgamal public key cryptosystem over an integer finite field (cf. [8], section 2.4), for each value of the encryption nonce, there exist q distinct such pairs producing the same cryptogram, where q is the generator order – so quite a lot, considering that a standard magnitude for q is 2^{256}!

Yet, cryptogram collisions are no use for the spy. In fact, given a cryptogram, he has to decrypt it with the correct private key to recover the plaintext, otherwise he would get nothing but garbage. Likewise, given a public key, he has to encrypt a correct plaintext to forge an unknown cryptogram, whose corresponding collisions are unknown as well. Of course, forging a known cryptogram would make sense just to detect the correct plaintext as the one resulting in that same cryptogram, but again, the correct public key alone can be used for this purpose, thus collisions are still useless. The moral is that modeling cryptograms via a type constructor does not entail any significant limitation on the spy's capabilities.

Another potential objection, this time targeting the very choice of modeling message syntax through such a type definition, is that the ranges of any two type constructors are disjoint, while messages of different kinds may match in practice. For instance, if their bit lengths are the same, there is a nonzero probability that a nonce generated at random will match an agent identifier. However, even in this case, both the sender and the recipient of such a message will interpret it unambiguously, based on the protocol rules: if, say, a nonce is expected, they will handle it as such all the same. Hence, notwithstanding its apparent shortcomings, this way of formalizing message syntax is in fact fully adequate, and then preferable to possible alternatives by virtue of its simplicity.

In any state machine, transitions between states are brought about by relevant *events*, so it comes as no surprise that this should happen for a state machine modeling a cryptographic protocol, too. The model described in [22] and [17] uses a single kind of event, namely the sending of a message X by an agent A to an intended recipient B, represented as event *Says* A B X according to the following type definition. Of course, if more kinds of event were needed, as many additional type constructors would have to be appended to this definition.

datatype *event* = *Says agent agent msg*

From events to states, it is a short step: states are simply modeled as *event traces*, namely event lists, which keep track of any occurred event in reverse chronological order. That is to say, the occurrence of event *ev* in state *evs*, if allowed, brings the state machine into the new state constructed by prefixing *ev* to list *evs*. As explained previously, Isabelle/HOL syntax uses infix operator # to denote the addition of an element to the front of a list; as a result, such new state can be written as *ev # evs*.

4.2.3 SPY'S CAPABILITIES

The spy's capabilities in learning and generating messages, respectively, are modeled through two inductively defined functions, *analz* and *synth*, both mapping a message set *H* to a larger one. In more detail, *analz H* and *synth H* contain every message *X* such that it is computationally feasible for one knowing every message in *H* to learn/generate *X*. Here below are their formal definitions, where *invKey* is a function that maps the numeric identifiers of public keys to those of the respective private keys.

inductive-set *analz* :: *msg set* ⇒ *msg set* **for** *H* :: *msg set* **where**
Inj: $X \in H \Longrightarrow X \in analz\ H$ |
Fst: $\{\!|X, Y|\!\} \in analz\ H \Longrightarrow X \in analz\ H$ |
Snd: $\{\!|X, Y|\!\} \in analz\ H \Longrightarrow Y \in analz\ H$ |
Decrypt: $[\![Crypt\ K\ X \in analz\ H;\ Key\ (invKey\ K) \in analz\ H]\!] \Longrightarrow X \in analz\ H$

inductive-set *synth* :: *msg set* ⇒ *msg set* **for** *H* :: *msg set* **where**
Inj: $X \in H \Longrightarrow X \in synth\ H$ |
Agent: $Agent\ A \in synth\ H$ |
MPair: $[\![X \in synth\ H;\ Y \in synth\ H]\!] \Longrightarrow \{\!|X, Y|\!\} \in synth\ H$ |
Crypt: $[\![X \in synth\ H;\ Key\ K \in H]\!] \Longrightarrow Crypt\ K\ X \in synth\ H$

Of course, the spy has learned and can generate any message he already knows (rules *analz.Inj*, *synth.Inj*). He can split compound messages into their components (rules *Fst*, *Snd*), as well as decrypt a cryptogram if he knows the private key associated with the public key used for encryption (rule *Decrypt*). Moreover, he can generate agent identifiers, all of which are supposed to be publicly known (rule *Agent*) and any combination of known messages (rule *MPair*), and he can encrypt any known message with any public key (rule *Crypt*).

A state is an event list, whereas functions *analz* and *synth* rather take a message set as input. Hence, there is a need for another function that takes an event list *evs* as input and returns the set of all the messages exchanged in *evs*, in turn usable as an input to *analz* and *synth*. In fact, all message exchanges are supposed to take place on a public, insecure network, so that any exchanged message is susceptible to be intercepted by the spy.

This is precisely the role played by function *knows*, defined by recursion on the input event list. The base case, namely the empty list one, refers to the *initial state* where no event has occurred yet. Thus, the corresponding function output contains any message that the spy knows in advance, either because it is public or by means other than attacking the protocol (for example, as a result of his own insider knowledge or of some social engineering attack). In the present case, the spy's initial knowledge comprises all public keys and the private keys of compromised agents, if any, which are modeled as the elements of agent set *bad*. As the spy, like any other agent, knows his own private key, this set will always contain at least the spy.

Given two functions *priK* and *pubK* mapping each agent to the numeric identifier of her private/public key, function *knows* can be defined as follows (notation *f* ‘ *A* denotes the image of set *A* under function *f*).

fun *initState* :: *agent* ⇒ *msg set* **where**
initState Spy = *Key* ‘ *range pubK* ∪ *Key* ‘ *priK* ‘ *bad*

fun *knows* :: *agent* ⇒ *event list* ⇒ *msg set* **where**
knows Spy [] = *initState Spy* |
knows Spy (*Says A B X* # *evs*) = *insert X* (*knows Spy evs*)

abbreviation *spies* :: *event list* ⇒ *msg set* **where**
spies ≡ *knows Spy*

The above header lines starting with keyword *fun* specify that two recursive functions, named *initState* and *knows*, are being defined. Both of them return a message set; the former one takes an agent as its sole input, whereas the latter one takes two inputs, an agent and an event list. Nonetheless, the only significant value for the parameter of type *agent*, at least in the case of this protocol, is *Spy* for both functions – in fact, abbreviation *spies* is introduced precisely as a shorter alias for function *knows Spy*.

The *fun* headings are followed by *equations* that, as happens in functional programming languages, specify the values taken by the functions in correspondence with their possible inputs, distinguished by pattern matching over type constructors. Since the only significant input agent is *Spy*, no equation is needed for the other constructor *Friend* of type *agent*. On the contrary, the definition of function *knows* contains an equation for each of the two constructors of polymorphic type *'a list*, namely *Nil* and *Cons*, respectively consisting of the empty list and the function adding an element to the front of a list, and denoted by symbol [] and infix operator #. The latter equation is recursive on the input event list and uses function *insert*, which adds the element taken as its first input to the set taken as its second input and returns the resulting, augmented set.

Based on the previous definitions, the sets of the messages that the spy can learn/generate in state *evs* are expressed as $analz\ (spies\ evs)$ and $synth\ (analz\ (spies\ evs))$, respectively.

The introduction of functions *analz* and *synth* goes in the direction of decoupling the formalization of the spy's capabilities from that of the protocol itself, consistently with the fact that they only depend on what is computationally feasible for the spy, and then are independent of how the protocol works, which just determines whether they are sufficient to breach the protocol security goals. In principle, this way of proceeding promises to provide a considerable benefit: one could define these functions and prove their properties just once, whereupon such definitions and properties could be reused over and over again to verify many different protocols, in the same way as a functionally independent software module can be reused in multiple software systems.

However, as both function definitions depend on type *msg*, this benefit only applies as long as the message formats used to model the target protocols remain unchanged. Indeed, the subsequent sections will show that in practice, when it comes to verifying a protocol making use of public key cryptography features more complex than mere encryption and decryption, some ad hoc message format, as well as some related spy's capability, are likely to be required. In any such case, applying the *analz-synth* approach would then result in facing the surplus complexity arising from two more inductive definitions, in addition to that of the protocol itself, without gains in terms of reuse.

Another disadvantage of this approach is the distinction between what the spy can learn and what he can generate. A real-world attacker is free to send any message he has learned to any other party, and conversely to use any message he has generated, such as nonces, session keys, or ephemeral key pairs, to further extend his knowledge. Hence, a fully realistic model should be such that in every protocol state, the sets of the messages that the spy can learn/generate should match. According to the current approach, property $H \subseteq synth\ H$, satisfied by function *synth* for any message set H, entails that $analz\ (spies\ evs) \subseteq synth\ (analz\ (spies\ evs))$, namely the spy can generate anything he can learn, but the converse fails to hold, since in general $H \subset synth\ H$.

This drawback is usually negligible for a protocol making use of static keys only, like the Needham-Schroeder-Lowe one, since such a protocol does not use any session key whose computation requires knowledge of other one-time messages generated on the fly, such as nonces, ephemeral key pairs, or one-time passwords. On the contrary, the drawback becomes serious if session keys are used, as happens in key establishment protocols, even in those making use of symmetric cryptography alone. One might suppose that once the spy has generated a one-time message, it will be contained in $analz\ (spies\ evs)$ for any subsequent event trace evs, but this may be true only if that message is exchanged, which is usually not the case for session keys or ephemeral private keys, and even so not necessarily. For instance, if the spy exchanges a fresh nonce after encrypting it with the public key of an uncompromised agent, the set $analz\ (spies\ evs)$ associated with the resulting event trace evs will not contain that nonce, since the decryption private key is unknown to the spy.

An additional problem emerges by observing that the definition of function *synth* does not include any introduction rule enabling the spy to generate a fresh nonce, something quite strange for a protocol where doing so is a breeze for legitimate participants. Indeed, unless this gap is filled in some way, they would turn out to have more computational power than the spy himself in this respect, which would be at odds with the requirement to allow for the spy doing all that is computationally feasible.

At first glance, this might be taken for nothing but an oversight, easily remediable by extending *synth*'s definition with an additional introduction rule such as $Nonce\ N \notin H \Longrightarrow Nonce\ N \in synth\ H$, where the freshness of the generated nonce should be ensured by assumption $Nonce\ N \notin H$. However, not only would this trick render function *synth* no longer monotone – namely, property $H \subseteq H' \Longrightarrow synth\ H \subseteq synth\ H'$ would cease to hold –, but it would badly miss its target. In fact, a nonce $Nonce\ N$ exchanged by any two uncompromised agents in the event trace evs will even so not be contained in $analz\ (spies\ evs)$, thus it will pass the alleged freshness test $Nonce\ N \notin analz\ (spies\ evs)$ though having already been used. While nonces may be encrypted with uncompromised keys when exchanged, ephemeral private keys are usually not exchanged at all, so that such a trick would be flawed for any kind of one-time private message.

The only working alternative would rather be to change the signature of function *synth*, for example by adding a second input message set deputed to contain *all used messages*, not only those known to the spy or being exchanged, or else by replacing the input message set H with event trace evs itself. Yet, as a sad result, the definition of function *synth*, as well as its properties and their proofs, would get much more convoluted – a heavy price to pay for something apparently so simple!

4.2.4 PROTOCOL RULES

Leaving aside the spy for the moment, those who cannot but have the capability to generate fresh nonces, considering how the protocol works, are its legitimate participants. Particularly, the model cannot do without an effective freshness test for their nonces, on pain of enabling the spy to break the protocol upon any reuse of some compromised nonce – a quite unrealistic attack pattern, as long as nonces are picked out at random from a sufficiently wide range of permitted values. In turn, such a test will consist of verifying that the alleged nonce is not included among the messages occurring in the current event trace *evs*, thus there is a need for some mechanism enabling to collect all those messages.

This is achieved by introducing a further inductively defined function, named *parts*. Like *analz* and *synth*, this function maps a message set H to a larger one, which in this case consists of all the components of the messages in H. Function *parts* is then employed in the recursive definition of still another function, named *used*, which takes an event list *evs* as input and returns the set of all the messages occurring in *evs*, whether as stand-alone messages, as components of compound messages, or within encrypted plaintexts.

The definitions of functions *parts* and *used* complete the formal toolbox required for the inductive definition of the set *ns-public* of event traces modeling the protocol itself, named after the Needham-Schroeder-Lowe public key protocol. The mentioned definitions are reported all together here below.

inductive-set *parts* :: *msg set* ⇒ *msg set* **for** H :: *msg set* **where**
Inj: $X \in H \Longrightarrow X \in parts\ H$ |
Fst: $\{\!|X, Y|\!\} \in parts\ H \Longrightarrow X \in parts\ H$ |
Snd: $\{\!|X, Y|\!\} \in parts\ H \Longrightarrow Y \in parts\ H$ |
Body: $Crypt\ K\ X \in parts\ H \Longrightarrow X \in parts\ H$

fun *used* :: *event list* ⇒ *msg set* **where**
$used\ [] = parts\ (initState\ Spy)$ |
$used\ (Says\ A\ B\ X\ \#\ evs) = parts\ \{X\} \cup used\ evs$

inductive-set *ns-public* :: *event list set* **where**

Nil:
$[] \in ns\text{-}public$ |

Fake:
  $[\![evsf \in ns\text{-}public; X \in synth\ (analz\ (spies\ evsf))]\!] \Longrightarrow$
    $Says\ Spy\ B\ X\ \#\ evsf \in ns\text{-}public$ |

NS1:
  $[\![evs1 \in ns\text{-}public; Nonce\ N_A \notin used\ evs1]\!] \Longrightarrow$
    $Says\ A\ B\ (Crypt\ (pubK\ B)\ \{\!|Nonce\ N_A, Agent\ A|\!\})\ \#\ evs1 \in ns\text{-}public$ |

NS2:
  $[\![evs2 \in ns\text{-}public; Nonce\ N_B \notin used\ evs2;$
    $Says\ A'\ B\ (Crypt\ (pubK\ B)\ \{\!|Nonce\ N_A, Agent\ A|\!\}) \in set\ evs2]\!] \Longrightarrow$
   $Says\ B\ A\ (Crypt\ (pubK\ A)\ \{\!|Nonce\ N_A, Nonce\ N_B, Agent\ B|\!\})$
    $\#\ evs2 \in ns\text{-}public$ |

NS3:
  $[\![evs3 \in ns\text{-}public;$
    $Says\ A\ B\ (Crypt\ (pubK\ B)\ \{\!|Nonce\ N_A, Agent\ A|\!\}) \in set\ evs3;$
    $Says\ B'\ A\ (Crypt\ (pubK\ A)\ \{\!|Nonce\ N_A, Nonce\ N_B, Agent\ B|\!\})$
     $\in set\ evs3]\!] \Longrightarrow$
   $Says\ A\ B\ (Crypt\ (pubK\ B)\ (Nonce\ N_B))\ \#\ evs3 \in ns\text{-}public$

Rule *Nil* states that the empty event trace, standing for the initial state where no event has occurred yet, is contained in *ns-public*. As *ns-public* is the least set closed under these rules, and all of them but *Nil* extend some event trace assumed to be contained in *ns-public*, without rule *Nil* this set would be empty!

Rule *Fake* enables the spy to generate a message *X* and send it to whatever agent (even to himself if he were egocentric enough, although this would hardly have any effect). The spy can pick *X* out of set *synth* (*analz* (*spies evsf*)), where *evsf* is the current state, so the bricks at his disposal to build *X* are limited to agent identifiers and to any message that he has managed to learn, recycled from past traffic. Messages containing fresh nonces, as observed previously, are out of scope.

Particularly, this rule gives the spy the capability to echo any message exchanged in past traffic, contained in *spies evsf* and then in the larger set *synth* (*analz* (*spies evsf*)) as well, by sending it to any other party. Consequently, every exchanged message can reach any agent, so that qualifying a message sending event with its intended recipient, such as *B* for event *Says A B X*, in fact provides needless information, exactly as gossip news in newspapers. In other words, qualifying message sending events with senders alone (which are relevant to verify message authenticity), implying that every message being issued can be broadcast to whatever agent, would result in an equally realistic, but simpler model.

Rules *NS1*, *NS2*, and *NS3* model how legitimate agents accomplish the first, the second, and the third step, respectively, of the Needham-Schroeder-Lowe protocol. The reader is encouraged to compare these rules with the previous informal description of the protocol steps, so as to grasp their correspondence with those steps. For example, according to rule *NS2*, if agent *B* has received step 1 cryptogram *Crypt* (*pubK B*) $\{\!|Nonce\ N_A,\ Agent\ A|\!\}$ (*set evs2* is the set of all the events contained in event list *evs2*), he can reply by generating a fresh nonce N_B ($Nonce\ N_B \notin used\ evs2$), computing step 2 cryptogram *Crypt* (*pubK A*) $\{\!|Nonce\ N_A, Nonce\ N_B, Agent\ B|\!\}$, and addressing this cryptogram to the alleged sender *A*. Strictly speaking, the probability of generating some nonce already used in past traffic does not exactly match zero. However, for cryptographically secure nonces it is so small, that restricting available nonces to fresh ones is a fully adequate approximation.

Interestingly enough, both of the rules assuming that some given message has been received, namely *NS2* and *NS3*, do not assume that such message has really been sent by the alleged, mentioned sender. In fact, both use distinct, independent variables for the real sender and the alleged one (A', *A* in rule *NS2* and B', *B* in rule *NS3*, respectively). This is a faithful formal counterpart of the fact that the recipient has no choice but to reply to the alleged sender, though not knowing in principle whether the received message really comes from that sender or has rather been sent by the spy. Indeed, as shown in what follows, an important purpose of the formal verification effort is precisely to prove that under appropriate assumptions, the real sender is guaranteed to match the alleged one.

Equally interesting is the observation that in rules *NS1*, *NS2*, and *NS3*, the agent sending the new message (namely *A*, *B*, and *A*, respectively) is not constrained to be a legitimate one, so that they may match the spy as well. This enables the spy to generate the same messages containing fresh nonces as honest agents, which compensates for the lack of fresh nonce generation in the definition of function *synth*. Moreover, no matter whether the spy plays the role of *A* or *B*, the newly generated nonce enters set *analz* (*spies evs*), namely gets known to the spy, upon the message sending itself if the recipient is compromised, or at the latest as soon as the recipient replies using the spy's own public key for encryption. However, such a trick would not work for any one-time private message not exchanged with other parties, such as an ephemeral private key, which would be left out of the spy's knowledge despite his capability to generate it.

Still another observation is that, even though protocol states are modeled as event lists, within the assumptions of *ns-public* introduction rules they are always converted either into event sets, using the library function *set*, or into message sets, using the custom recursive functions *spies* and *used*. Therefore, the truth values of those assumptions for a given protocol state *evs* depend only on the events contained in *evs*, rather than on the actual order in which they occur. Moreover, the same conversions are invariably used also in the formal statements of protocol properties, as shown in what follows, which should come as no surprise because such properties derive from protocol rules. This is rightly so, as a realistic model cannot rely upon fixed sequences of message exchanges given that the spy can freely intercept and forward messages, or even generate and send his own ones, nor could attacks working by interleaving messages pertaining to different concurrent protocol runs be safely ruled out a priori. But then, since the required protocol states are event sets, their representation as event lists just provides needless additional information (namely, event chronological order), at the cost of model simplicity due to the need for conversion functions (particularly, for custom recursive ones).

4.2.5 PROTOCOL PROPERTIES

As with any inductively defined set, Isabelle/HOL automatically generates an induction rule for set *ns-public*, named *ns-public.induct* after the set's name. For a statement of the form $[\![evs \in ns\text{-}public; P\ evs]\!] \Longrightarrow Q\ evs$, such that some conclusion $Q\ evs$ is drawn about any protocol state *evs* satisfying some appropriate assumption $P\ evs$ – which is the form of the statements used to formalize protocol properties –, this induction rule looks as follows.

$$
\begin{aligned}
&[\![evs \in ns\text{-}public;\ P\ [\,] \Longrightarrow Q\ [\,];\\
&\ \textstyle\bigwedge evsf\ X\ B.\\
&\quad [\![evsf \in ns\text{-}public;\ P\ evsf \Longrightarrow Q\ evsf;\ X \in synth\ (analz\ (spies\ evsf));\\
&\quad P\ (Says\ Spy\ B\ X\ \#\ evsf)]\!]\\
&\quad \Longrightarrow Q\ (Says\ Spy\ B\ X\ \#\ evsf);\\
&\ \textstyle\bigwedge evs1\ N_A\ A\ B.\\
&\quad [\![evs1 \in ns\text{-}public;\ P\ evs1 \Longrightarrow Q\ evs1;\ Nonce\ N_A \notin used\ evs1;\\
&\quad P\ (Says\ A\ B\ (Crypt\ (pubK\ B)\ \{\!|Nonce\ N_A, Agent\ A|\!\})\ \#\ evs1)]\!]\\
&\quad \Longrightarrow Q\ (Says\ A\ B\ (Crypt\ (pubK\ B)\ \{\!|Nonce\ N_A, Agent\ A|\!\})\ \#\ evs1);\\
&\ \textstyle\bigwedge evs2\ N_B\ A'\ B\ N_A\ A.\\
&\quad [\![evs2 \in ns\text{-}public;\ P\ evs2 \Longrightarrow Q\ evs2;\ Nonce\ N_B \notin used\ evs2;\\
&\quad Says\ A'\ B\ (Crypt\ (pubK\ B)\ \{\!|Nonce\ N_A, Agent\ A|\!\}) \in set\ evs2;\\
&\quad P\ (Says\ B\ A\ (Crypt\ (pubK\ A)\ \{\!|Nonce\ N_A, Nonce\ N_B, Agent\ B|\!\})\ \#\ evs2)]\!]\\
&\quad \Longrightarrow Q\ (Says\ B\ A\ (Crypt\ (pubK\ A)\ \{\!|Nonce\ N_A, Nonce\ N_B, Agent\ B|\!\})\ \#\ evs2);\\
&\ \textstyle\bigwedge evs3\ A\ B\ N_A\ B'\ N_B.\\
&\quad [\![evs3 \in ns\text{-}public;\ P\ evs3 \Longrightarrow Q\ evs3;\\
&\quad Says\ A\ B\ (Crypt\ (pubK\ B)\ \{\!|Nonce\ N_A, Agent\ A|\!\}) \in set\ evs3;\\
&\quad Says\ B'\ A\ (Crypt\ (pubK\ A)\ \{\!|Nonce\ N_A, Nonce\ N_B, Agent\ B|\!\}) \in set\ evs3;
\end{aligned}
$$

$P\ (Says\ A\ B\ (Crypt\ (pubK\ B)\ (Nonce\ N_B))\ \#\ evs3)]\!]$
$\Longrightarrow Q\ (Says\ A\ B\ (Crypt\ (pubK\ B)\ (Nonce\ N_B))\ \#\ evs3);$
$P\ evs]\!]$
$\Longrightarrow Q\ evs$

Unsurprisingly, in light of the previous considerations, rule induction with this induction rule is the primary means how protocol properties can be proven. Although such a lengthy rule might have an unpromising aspect, what it states is rather simple: apart from the empty initial state, which is trivial, the above statement can be proven by verifying that $Q\ (ev\ \#\ evs)$ holds for every new state $ev\ \#\ evs$ produced by *ns-public* introduction rules, given assumptions $P\ evs \Longrightarrow Q\ evs$ (the *induction hypothesis*) and $P\ (ev\ \#\ evs)$, along with any further assumption satisfied by state *evs* according to those rules. Whenever rule *ns-public.induct* is applied to prove a protocol property, Isabelle instantiates its free variables P and Q to the properties of protocol state *evs* occurring in the protocol property's assumptions and in its conclusion, respectively.

As explained above, the application of rule *ns-public.induct* leaves each assumption following its major premise $evs \in ns\text{-}public$, one for each *ns-public* introduction rule, as a distinct subgoal to be proven. Thanks to the idea of using different names for the free *event list* variables contained in *ns-public* introduction rules, each subgoal can easily be traced back to the introduction rule from which it originates, as Isabelle keeps the same name for the bound *event list* variable contained in the subgoal. For example, the subgoal arising from introduction rule *NS2* is the one containing variable *evs2*; it demands to prove the conclusion about *evs2* to the right of the $\Longrightarrow$ arrow, given the assumptions about *evs2* to the left, enclosed within brackets $[\![$ and $]\!]$ and separated by semicolons.

Confidentiality properties state that under proper assumptions, a given message X is unknown to the spy in a protocol state *evs*. Correspondingly, they have a conclusion of the form $X \notin analz\ (spies\ evs)$. An example is the following property, which states that if agent B addresses nonce N_B to agent A, and both agents are uncompromised, then nonce N_B is guaranteed to remain unknown to the spy.

theorem *Spy-not-see-NB*:
$[\![Says\ B\ A\ (Crypt\ (pubK\ A)\ \{\!|Nonce\ N_A, Nonce\ N_B, Agent\ B|\!\})\in set\ evs;$
$A \notin bad;\ B \notin bad;\ evs \in ns\text{-}public]\!] \Longrightarrow$
$Nonce\ N_B \notin analz\ (spies\ evs)$
$\langle proof \rangle$

Rule induction generates subgoals with assumptions applying to state *evs* and conclusions of the form *Nonce* $N_B \notin$ *analz* (*spies* (*ev* # *evs*)), so set *analz* (*spies* (*ev* # *evs*)) has to be expressed in terms of set *analz* (*spies evs*) and of the components of the message exchanged upon event *ev*. For this reason, some lemmas about function *analz* like the following one, applying to cryptograms, have to be proven for each constructor of type *msg* (unsurprisingly, by rule induction with *analz* induction rule *analz.induct*).

lemma *analz-Crypt-if*:
analz (*insert* (*Crypt K X*) *H*) =
  (*if* (*Key* (*invKey K*) $\in$ *analz H*)
   then insert (*Crypt K X*) (*analz* (*insert X H*))
   else insert (*Crypt K X*) (*analz H*))
⟨*proof*⟩

Furthermore, the following intermediate lemmas about the protocol itself, as such proven by rule induction with rule *ns-public.induct*, are also needed. They are *regularity lemmas*, namely lemmas applying to some message *X* such that $X \in$ *parts* (*spies evs*), where *evs* is a protocol state. Particularly, the last two ones are *unicity lemmas*, namely lemmas stating that some message may occur only once in *evs* unless it is compromised.

Actually, such lemmas are required for the proof of the main protocol properties so often, that they deserve a full-fledged classification. Curiously, function *parts*, introduced to model nonce freshness by collecting all the messages occurring in past traffic, turns out to be also a valuable tool for the proof of protocol properties. In the context of the inductive method, the fact that it proves easier to reason about *parts* than about *analz* certainly contributes to the success of *parts* in this second role. Nonetheless, the subsequent sections will show that *parts* keeps being an essential proof tool even if message freshness and spy's capabilities are rather modeled by means other than *parts*, *analz*, and *synth*.

In more detail, the first lemma states that a private key may occur in *evs* just in case it is compromised, the second one that the same nonce may be used both as N_A and as N_B in *evs* only if it is compromised, and the third one that an uncompromised nonce may be used just once as N_B in *evs*.

lemma *Spy-see-priK*:
evs $\in$ *ns-public* $\Longrightarrow$ (*Key* (*priK A*) $\in$ *parts* (*spies evs*)) = (*A* $\in$ *bad*)
⟨*proof*⟩

lemma *no-nonce-NS1-NS2*:

$[\![evs \in \textit{ns-public};$
$\quad Crypt\ (pubK\ A')\ \{\!|X, Nonce\ N, Agent\ B'|\!\} \in parts\ (spies\ evs);$
$\quad Crypt\ (pubK\ B)\ \{\!|Nonce\ N, Agent\ A|\!\} \in parts\ (spies\ evs)]\!] \Longrightarrow$
$Nonce\ N \in analz\ (spies\ evs)$
$\langle proof \rangle$

lemma *unique-NB*:
$[\![Crypt\ (pubK\ A)\ \{\!|Nonce\ N_A, Nonce\ N_B, Agent\ B|\!\} \in parts\ (spies\ evs);$
$\quad Crypt\ (pubK\ A')\ \{\!|Nonce\ N_{A'}, Nonce\ N_B, Agent\ B'|\!\} \in parts\ (spies\ evs);$
$\quad Nonce\ N_B \notin analz\ (spies\ evs);\ evs \in \textit{ns-public}]\!] \Longrightarrow$
$A = A' \land N_A = N_{A'} \land B = B'$
$\langle proof \rangle$

Confidentiality property *Spy-not-see-NB* can be proven by applying rule induction and then proving the resulting subgoals by means of the above lemmas, as well as of some other properties of functions *parts*, *analz*, and *synth*. Here below is an outline of the proofs of the nontrivial subgoals. Each subgoal is qualified with the *ns-public* introduction rule from which it originates, which is all that is required to discern its full structure based on the *ns-public.induct* induction rule reported previously.

Fake

Being $B \neq Spy$ as $B \notin bad$, the issue of the step 2 cryptogram containing nonce N_B cannot match the event extending *evsf*, so it must already occur in *evsf*. Thus, $Nonce\ N_B \notin analz\ H$, where $H = spies\ evsf$, by the induction hypothesis, so $Nonce\ N_B \notin synth\ (analz\ H)$ as well. But $analz\ (insert\ X\ H) \subseteq synth\ (analz\ H)$, therefore $Nonce\ N_B \notin analz\ (insert\ X\ H)$, QED.

NS1

Due to the mismatch between step 1 and 2 cryptograms, the issue of the step 2 one containing nonce N_B cannot match the event extending *evs1*, so it must already occur in *evs1*. Thus, $Nonce\ N_B \notin analz\ H$, where $H = spies\ evs1$, by the induction hypothesis. Hence, by lemma *analz-Crypt-if*, nonce N_B can be disclosed only if it matches the nonce $N_{A'}$ contained in the new step 1 cryptogram. But being $Nonce\ N_B \in used\ evs1$, $Nonce\ N_{A'} \notin used\ evs1$, this is not the case, QED.

Since the nonce contained in the new step 1 cryptogram is referred to as N_A within rule *ns-public.induct*, the reader might be wondering why it is assigned a different name, such as $N_{A'}$, within the present proof. The reason is that the statement of theorem *Spy-not-see-NB* already contains a free variable N_A, so the bound variable having the same name in rule *ns-public.induct* has to be renamed. Of course, the same happens whenever there are a free variable and a bound one with identical names, as the reader can notice in the proofs of the subsequent subgoals, too.

NS2

The sending of the step 2 cryptogram containing nonce N_B either matches the event extending *evs2*, or already occurs in *evs2*.
In the former case, *Nonce* $N_B \notin$ *used evs2*, so that *Nonce* $N_B \notin$ *analz H*, where $H =$ *spies evs2*. Hence, by lemmas *analz-Crypt-if* and *Spy-see-priK*, nonce N_B can be disclosed only if recipient A is compromised, which is not the case since $A \notin$ *bad*.
In the latter case, *Nonce* $N_B \notin$ *analz H* by the induction hypothesis. Thus, again by lemma *analz-Crypt-if*, nonce N_B can be captured only if it matches either nonce $N_{A'}$ or $N_{B'}$ contained in the new step 2 cryptogram. But being *Nonce* $N_B \notin$ *analz H*, it is $N_B \neq N_{A'}$ by lemma *no-nonce-NS1-NS2*, and since *Nonce* $N_B \in$ *used evs2*, *Nonce* $N_{B'} \notin$ *used evs2*, it is also $N_B \neq N_{B'}$, QED.

NS3

Due to the mismatch between step 2 and 3 cryptograms, the issue of the step 2 one containing nonce N_B cannot match the event extending *evs3*, so it must already occur in *evs3*. Thus, *Nonce* $N_B \notin$ *analz H*, where $H =$ *spies evs3*, by the induction hypothesis. Hence, by lemmas *analz-Crypt-if* and *Spy-see-priK*, nonce N_B can be disclosed only if it matches the nonce $N_{B'}$ contained in the new step 3 cryptogram and the recipient B' of this message is compromised. However, being *Nonce* $N_B \notin$ *analz H*, assumption $N_B = N_{B'}$ entails that $B = B'$ by lemma *unique-NB*, so $B' \notin$ *bad*, QED.

Authenticity properties state that under proper assumptions, a given message X exchanged in a protocol state *evs* has been generated by its expected, legitimate sender A, so that A has authenticated herself toward its intended recipient B. Therefore, these protocol properties have a conclusion of the form *Says A B X* $\in$ *set evs*.

An example of such a property is the following one, which states that if agent B has sent a step 2 cryptogram to agent A, has received the expected step 3 cryptogram in response, and both agents are uncompromised, then that reply is guaranteed to have been generated by A. Altogether, this property and the confidentiality one *Spy-not-see-NB* demonstrate that the Needham-Schroeder-Lowe protocol succeeds in achieving its security goals as regards Bob, who was precisely the harmed participant in the original, flawed version of the protocol. Particularly, they confirm that Lowe's attack is no longer applicable. Similar properties can be proven also for Alice.

theorem *B-trusts-NS3*:
$[\![$*Says B A* (*Crypt* (*pubK A*) $\{\!|$*Nonce* N_A, *Nonce* N_B, *Agent B*$|\!\}$) $\in$ *set evs*;
Says A' B (*Crypt* (*pubK B*) (*Nonce* N_B)) $\in$ *set evs*;
$A \notin$ *bad*; $B \notin$ *bad*; *evs* $\in$ *ns-public*$]\!]$ $\Longrightarrow$
Says A B (*Crypt* (*pubK B*) (*Nonce* N_B)) $\in$ *set evs*
$\langle$*proof*$\rangle$

This property can be proven as a direct consequence of a lemma obtained by replacing the assumption on the step 3 cryptogram with a weaker one, *Crypt* (*pubK B*) (*Nonce* N_B) $\in$ *parts* (*spies evs*). After applying rule induction, the proof of this lemma gives rise to four nontrivial subgoals, which can be proven as follows. Again, each subgoal is qualified with the *ns-public* introduction rule from which it originates.

Fake

Since $B \notin$ *bad*, the sending of the step 2 cryptogram cannot match the event extending *evsf*, so it must already occur in *evsf*. Thus, *Nonce* $N_B \notin$ *analz H* by theorem *Spy-not-see-NB*, where $H =$ *spies evsf*. Being *parts* (*insert X H*) $\subseteq$ *synth* (*analz H*) $\cup$ *parts H*, the step 3 cryptogram is contained either in *synth* (*analz H*) or in *parts H*. But in the former case, being *Nonce* N_B $\notin$ *analz H*, the cryptogram is contained in *analz H*, and then in *parts H* since *analz H* $\subseteq$ *parts H*. Hence, the step 3 cryptogram turns out to be contained in *parts H* in both possible cases, so that the conclusion follows from the induction hypothesis, QED.

NS1

As a result of the mismatch between step 1 and 2 cryptograms, the issue of the step 2 one cannot match the event extending *evs1*, so it must already occur in *evs1*. Likewise, being *parts* (*insert X H*) $=$ *parts* $\{X\}$ $\cup$ *parts H*, the mismatch between step 1 and 3 cryptograms entails that the step 3 one is contained in *parts* (*spies evs1*). Therefore, the conclusion follows from the induction hypothesis, QED.

NS2

As a result of the mismatch between step 2 and 3 cryptograms, the step 3 one must be contained in *parts* (*spies evs2*), so nonce N_B is, too. Thus, the issue of the step 2 cryptogram cannot match the event extending *evs2*, in which case *Nonce* $N_B \notin$ *used evs2*. Hence, the step 2 cryptogram's sending already occurs in *evs2*, so that the conclusion follows from the induction hypothesis, QED.

NS3

As a result of the mismatch between step 2 and 3 cryptograms, the issue of the step 2 one cannot match the event extending *evs3*, so it must already occur in *evs3*. Consequently, *Nonce* $N_B \notin$ *analz H* by theorem *Spy-not-see-NB*, where $H =$ *spies evs3*.
Moreover, the step 3 cryptogram either matches the new one, or is contained in *parts H*.
In the former case, being *Nonce* $N_B \notin$ *analz H*, lemma *unique-NB* implies that $A = A'$ and $B = B'$, so that the event extending *evs3* matches precisely the one required for the conclusion to hold.
In the latter case, the conclusion follows from the induction hypothesis, QED.

Why does the assumption on the step 3 cryptogram need to be weakened? Using the original one, namely *Says* A' B (*Crypt* (*pubK* B) (*Nonce* N_B)) $\in$ *set evs* (a similar drawback would arise using the alternative one *Crypt* (*pubK* B) (*Nonce* N_B) $\in$ *spies evs*), even by generalizing A' to an arbitrary agent, the induction hypothesis of the subgoal deriving from rule *Fake* would look as follows.

$$\bigwedge A''. \; [\![Says\ B\ A\ (Crypt\ (pubK\ A)\ \{\!|Nonce\ N_A, Nonce\ N_B, Agent\ B|\!\}) \in set\ evsf; \; Says\ A''\ B\ (Crypt\ (pubK\ B)\ (Nonce\ N_B)) \in set\ evsf]\!] \Longrightarrow Says\ A\ B\ (Crypt\ (pubK\ B)\ (Nonce\ N_B)) \in set\ evsf$$

Assumption *Says* A' B (*Crypt* (*pubK* B) (*Nonce* N_B)) $\in$ *set* (*Says Spy* B' X # *evsf*) implies that either *Says* A' B (*Crypt* (*pubK* B) (*Nonce* N_B)) $\in$ *set evsf*, in which case the conclusion follows from the induction hypothesis, or $A' = Spy \wedge Crypt\ (pubK\ B)\ (Nonce\ N_B) = X$. As no assumption prevents agent A' from matching the spy, an appeal to the induction hypothesis is also required in the latter case. However, assumption $X \in synth$ $(analz\ (spies\ evsf))$ does not entail the existence of some event *Says* A'' B $(Crypt\ (pubK\ B)\ (Nonce\ N_B)) \in set\ evsf$, but only (as in the above proof) that $Crypt\ (pubK\ B)\ (Nonce\ N_B)$ is included, possibly as a component of a larger message, in the traffic occurred up to state *evsf*. This is too little to resort to the induction hypothesis, so that the subgoal proof is doomed to failure.

Thus, weakening the assumption on the step 3 cryptogram is necessary for the proof by rule induction of theorem *B-trusts-NS3* to succeed. In fact, it would fail to be an authenticity property if the issue were rather solved by adding assumption $A' \notin bad$. However, unlike authenticity properties, nothing in principle prevents unicity lemmas from dealing with uncompromised senders only. So, why do unicity lemmas *no-nonce-NS1-NS2* and *unique-NB* contain assumptions of the form $X \in parts\ (spies\ evs)$, rather than of the form $Says\ C\ D\ X \in set\ evs$, where $C \notin bad$? Indeed, the latter form would in principle allow for simpler proofs, since those lemmas would no longer refer to the inductively defined function *parts*.

Once more, the reason lies in the inductive method's limitations in modeling spy's fresh message generation, which make it necessary not to restrict message senders in rules *NS1*, *NS2*, and *NS3* to legitimate agents, as explained previously. Hence, the same holds for the respective subgoals arising from rule induction, which then would not allow for the use of unicity lemmas applying to uncompromised senders only. Consequently, such lemmas have to apply to fake messages as well, so that assumptions of the form $X \in parts\ (spies\ evs)$ are required for proving them by rule induction, precisely as happens with authenticity properties.

Formal proofs for each protocol property cited in this section can be found in [21]. However, in case the reader still had doubts, the above informal proof sketches should suffice as a confirmation of what had been anticipated previously, namely that a proof assistant is required given the number and the subtlety of the cases arising in these proofs, which often also make it necessary to prove suitable intermediate lemmas before the main protocol properties can be proven. Particularly, as shown above, many such lemmas consist of properties of the inductively defined functions *analz* and *synth*, which renders it still more desirable to model the spy's attack capabilities in an alternative, cheaper way.

4.3 THE RELATIONAL METHOD

This section illustrates the relational method for formal protocol verification as described in [18].

4.3.1 BINARY RELATIONS AND THE RELATIONAL METHOD

Without doubt, the development of the inductive method by Paulson was a remarkable accomplishment. In fact, as shown in the previous section, it offered a general, elegant solution, working well with many different cryptographic protocols, to the problem of modeling them as infinite state machines, as required to build fully realistic models, and formally verifying their properties. On the other hand, the above description also highlighted some aspects of the method for which there is room for further improvement, specially when it comes to verifying protocols using public key cryptography features more complex than mere encryption and decryption. Indeed, this section will address a more recent method for formal protocol verification described in [18], the *relational method*, whose purpose is exactly to enhance those aspects while retaining the basic insights and the strengths of the inductive method.

As the main reasoning tool for formal protocol verification is generalized mathematical induction, the quest for a further enhanced method cannot but start with a related question: does there exist any such form of induction more convenient than induction over inductively defined sets? Since the possible alternatives derive from the same origin, namely induction over a well-founded relation, this might actually seem a rather silly question: ultimately, all these forms of induction differ just in the mathematical objects to which they apply. Yet, this very consideration suggests a sense in which the previous question is in fact quite meaningful: do there exist any mathematical objects supporting not only some extended induction rule, but also some other inference rule lending itself to reasoning about protocols? Such objects, provided that there really were any, would indeed be the best possible candidates for representing protocol states.

Following induction, the other major rule applying to inductively defined sets is case analysis, whose application is named *rule inversion* in the Isabelle jargon (cf. [17], section 7.1.5). However, case analysis proves to be of little use for formal protocol verification (as opposed to, say, programming language semantics). In fact, the capabilities of a realistic spy are so extensive, that whatever is the format of a message possibly generated by a legitimate agent, the spy can generate messages in the same format, too. Therefore, the assumptions of the subgoals arising from rule inversion are too weak to rule out the possibility that the spy has learned or generated some given message, which is precisely what is required to prove protocol properties, so there remains little choice but to resort to the stronger assumptions resulting from rule induction. Consequently, the elements of an inductively defined set do not fulfil the previous, more ambitious selection criterion.

A promising class of mathematical objects is the one consisting of any type of objects on which a *binary relation*, namely a set of ordered pairs, is defined. In fact, being of whatever type, such objects are generic enough to lend themselves to modeling protocol states, and what is more, the *reflexive transitive closure* r^* of any binary relation r supports the following extended induction rule, named *rtrancl-induct* in the Isabelle/HOL library.

$$[\![(a, b) \in r^*; P\, a; \bigwedge y\, z.\ [\![(a, y) \in r^*; (y, z) \in r; P\, y]\!] \Longrightarrow P\, z]\!] \Longrightarrow P\, b$$

Both the context where this rule is applicable and the outcome of its application resemble what happens with rule induction. In more detail, rule *rtrancl-induct* can be applied to infer $P\ b$ for any object b satisfying its major premise $(a, b) \in r^*$, where a is some other object such that proving P a is easier. As a result of its application, its remaining assumptions, namely $P\ a$ (the *base case*) and $\bigwedge y\, z.\ [\![(a, y) \in r^*; (y, z) \in r; P\, y]\!] \Longrightarrow P\, z$ (the *induction step*), constitute as many subgoals left to be proven.

This rule states that for any two objects a, b such that $(a, b) \in r^*$, property P holds for b if it holds for a, as well as for any z such that $(a, y) \in r^*$ and $(y, z) \in r$ for some y, assuming that it holds for y. Once more, this rule can easily be viewed as just another specialization of the previous "dominoes" induction rule, where the property of having fallen is replaced by P and the arbitrary consecutive dominoes x, y by y and z, respectively.

Typically, a's role is played by some "initial" object whose very definition directly entails $P\ a$, whereas b is a generic object, for which no such definition is available, "reachable" from a via relation r^*. Therefore, if relation r models the state transitions permitted for a given protocol, and a is the initial protocol state, rule *rtrancl-induct* enables to make inferences about every protocol state b reachable from a, which is precisely what is needed to prove protocol properties.

Using rule *rtrancl-induct*, it is possible to prove another rule looking as follows.

$$[\![(a, b) \in r^*; P\, b; \neg\, P\, a]\!]$$
$$\Longrightarrow \exists u\, v.\ (a, u) \in r^* \wedge (u, v) \in r \wedge (v, b) \in r^* \wedge \neg\, P\, u \wedge P\, v$$

This rule, referred to as *rtrancl-start* in [18], states that for any two objects a, b such that $(a, b) \in r^*$, if property P holds for b but not for a, there must exist two intermediate, consecutive objects u, v such that P does not hold for u but holds for v. Indeed, it can be used to prove the existence of two such objects for any a and b satisfying its major premise $(a, b) \in r^*$, and as a result of its application, its remaining assumptions, namely $P\ b$ and $\neg\, P\ a$, are left as subgoals to be proven.

At first glance, this rule seems as trivial as saying that a turkey in shape on December 1 and dead by New Year's Eve must have passed away on an intermediate day (probably on Christmas Eve). Yet, if relation r models the permitted state transitions, a is the initial protocol state, and b is a protocol state reachable from a, this rule enables to make inferences *based on the state transitions allowing P to start to hold.* This interpretation for free variables r, a, and b luckily matches the above one applying to rule *rtrancl-induct*, which enables to use either rule interchangeably for the proof of protocol properties.

The italicized phrase implies a salient difference from rule *rtrancl-induct*, so it deserves some word of explanation. If rule *rtrancl-induct* is applied, the induction step involves two protocol states y, z such that $(y, z) \in r$, namely linked by a single state transition, and such that property P holds in both states; indeed, the induction step demands to deduce $P\ z$ from $P\ y$ (the *induction hypothesis*). If rule *rtrancl-start* is used, its conclusion rather introduces two protocol states u, v as well linked by a single state transition, but such that P starts to hold in v. As a result, the transition leading from u to v may only be one of those, if any, allowing P to begin to hold, which enables to make inferences based on the features of such transitions.

Rule *rtrancl-start* differs from *rtrancl-induct* in still another significant respect. If the latter rule is used, both subgoals left to be proven do not refer to state b any longer. Particularly, nothing is known about the states y, z occurring in the induction step but its assumptions, which do not mention b. Moreover, the induction step demands to deduce something about z (P z) from an assumption about y ($P\ y$), whereas no inference can be made the other way around. On the contrary, if rule *rtrancl-start* is applied, the states u, v introduced by its conclusion are linked to state b by property $(v, b) \in r^*$. This enables to make inferences about b based on what is known about u and v, or even vice versa. Particularly, inferences the other way around can be used to build *proofs by contradiction* such that some contradiction about u or v is deduced from what is known about b.

Thus, as opposed to the elements of an inductively defined set, the objects bound by the reflexive transitive closure of a relation make available a further powerful tool besides induction to prove protocol properties, but there is more to it than that. In fact, those objects turn out to provide still one more such tool, which would enable to reason about the state transitions possibly leading from one protocol state to another at a finer-grained level of detail, namely by also considering their precise sequence.

In more detail, this tool would consist of a predicate defined by recursion over an input list of states, and returning *True* just in case any two consecutive states x, y in the list were such that $(x, y) \in r$, namely were bound by some allowed state transition – that is, if and only if the input list were a valid *small-step protocol execution*. A list complying with this condition, and having two given states a, b as its first and last item, could easily be proven to exist just in case a and b were such that $(a, b) \in r^*$, namely were linked by a *big-step protocol execution*.

Therefore, the properties of big-step executions could also be investigated via small-step ones, using the induction rule automatically generated by Isabelle/HOL as a result of that predicate's definition, as with any recursive function, or else using any one of the rules in the Isabelle/HOL library allowing for induction over lists (needless to say, all such induction rules derive from generalized mathematical induction, too). This approach, *mutatis mutandis*, has indeed been successfully applied in the field of programming language semantics as a strategy to shorten formal compiler correctness proofs [19].

On these (strong) grounds, in the relational method a cryptographic protocol is modeled precisely as the reflexive transitive closure of the set-theoretic union of its state transitions, represented as binary relations defined over states, which is the feature that originates the method's very name. The method's application described in [18] makes an extensive use of both induction rule *rtrancl-induct* and rule *rtrancl-start*, while small-step protocol executions are not employed. Actually, the fact that two protocols using nontrivial public key cryptography features are verified without even making use of such a powerful reasoning tool is a clear hint of the potential of the relational method.

Following [18], this section addresses the problem of formal cryptographic protocol verification by applying the relational method via interactive theorem proving with Isabelle/HOL. Other approaches to this problem investigated by recent research include extensions to Paulson's inductive method with Isabelle/HOL [3] (particularly, this paper considers extensions aimed at modeling anonymity in electronic voting protocols), as well as methods for combining the benefits of fully automated verification tools (high level of automation/productivity) with those of interactive theorem proving (high assurance), using Isabelle [2] [12] [7], other general-purpose proof assistants such as Coq [5], or ad hoc tools [13].

4.3.2 CASE STUDY: A PAKE PROTOCOL

The remainder of this section is dedicated to illustrating the relational method by considering its application to the *Password-Authenticated Key Agreement (PAKE)* public key protocol addressed in [18]. In this protocol, *Password Authenticated Connection Establishment (PACE) with Chip*

Authentication Mapping [10] is used by an asset *owner* (for example, a cardholder) to establish a secure channel with her own *asset* (for instance, her smart card). Using Chip Authentication Mapping, the asset also authenticates itself to the owner via this PACE run. Then, the secure channel is used by the owner to authenticate herself to the asset by sending it a password, independent of the one from which the PACE key is derived and assigned to that owner only. In turn, the asset replies with a final acknowledgment, sent over the same secure channel.

The main distinguishing feature of this protocol is that the use of Chip Authentication Mapping and the sending of the owner's unique password enable the asset and the owner, respectively, to reliably authenticate themselves to each other even if the PACE key is shared by multiple owners or is weak, as happens in electronic passports [10]. The Diffie-Hellman key agreements provided for by PACE are performed using a cryptographically secure elliptic curve group [10].

In more detail, this protocol consists of the following steps.

1. Asset $\rightarrow$ Owner: $\{\widetilde{SK_S}\}_{K_\pi}$

   The asset generates an ephemeral private key $\widetilde{SK_S}$, encrypts it with the PACE key K_π, and sends the resulting *encrypted nonce* to the owner.

2. Owner $\rightarrow$ Asset: $\{1, \widetilde{PK_A}\}$

   The owner generates an ephemeral key pair $(\widetilde{SK_A}, \widetilde{PK_A})$ and sends the ephemeral public key $\widetilde{PK_A}$ to the asset, along with a numeric label (used to denote the protocol step to which the exchanged public key pertains).

3. Asset $\rightarrow$ Owner: $\{2, \widetilde{PK_B}\}$

   The asset also generates an ephemeral key pair $(\widetilde{SK_B}, \widetilde{PK_B})$ and sends the ephemeral public key $\widetilde{PK_B}$ to the owner, along with a numeric label.

4. Owner $\rightarrow$ Asset: $\{3, \widetilde{PK_C}\}$

   The owner retrieves $\widetilde{SK_S}$ by decrypting the asset's encrypted nonce with the PACE key K_π. As the plaintext does not contain either padding or any other check data, the owner cannot detect the garbage resulting from decrypting a wrong message instead. In this case, the

protocol will go on all the same, only to end up with the owner and the other party not agreeing on the same session key.
Then, the owner maps the public group generator G to a shared *ephemeral generator* $\tilde{G}$ via a Diffie-Hellman key agreement, by computing $\tilde{G} = [\widetilde{SK_S}]G + [\widetilde{SK_A}]\widetilde{PK_B}$.
Finally, she generates another ephemeral key pair $(\widetilde{SK_C}, \widetilde{PK_C})$ using $\tilde{G}$ as group generator (namely, $\widetilde{PK_C} = [\widetilde{SK_C}]\tilde{G}$), and sends the ephemeral public key $\widetilde{PK_C}$ to the asset, along with a numeric label.

5. Asset $\rightarrow$ Owner: $\{4, \widetilde{PK_D}\}$

   The asset agrees on the same ephemeral generator by computing $\tilde{G} = [\widetilde{SK_S}]G + [\widetilde{SK_B}]\widetilde{PK_A}$.
   Then, the asset generates another ephemeral key pair $(\widetilde{SK_D}, \widetilde{PK_D})$ using generator $\tilde{G}$ (namely, $\widetilde{PK_D} = [\widetilde{SK_D}]\tilde{G}$), and sends the ephemeral public key $\widetilde{PK_D}$ to the owner, along with a numeric label.

6. Owner $\rightarrow$ Asset: $\{\widetilde{PK_D}\}_{\tilde{K}}$

   The owner computes the shared secret $[\widetilde{SK_C}]\widetilde{PK_D}$ by means of another Diffie-Hellman key agreement, and uses it to derive a shared session key $\tilde{K}$.
   Then, she encrypts $\widetilde{PK_D}$ with $\tilde{K}$ and sends the resulting cryptogram to the asset.

7. Asset $\rightarrow$ Owner: $\{\{\widetilde{PK_C}\}_{\tilde{K}}, \{(SK_{Auth})^{-1} \times \widetilde{SK_B}\}_{\tilde{K}},$
   $\{\{H(Owner), H(PK_{Auth})\}_{SK_{Sign}}\}_{\tilde{K}}\}$

   The asset agrees on the same secret by computing $[\widetilde{SK_D}]\widetilde{PK_C}$, and uses it to derive the same session key $\tilde{K}$, which is used to decrypt the owner's cryptogram and verify that the resulting plaintext matches $\widetilde{PK_D}$.
   Then, the asset computes the *Chip Authentication Data CA* as the product of the inverse of its *authentication private key* SK_{Auth}, a static key, and $\widetilde{SK_B}$ (modulo the order of the elliptic curve group).
   Finally, the asset uses $\tilde{K}$ to encrypt $\widetilde{PK_C}$, the Chip Authentication Data, and a precomputed trusted party's signature, generated over a pre-shared hash of some owner's unambiguous data and a hash of the asset's *authentication public key* PK_{Auth} (where $PK_{Auth} = [SK_{Auth}]G$), and sends the concatenation of the resulting cryptograms, serving as the asset's *authentication token*, to the owner.

8. Owner $\rightarrow$ Asset: $\{Password\}_{\tilde{K}}$

   The owner decrypts the fields of the asset's authentication token with $\tilde{K}$ and verifies that the first plaintext matches $\widetilde{PK_C}$, the Chip Authentication Data CA are such that $[CA]PK_{Auth} = \widetilde{PK_B}$, and the last plaintext is a valid signature of message $\{H(Owner), H(PK_{Auth})\}$.
   Then, she encrypts her unique password with $\tilde{K}$ and sends the resulting cryptogram to the asset.

9. Asset $\rightarrow$ Owner: $\{0\}_{\tilde{K}}$

   The asset decrypts the owner's latest cryptogram with $\tilde{K}$ and verifies that the obtained plaintext matches the owner's unique password.
   Then, the asset encrypts a success code with $\tilde{K}$ and sends the resulting cryptogram, which serves as a final acknowledgment concluding the protocol, to the owner.

In steps 1 to 5, the asset and the owner set up their secure channel. In step 6, the owner provides the asset with *key confirmation*, namely with assurance that the other party shares the same session key. In step 7, the asset authenticates itself to the owner and provides her with key confirmation, too. The owner in turn authenticates herself to the asset in step 8, and the asset sends its final acknowledgment to the owner in step 9.

According to [10], the PACE run should start with the owner sending a plain request to the asset, and the trusted party's signature should be retrieved by the owner as a distinct message, instead of being included in the asset's authentication token, but these details are not reflected by the model since they are irrelevant for protocol security. Similarly, the owner should read the asset's authentication public key over the established secure channel as an additional step, given that this key is required to verify the asset's Chip Authentication Data, but this detail turns out to be negligible, too, as shown in what follows.

The only relevant difference from [10] is that in the PAKE protocol considered in this chapter, the trusted party's signature binds the asset's authentication public key to the owner's identity, whereas according to [10], the signed file storing the chip's authentication public key for Chip Authentication Mapping on an electronic passport is not required to contain any link to the holder's identity. As a matter of fact, this is a substantial difference in terms of security. In the considered PAKE protocol, the spy cannot masquerade as an asset but knowing the authentication private key of that specific asset. On the contrary, in the case of PACE with Chip

Authentication Mapping as supported by an electronic passport, the spy would suffice to know the authentication private key of whatever genuine chip whose authentication public key is signed by the same trusted party.

4.3.3 AGENTS AND MESSAGES

Agents consist of an infinite population of assets and owners, plus the spy, while the trusted party need not be included, as it does not exchange any message. Assets and owners are identified through natural numbers, implying that for each *n*, *Asset n* is *Owner n*'s asset.

type-synonym *agent-id* = *nat*

datatype *agent* =
Asset agent-id |
Owner agent-id |
Spy

Keys comprise private keys, public keys, static shared keys (namely, PACE keys) – all of which are identified by natural numbers –, session keys, and a single trusted party's key pair for signature generation/verification. Notably, the type constructor for session keys takes an input of a rather complex type, defined as follows.

type-synonym *key-id* = *nat*

type-synonym *seskey-in* = *key-id option* × *key-id set* × *key-id set*

As type *key-id* is just an alias for natural numbers, the important part of this definition is the polymorphic types *'a option* and *'a set*, whose type variable *'a* is here filled with *key-id*. The latter is the type of sets, and has already occurred many times in the previous section, whereas the former type provides for two constructors: *None*, taking no input, and *Some*, taking a single input of type *'a* (namely, of type *key-id* in this case).

This type definition enables to use a 3-tuple of either form $(Some\ S, \{A, B\}, \{C, D\})$ or $(None, \{A, B\}, \{C, D\})$ to model the session key produced by a PACE run. Here, S, A, B, C, D are the numeric identifiers of the ephemeral private keys $\widetilde{SK_S}, \widetilde{SK_A}, \widetilde{SK_B}, \widetilde{SK_C}, \widetilde{SK_D}$ generated by either party. The reason why they all have to be mentioned is that the session key is computed as a function of all of them: for instance, as seen above, the owner derives the session key from the shared secret $[\widetilde{SK_C}]\widetilde{PK_D} = [\widetilde{SK_C} \times \widetilde{SK_D}]\tilde{G} = [\widetilde{SK_C} \times \widetilde{SK_D}]([\widetilde{SK_S} + \widetilde{SK_A} \times \widetilde{SK_B}]G)$.

The use of unordered sets for the second and the third component of such 3-tuples models the fact that $[SK_1]PK_2 = [SK_2]PK_1$ for any two key pairs (SK_1, PK_1), (SK_2, PK_2). Furthermore, term *None* is used as first component in place of term *Some S* to represent the junk session key obtained by an owner who decrypts a wrong message instead of a valid encrypted nonce.

According to this approach, if the protocol had provided for a single Diffie-Hellman key agreement without any encrypted nonce, the resulting session key would have rather been identified just by a single set of private key numeric identifiers, such as $\{A, B\}$. This is a good confirmation of a drawback of the *analz-synth* approach mentioned in the previous section, namely that in the face of increased complexity, reuse is hindered by the need for ad hoc message formats likely arising in case of protocols using nontrivial public key cryptography.

Message syntax is modeled by means of the following type definition.

datatype *msg* =
Num nat |
Agent agent-id |
Pwd agent-id |
Key key |
Mult key-id key-id (**infixl** $\otimes$ *70*) |
Hash msg |
Crypt key msg |
MPair msg msg |
IDInfo agent-id msg |
Log msg

Constructors *Agent*, *Key*, *Crypt*, and *MPair* have the same meanings as before. *Num n* is numeric constant *n*, used to model message labels and success codes, *Pwd n* is *Owner n*'s unique password, and *Hash X* is a hash of message *X*. Notation $A \otimes B$ stands for *Mult A B*, and represents the Chip Authentication Data resulting from the private keys with numeric identifiers *A* and *B*, namely $(SK_A)^{-1} \times SK_B$ (another ad hoc message format).

As with cryptograms, formalizing hash values and Chip Authentication Data via as many injective type constructors has the effect of dismissing real-world collisions from the model. Nonetheless, similar considerations as with cryptograms also apply to Chip Authentication Data, whereas the exclusion of hash collisions formalizes the fact that meaningful ones are highly unlikely to be detected for a cryptographically secure hash function, exactly as condition *Nonce N* $\notin$ *used evs* modeled the high unlikelihood of nonce repetitions. Therefore, again, type constructors are fully adequate to model message formats.

What about the remaining constructors, *IDInfo* and *Log*? Well, they have been added on the grounds that this formal protocol verification not only addresses message confidentiality and authenticity, but also a further security property, *message anonymity*. In this respect, the model targets the problem of verifying that under some proper conditions, the spy cannot map a given message X_A, pertaining to agent A, to this agent by running into some other message Y_A mapped to A within a compound message also containing X_A. In other words, the purpose is to verify that message X_A remains *anonymous*, although the spy can log all the messages generated or accepted by some legitimate agent and map any two observable messages contained in any such message to the same agent. Actually, the present model focuses on the specific case where A is some given asset or owner and X_A is their password, authentication private key, or PACE key. However, as shown in what follows, the approach to modeling and verifying *logging-dependent message anonymity* introduced through this case study is fairly general and can be applied to whatever protocol.

After this preamble, the purposes of constructors *IDInfo* and *Log* should sound less arcane. Notation $\langle n, X \rangle$ stands for message *IDInfo n X*, whose meaning is "message X is mapped to *Asset n*/*Owner n*" (distinguishing between the two agents is unnecessary). So, a message X_n related to *Asset n*/*Owner n* is anonymous just in case the spy does not know message $\langle n, X_n \rangle$. Message *Log X* means "message X has been generated or accepted by some legitimate agent", so that a given message X has been logged by the spy just in case he knows *Log X*.

Of course, the spy will be able to observe a message X within a logged message X', and so to map it to some agent, if and only if he can look inside any cryptogram that encapsulates X in X', namely if he knows all the related decryption keys. To understand how this condition can be formalized, consider the sample logged message $X' = \{H(\{X\}_{K_2}), Y\}_{K_1}$. If function *InvKey* associates each encryption key K to message *Key* K', where K' is the corresponding decryption key, the above message X will be observable just in case the spy knows both messages *InvKey* K_1 and *InvKey* K_2. However, the spy has not logged message $\{\{X\}_{K_2}\}_{K_1}$; he has just logged X', where X is input to a hash function, whose output is in turn part of a compound message.

So, some function is needed that can take a set of logged messages as input, and returns a larger set also containing the encryption of any inner message nested within hash functions or compound messages. In fact, if this function were input a message set containing X', it would return a larger set also containing $\{\{X\}_{K_2}\}_{K_1}$, and this would render it a breeze to model the fact that the spy should know *InvKey* K_1 and *InvKey* K_2 to observe *X*.

Such a function can be defined inductively as follows (for any list of keys $KS = [K_1, ..., K_n]$, *foldr Crypt KS X* matches *Crypt* K_1 $(...(Crypt\ K_n X)))$.

inductive-set *crypts* :: *msg set* ⇒ *msg set* **for** *H* :: *msg set* **where**

crypts-used:
 $X \in H \Longrightarrow X \in$ *crypts H* |

crypts-hash:
 foldr Crypt KS (*Hash X*) ∈ *crypts H* ⟹
  foldr Crypt KS X ∈ *crypts H* |

crypts-fst:
 foldr Crypt KS ⦃*X*, *Y*⦄ ∈ *crypts H* ⟹ *foldr Crypt KS X* ∈ *crypts H* |

crypts-snd:
 foldr Crypt KS ⦃*X*, *Y*⦄ ∈ *crypts H* ⟹ *foldr Crypt KS Y* ∈ *crypts H*

The missing link is another function *key-sets* that takes a message *X* and a set of messages *H* as inputs, and returns a set containing any set of decryption keys, encoded as messages, such that the corresponding encryption keys are used to encrypt *X* in *H*. In this way, if *H* is a set of messages logged by the spy, the spy can observe *X* within some of them just in case there exists a message set *U* of keys known to the spy and such that *U* ∈ *key-sets X* (*crypts H*). For example, if this function were input the above message *X* and message set *crypts H*, containing $\{\{X\}_{K_2}\}_{K_1}$, it would return a set containing message set $\{InvKey\ K_1, InvKey\ K_2\}$ as an element, according to expectations.

Such a function can be defined by means of a single equation, as follows.

definition *key-sets* :: *msg* ⇒ *msg set* ⇒ *msg set set* **where**
key-sets X H ≡ {*InvKey* ‘ *set KS* | *KS*. *foldr Crypt KS X* ∈ *H*}

In general, encrypted messages can be mapped to agents not only by decrypting them with decryption keys, but also by recomputing the cryptograms with encryption keys, as long as the plaintexts are known, and checking whether they match the logged ones. Since function *key-sets* collects decryption keys only, it just models the former attack pattern. In fact, for the PAKE protocol under consideration, the latter one need not be considered as it would not be exploitable, on the grounds that PACE keys and session keys are symmetric, so that encryption and decryption keys are the same, while the only asymmetric encryption key being used, namely the trusted party's signature generation key, is unknown to the spy. On the contrary, in case of a protocol that uses (also) public keys for encryption, like the Needham-Schroeder-Lowe one, encryption keys (also) have to be collected, as the corresponding decryption keys, namely private keys, are unknown to the spy, unless they are already compromised in the initial state or can get disclosed in some way.

Logging-independent message anonymity rather is the property for a message to be anonymous notwithstanding the spy's capability of mapping messages of that sort to agents without resorting to message logging, namely via some intrinsic feature of such messages. If the only kind of anonymity of interest for a given protocol is the logging-independent one, the sole anonymity-related constant to be included in the model is constructor *IDInfo*, whereas constructor *Log* and functions *crypts*, *key-sets* can be left out. It is also possible to include both kinds of anonymity in the model, in which case there will be some protocol rule enabling the spy to map messages of some sort to agents only if they are observable within logged messages, and some other protocol rule enabling the spy to do so independently of this condition.

Protocol rules are discussed in what follows, while an application of logging-independent anonymity to a real-world protocol is described in [20].

4.3.4 EVENTS AND STATES

In the relational method, events are modeled as *ordered pairs* of the form (A, X), where A is an agent and X is a message, with the following possible interpretations.

1. *The spy has learned/generated X and may have sent X to any other agent*, in which case $A = Spy$.
   According to the observations made in the previous section, this interpretation of spy's events abolishes any distinction between what the spy can learn and what he can generate and send to other agents.
2. *A has sent X to another agent*, where A is a legitimate agent.
   For any state transition triggered by such an event (A, X), event (Spy, X) will also occur. Based on the above interpretation of spy's events,

this companion event models the spy's interception of X over the public, insecure network on which all messages are supposed to be exchanged.
As observed before, the omission of the intended recipient reflects the fact that whoever they are, message X can actually reach whatever agent because of the spy.
3. *A has generated private message X without sending it*, where A is a legitimate agent.
Unlike the previous case, no companion event (Spy, X) will occur in this one, unless the spy has hacked A to steal her private messages.
Of course, this interpretation does not imply that X cannot be sent to some other agent upon a subsequent state transition, or as part of another message; it just means that X is not exchanged as is in the current state transition.
Such events typically express the fact that a given one-time private message (such as a nonce, an ephemeral private key, or a one-time password) has already been generated, so that it cannot occur again (which models the high unlikelihood of this eventuality for cryptographically secure one-time messages).

In addition to the aforesaid ones, further context-specific interpretations, dependent on the state transition brought about by the event, are also possible. An example will be supplied by one of the protocol rules reported here below, which provides for an event modeling the disposal of a message by its originator after using it once.

States are modeled as *sets of events*, or equivalently, as *binary relations* between agents and messages – in fact, as explained previously, binary relations are sets of ordered pairs, and events are agent-message pairs. Thus, an event has occurred in a given state just in case it is an element of that state. As suggested in the previous section, an event set is all that is needed as a state, since keeping track of the chronological order of events is unnecessary.

Two important functions, *used* and *spied*, both taking a state s as input and returning a message set, are introduced as follows.

- *used s* is the set of all the messages already used in state s.
  It is defined as the *range* of relation s, namely as the set comprising every message X such that, for some agent A, event (A, X) is contained in set s.
- *spied s* is the set of all the messages that the spy has learned/generated and possibly sent to any agent in state s.
  It is defined as the *image* of singleton $\{Spy\}$ under relation s,

namely as the set comprising every message X such that event (Spy, X) is contained in set s.

So, two simple expressions in terms of as many library functions are enough, whereas defining these message sets in the inductive method required three inductively defined functions (*parts*, *analz*, and *synth*) and two recursively defined ones (*used* and *spies*), whose custom properties had to be proven, too – this looks like a good deal!

The initial state s_0 defines the spy's starting knowledge, which consists of the following messages.

- All public ones, namely the signature verification key and all numeric constants, authentication public keys, and authentication signatures.
- Disclosed signed unambiguous data, passwords, authentication private keys, and PACE keys.
- Mappings of anonymity-compromised passwords, authentication private/public keys, and PACE keys to their respective agents.

The sets of anonymity-compromised agents referred to by s_0 are defined by adhering to two constraints. The first one, also applying to protocol rules, is that the spy can map known messages alone to agents. This is consistent with the principle that he can map anything logged only if it is observable, and indeed, mapping unknown messages to agents would be too much even for a so powerful attacker, like for anyone else! Thus, for instance, the set *bad-id-prikey* of the agents with anonymity-compromised private keys is a subset of the set of the agents with disclosed private keys. The second constraint is to extend those sets with any mapping that the spy can perform based on his sole initial knowledge, which enables to simplify the assumptions of anonymity properties (and then their proofs, too). So, for example, the set *bad-id-pubkey* of the agents whose public keys are anonymity-compromised also contains any agent with disclosed signed unambiguous data or having an anonymity-compromised private key.

Moreover, the range of relation s_0 also comprises all authentication private keys, so as to mark them as having already been used from the very beginning. This is necessary to prevent any one of them from occurring again as an ephemeral private key during protocol execution.

Here below are the formal definitions of *used*, *spied*, and s_0 (notations f ` A and r `` B denote the images of sets A and B under function f and relation r, respectively).

abbreviation *used* :: *state* $\Rightarrow$ *msg set* **where**
used s $\equiv$ *Range s*

abbreviation *spied* :: *state* $\Rightarrow$ *msg set* **where**
spied s $\equiv$ *s* '' {*Spy*}

abbreviation s_0 :: *state* **where**
$s_0 \equiv$ *range* ($\lambda n.$ (*Asset n*, *Auth-PriKey n*)) $\cup$ {*Spy*} $\times$ *insert VerKey*
(*range Num* $\cup$ *range Auth-PubKey* $\cup$ *range* ($\lambda n.$ *Sign n* (*Auth-PriK n*)) $\cup$
Agent ' *bad-agent* $\cup$ *Pwd* ' *bad-pwd* $\cup$ *PriKey* ' *bad-prik* $\cup$ *ShaKey* ' *bad-shak* $\cup$
($\lambda n.$ $\langle n$, *Pwd n*$\rangle$) ' *bad-id-password* $\cup$
($\lambda n.$ $\langle n$, *Auth-PriKey n*$\rangle$) ' *bad-id-prikey* $\cup$
($\lambda n.$ $\langle n$, *Auth-PubKey n*$\rangle$) ' *bad-id-pubkey* $\cup$
($\lambda n.$ $\langle n$, *Key* (*Auth-ShaKey n*)$\rangle$) ' *bad-id-shakey*)

4.3.5 PROTOCOL RULES

As anticipated before, state transitions are defined as binary relations defined over states, namely as sets of state pairs (s, s'), so that such is also their set-theoretic union. Thus, for any two states s, s', there exists a transition leading from s to s' just in case the ordered pair (s, s') is contained in the union of all transitions, which is denoted using infix notation $s \vdash s'$. More generally, s' may be reached from s compatibly with protocol rules just in case pair (s, s') is contained in the reflexive transitive closure of that union, which is denoted using infix notation $s \models s'$. Particularly, any given state s is an allowed protocol state, namely may be reached from initial state s_0 compatibly with protocol rules, if and only if $s_0 \models s$.

The choice of using operators $\vdash$ and $\models$ in place of more common ones for transition systems, such as $\rightarrow$ and $\rightarrow*$, has been made to avoid confusion with the implication arrow $\longrightarrow$. However, of course, it is just a matter of convention, and the reader would be free to replace $\vdash$ and $\models$ with other operators, say $\rightarrow$ and $\rightarrow*$, in their own applications of the relational method.

Every state transition is defined so that for any element (s, s'), s' contains all the events in s, as well as one or more additional events. In other words, protocol rules consist of rules allowing states that comply with some proper conditions to be extended with further suitable events, which is the reason why the following properties hold.

proposition *state-subset*:
$s \models s' \Longrightarrow s \subseteq s'$
$\langle$*proof*$\rangle$

proposition *used-subset*:
$s \models s' \Longrightarrow used\ s \subseteq used\ s'$
⟨*proof*⟩

proposition *spied-subset*:
$s \models s' \Longrightarrow spied\ s \subseteq spied\ s'$
⟨*proof*⟩

Interestingly enough, these properties are essential to prove protocol properties by means of rule *rtrancl-start*. In fact, if this rule is applied to prove something about a state b, what it allows to do is to reason about another state v such that $v \models b$. Thus, v and b are not necessarily linked by a single state transition, as opposed to the states y and z in the induction step of rule *rtrancl-induct*, which are such that $y \vdash z$. The latter, stronger link allows to draw conclusions about z from the induction hypothesis about y based on protocol rules. This is not possible for v and b, in which case the previous properties can rather be used to make inferences about b based on what is known about v, or even vice versa, as explained above.

Protocol rules can be parted into three categories depending on the modeled agents' capabilities, as follows.

1. Legitimate agents' capabilities, as provided for by protocol steps.
2. Spy's capabilities enabling the spy to learn/generate messages.
3. Spy's capabilities enabling the spy to map known messages to agents.

Here below is an example from the first category, which formalizes the second protocol step, namely the owner's generation and sending of her former ephemeral public key.

abbreviation *rel-owner-ii* :: (*state* × *state*) *set* **where**
$rel\text{-}owner\text{-}ii \equiv \{(s, s') \mid s\ s'\ n\ S\ A\ K.$
$\quad s' = insert\ (Owner\ n,\ PriKey\ A)\ s \cup$
$\quad\quad \{Owner\ n,\ Spy\} \times \{\{\!|Num\ 1,\ PubKey\ A|\!\}\} \cup$
$\quad\quad \{Spy\} \times Log\ `\ \{Crypt\ K\ (PriKey\ S),\ \{\!|Num\ 1,\ PubKey\ A|\!\}\} \wedge$
$\quad Crypt\ K\ (PriKey\ S) \in used\ s\ \wedge$
$\quad PriKey\ A \notin used\ s\}$

Being a binary relation between states, protocol rule *rel-owner-ii* is defined as a set of state pairs (s, s'), where state s' extends state s by five events, as follows.

- According to the previous classification of events, $(Owner\ n, PriKey\ A)$ is a type 3 event, standing for the generation of private key $\widetilde{SK_A}$ by *Owner n* (as explained above, *insert* $(Owner\ n, PriKey\ A)\ s = s \cup \{(Owner\ n, PriKey\ A)\}$). Since no owner is supposed to have been hacked by the spy, no event $(Spy, PriKey\ A)$ occurs.
  Condition $PriKey\ A \notin used\ s$ ensures key freshness, while the addition of event $(Owner\ n, PriKey\ A)$ to state s' entails that $PriKey\ A \in used\ t$ for any state t such that $s' \models t$, by virtue of property *used-subset*. This ensures that the key cannot occur again, as a condition of the form $PriKey\ N \notin used\ s$ is included in every protocol rule providing for the generation of some ephemeral private key $\widetilde{SK_N}$.
- Event $(Owner\ n, \{\!|Num\ 1, PubKey\ A|\!\})$ is a type 2 one, standing for *Owner n*'s sending of public key $\widetilde{PK_A}$, so it comes along with type 1 event $(Spy, \{\!|Num\ 1, PubKey\ A|\!\})$, which models the spy's interception of the exchanged message (according to ordinary mathematical notation, $\{Owner\ n, Spy\} \times \{\{\!|Num\ 1, PubKey\ A|\!\}\} = \{(Owner\ n, \{\!|Num\ 1, PubKey\ A|\!\}), (Spy, \{\!|Num\ 1, PubKey\ A|\!\})\}$). Message $\{\!|Num\ 1, PubKey\ A|\!\}$ is *Owner n*'s reply to the alleged encrypted nonce $Crypt\ K\ (PriKey\ S)$ whoever is its actual sender, as expressed by condition $Crypt\ K\ (PriKey\ S) \in used\ s$.
  Condition $Crypt\ (Auth\text{-}ShaKey\ n)\ (PriKey\ S) \in used\ s$ should rather have been used, where *Auth-ShaKey n* is *Owner n*'s PACE key, if *Owner n* were able to validate the received message, but this is not the case, as remarked previously. Provided that the message format and size are correct, the protocol goes on, which is precisely what the weaker condition $Crypt\ K\ (PriKey\ S) \in used\ s$ expresses.
- Two more type 1 events are added to state s', $(Spy, Log\ (Crypt\ K\ (PriKey\ S)))$ and $(Spy, Log\ \{\!|Num\ 1, PubKey\ A|\!\})$, modeling the spy's logging of the alleged encrypted nonce and of the message containing $\widetilde{PK_A}$, respectively accepted and generated by legitimate agent *Owner n* (as explained previously, $Log\ `\ \{Crypt\ K\ (PriKey\ S), \{\!|Num\ 1, PubKey\ A|\!\}\} = \{Log\ (Crypt\ K\ (PriKey\ S)), Log\ \{\!|Num\ 1, PubKey\ A|\!\}\}$).
  Particularly, the addition of the former event is necessary because the alleged encrypted nonce could have been generated by the spy as well, in which case it would have not been logged upon its generation.

Here below are two further protocol rules from the first category, respectively modeling the seventh and the eighth protocol step, namely the asset's sending of its authentication token and the owner's sending of her unique password.

abbreviation *rel-asset-iv* :: (*state* × *state*) *set* **where**
rel-asset-iv ≡ {(*s*, *s'*) | *s s' n S A B C D SK*.
  s' = *s* ∪ {*Asset n*} × {*SesKey SK*, *PubKey B*} ∪
   {*Asset n*, *Spy*} × {*Token n* (*Auth-PriK n*) *B C SK*} ∪
   {*Spy*} × *Log* ‘ {*Crypt* (*SesK SK*) (*PubKey D*),
    Token n (*Auth-PriK n*) *B C SK*} ∧
  {*Asset n*} × {*Crypt* (*Auth-ShaKey n*) (*PriKey S*),
   ⦃*Num 2*, *PubKey B*⦄, ⦃*Num 4*, *PubKey D*⦄} ⊆ *s* ∧
  {⦃*Num 1*, *PubKey A*⦄, ⦃*Num 3*, *PubKey C*⦄,
   Crypt (*SesK SK*) (*PubKey D*)} ⊆ *used s* ∧
  (*Asset n*, *PubKey B*) ∉ *s* ∧
  SK = (*Some S*, {*A*, *B*}, {*C*, *D*})}

abbreviation *rel-owner-v* :: (*state* × *state*) *set* **where**
rel-owner-v ≡ {(*s*, *s'*) | *s s' n A B C SK*.
  s' = *s* ∪ {*Owner n*, *Spy*} × {*Crypt* (*SesK SK*) (*Pwd n*)} ∪
   {*Spy*} × *Log* ‘ {*Token n A B C SK*, *Crypt* (*SesK SK*) (*Pwd n*)} ∧
  Token n A B C SK ∈ *used s* ∧
  (*Owner n*, *SesKey SK*) ∈ *s* ∧
  B ∈ *fst* (*snd SK*)}

The former rule provides an example of an event, (*Asset n*, *PubKey B*), that is assigned a context-specific interpretation. The rule is guarded by condition (*Asset n*, *PubKey B*) ∉ *s* and then, if this condition and the other ones are all fulfilled, it brings about the addition of event (*Asset n*, *PubKey B*) to state *s'*. Thus, *Asset n* may use just once the ephemeral private key $\widetilde{SK_B}$, generated by *Asset n* in the third protocol step (as stated by condition (*Asset n*, ⦃*Num 2*, *PubKey B*⦄) ∈ *s*), in the computation of its authentication token, after which the key is discarded. Event (*Asset n*, *PubKey B*), rather than (*Asset n*, *PriKey B*), is employed for this purpose since the latter event is already used as a type 3 one in the rule modeling the third protocol step. This example shows how flexible events are in the relational method – indeed, they can be used to keep track of whatever one wants!

Although all ephemeral key pairs are thrown away after a single use in real-world PACE, this is not generally the case in the model, simply because the protocol properties of interest are independent of this good practice. The only exception is precisely the asset's former ephemeral key pair, since limiting its use to at most one asset's authentication token computation is required to establish an important unicity property, as shown in what follows.

What is above all interesting in the latter rule is how the validation of the asset's authentication token is formalized – actually, via a condition as simple as $Token\ n\ A\ B\ C\ SK \in used\ s$. This token differs from the one generated by *Asset n* in rule *rel-asset-iv* in that the numeric identifier *Auth-PriK n* of *Asset n*'s authentication private key is replaced by a generic one, A. The reason is that *Owner n* does not know that key, so she cannot verify that the token was generated using precisely that key, but only that it was generated with a private key corresponding to a public key signed along with *Owner n*'s unambiguous data. This is what condition $Token\ n\ A\ B\ C\ SK \in used\ s$ expresses based on the equations used to define function *Token*, which look as follows.

$$
\begin{aligned}
&Token\ n\ A\ B\ C\ SK \equiv \\
&\{\!|Crypt\ (SesK\ SK)\ (PubKey\ C),\ Crypt\ (SesK\ SK)\ (A \otimes B), \\
&\ Crypt\ (SesK\ SK)\ (Sign\ n\ A)|\!\}
\end{aligned}
$$

$$
Sign\ n\ A \equiv Crypt\ SigK\ \{\!|Hash\ (Agent\ n),\ Hash\ (PubKey\ A)|\!\}
$$

Thus, although the actual operations executed by an owner to verify the received Chip Authentication Data presuppose the exchange of the authentication public key PK_{Auth}, they can be formalized doing without this detail, which is one of the two reasons why the model can leave it out at all. The other one is that the spy can already observe PK_{Auth} within any trusted party's signature computed over that key under the same condition, namely by knowing the session key (as the signature verification key is public knowledge), so that the spy's attack patterns compromising anonymity are not affected by omitting PK_{Auth} exchange.

Spy's capabilities can be modeled as protocol rules extending a state with one or more events of type 1, namely of the form (*Spy*, *X*) for some given message *X*. Here below are two such protocol rules from the second above category, respectively modeling the generation of a fresh private key and the computation of a session key on the part of the spy.

$$
\begin{aligned}
&\textbf{abbreviation}\ \textit{rel-prik} :: (state \times state)\ set\ \textbf{where} \\
&\textit{rel-prik} \equiv \{(s, s')\ |\ s\ s'\ A. \\
&\ s' = insert\ (Spy, PriKey\ A)\ s\ \wedge \\
&\ PriKey\ A \notin used\ s\}
\end{aligned}
$$

$$
\begin{aligned}
&\textbf{abbreviation}\ \textit{rel-sesk} :: (state \times state)\ set\ \textbf{where} \\
&\textit{rel-sesk} \equiv \{(s, s')\ |\ s\ s'\ A\ B\ C\ D\ S.
\end{aligned}
$$

$s' = insert\ (Spy,\ SesKey\ (Some\ S,\ \{A, B\},\ \{C, D\}))\ s\ \wedge$
$\{PriKey\ S,\ PriKey\ A,\ PubKey\ B,\ PriKey\ C,\ PubKey\ D\} \subseteq spied\ s\}$

The former rule shows how simply the inductive method's limitations in modeling spy's fresh message generation are overcome by the relational method, whereas the latter rule reflects the fact that one can compute a session key just in case they know the initial nonce, as well as both a private and a public key for each key agreement.

Finally, here below are two protocol rules from the third above category. Based on the previous discussion on message anonymity, it comes as no surprise that every rule in this category extends a state with one or more events of the form $(Spy, \langle n, X\rangle)$ for some given agent identifier n and message X. Particularly, the following rules enable the spy to map any known private keys used in anonymity-compromised Chip Authentication Data to their same agent, and conversely to map any known Chip Authentication Data computed with anonymity-compromised private keys to their same agent. Notation f $-`$ $A$ denotes the inverse image of set $A$ under function $f$, so *Log* $-`$ *spied s* is the set of all the messages logged by the spy in state s.

abbreviation *rel-id-fact* :: $(state \times state)\ set$ **where**
$rel\text{-}id\text{-}fact \equiv \{(s, s') \mid s\ s'\ n\ A\ B.$
$s' = s \cup \{Spy\} \times \{\langle n, PriKey\ A\rangle,\ \langle n, PriKey\ B\rangle\} \wedge$
$\{PriKey\ A,\ PriKey\ B,\ \langle n, A \otimes B\rangle\} \subseteq spied\ s\}$

abbreviation *rel-id-mult* :: $(state \times state)\ set$ **where**
$rel\text{-}id\text{-}mult \equiv \{(s, s') \mid s\ s'\ n\ A\ B\ U.$
$s' = insert\ (Spy,\ \langle n, A \otimes B\rangle)\ s\ \wedge$
$U \cup \{PriKey\ A,\ PriKey\ B,\ A \otimes B\} \subseteq spied\ s\ \wedge$
$(\langle n, PriKey\ A\rangle \in spied\ s \vee \langle n, PriKey\ B\rangle \in spied\ s)\ \wedge$
$U \in key\text{-}sets\ (A \otimes B)\ (crypts\ (Log\ -`\ spied\ s))\}$

Both rules enforce the constraint that only known messages can be mapped to agents by providing for condition $X \in spied\ s$ for each message X such that event $(Spy, \langle n, X\rangle)$ is added to state s'. However, the latter rule alone reflects the requirement for the spy to be able to observe any such X in some logged message, modeled by the existence of a message set U satisfying both conditions $U \subseteq spied\ s$ and $U \in key\text{-}sets\ X\ (crypts\ (Log\ -`\ spied\ s))$.

In fact, applying this requirement to the components of an anonymity-compromised compound message is not necessary if they are observable within that message whenever they are known, as happens with components *PriKey A* and *PriKey B* of message $A \otimes B$ in the former rule. In this case, the enforcement of the requirement for that compound message alone, as happens with $A \otimes B$ in the latter rule, is sufficient to ensure that its components also are observable if they are known to the spy.

4.3.6 PROTOCOL PROPERTIES

As explained above, the primary means to prove protocol properties consist of the instantiations of rules *rtrancl-induct* and *rtrancl-start* in terms of operators $\vdash$ and $\models$, which look as follows.

$$[\![a \models b; P\ a; \bigwedge y\ z.\ [\![a \models y; y \vdash z; P\ y]\!] \Longrightarrow P\ z]\!] \Longrightarrow P\ b$$

$$[\![a \models b; P\ b; \neg\ P\ a]\!] \Longrightarrow \exists u\ v.\ a \models u \wedge u \vdash v \wedge v \models b \wedge \neg\ P\ u \wedge P\ v$$

Particularly, both rules have the same major premise $a \models b$. In turn, protocol properties are formalized as statements of the form $[\![s_0 \models s; P\ s]\!] \Longrightarrow Q\ s$, such that some conclusion $Q\ s$ is drawn about any protocol state s satisfying some appropriate assumption $P\ s$. Hence, to prove a protocol property, the above rules can be applied (although this is not the only available option in the case of rule *rtrancl-start*, as explained here below) by satisfying their major premise via the property's own major premise $s_0 \models s$. In this case, the free variables a and b occurring in the rules are instantiated to s_0 and s, respectively.

In more detail, to prove a protocol property, rule *rtrancl-induct* is typically used in the same way as happens with rule induction in the inductive method; that is, it is applied to turn the initial statement of the property into a few tractable subgoals left to be proven. Such subgoals match the two assumptions following the major premise $s_0 \models s$ in the resulting instantiation of the rule, which looks as follows.

$$[\![s_0 \models s; P\ s_0 \Longrightarrow Q\ s_0; \bigwedge y\ z.\ [\![s_0 \models y; y \vdash z; P\ y \Longrightarrow Q\ y; P\ z]\!] \Longrightarrow Q\ z; P\ s]\!] \Longrightarrow Q\ s$$

In turn, in the proof of the latter subgoal, the assumption $y \vdash z$ gives rise to a distinct subgoal for any protocol rule such that the conclusion $Q\ z$ is not an immediate consequence of assumptions $P\ y \Longrightarrow Q\ y$ (the *induction hypothesis*) and $P\ z$.

Rule *rtrancl-start* can be used in the same way, namely to transform the initial statement of a protocol property into tractable subgoals, by applying it as a *destruction rule* (cf. [17], section 5.7) to the property's major premise $s_0 \models s$. This option is viable for any protocol property of the form $[\![s_0 \models s; P\ s[; R]]\!] \Longrightarrow Q\ s$, where predicate P is such that $\neg\ P\ s_0$ and some further assumption R may be present, too. If rule *rtrancl-start* is applied to prove such a property, the following three subgoals are left to be proven.

- $[\![P\ s[;\ R]]\!] \Longrightarrow P\ s$, which is trivial as the conclusion matches an assumption.
- $[\![P\ s[;\ R]]\!] \Longrightarrow \neg\ P\ s_0$, which asks for proving that $\neg\ P\ s_0$.
- $[\![P\ s[;\ R];\ \exists u\ v.\ s_0 \models u \land u \vdash v \land v \models s \land \neg\ P\ u \land P\ v]\!] \Longrightarrow Q\ s$, which resembles the original statement of the property, but contains the conclusion of rule *rtrancl-start* as an additional assumption.

Yet, rule *rtrancl-start* can also be applied in a different way, that is, as an intermediate step within a structured proof, by instantiating its free variables a and b to any two protocol states satisfying its assumptions in the current proof state. As explained previously, its application enables to make either *forward* inferences about b, based on what is known about u and v, or *backward* inferences, proceeding the other way around.

Confidentiality properties have a conclusion of the form $X \notin spied\ s$ for a given message X. Two such properties are reported here. The former one states that the session keys used by *Owner n* to encrypt her unique password are guaranteed to remain unknown to the spy, unless *Asset n*'s PACE key and authentication private key are both compromised. The latter property provides a similar guarantee for the session keys used by *Asset n* to encrypt the final acknowledgment.

theorem *owner-seskey-secret*:
  assumes
    A: $s_0 \models s$ **and**
    B: $n \notin bad\text{-}shakey \cap bad\text{-}prikey$ **and**
    C: $(Owner\ n,\ Crypt\ (SesK\ SK)\ (Pwd\ n)) \in s$
  shows $SesKey\ SK \notin spied\ s$
⟨*proof*⟩

theorem *asset-seskey-secret*:
  assumes
    A: $s_0 \models s$ **and**
    B: $n \notin bad\text{-}shakey \cap (bad\text{-}pwd \cup bad\text{-}prikey)$ **and**
    C: $(Asset\ n,\ Crypt\ (SesK\ SK)\ (Num\ 0)) \in s$
  shows $SesKey\ SK \notin spied\ s$
⟨*proof*⟩

Authenticity properties have a conclusion of the form $(A, X) \in s$, where X is an exchanged message and A is its expected, legitimate sender. Again, two such properties are reported here. The former one states that if *Owner n* has sent her unique password and the corresponding final acknowledgment has been exchanged in response, then that reply is guaranteed to have been generated by *Asset n*, unless *Asset n*'s PACE key and authentication private key are both compromised. The latter property provides *Asset n* with the similar guarantee that under proper assumptions, if *Asset n* has sent the final acknowledgment, *Owner n* herself must have sent her unique password.

theorem *owner-num-genuine*:
  assumes
    A: $s_0 \models s$ **and**
    B: $n \notin$ *bad-shakey* $\cap$ *bad-prikey* **and**
    C: (*Owner n*, *Crypt* (*SesK SK*) (*Pwd n*)) $\in s$ **and**
    D: *Crypt* (*SesK SK*) (*Num 0*) $\in$ *used s*
  shows (*Asset n*, *Crypt* (*SesK SK*) (*Num 0*)) $\in s$
⟨*proof*⟩

theorem *asset-pwd-genuine*:
  assumes
    A: $s_0 \models s$ **and**
    B: $n \notin$ *bad-shakey* $\cap$ (*bad-pwd* $\cup$ *bad-prikey*) **and**
    C: (*Asset n*, *Crypt* (*SesK SK*) (*Num 0*)) $\in s$
  shows (*Owner n*, *Crypt* (*SesK SK*) (*Pwd n*)) $\in s$
⟨*proof*⟩

Notably, property *owner-num-genuine* contains an assumption of the form $X \in$ *used s*, whereas property *asset-pwd-genuine* does not. In fact, according to the protocol rules, the message proven to be genuine in *owner-num-genuine*, namely the asset's final acknowledgment, is sent *after* the one assumed to have been sent by its intended recipient, namely the owner's unique password. Hence, the occurrence of the latter message in past traffic is not sufficient to infer that of the former message, which then has to be included as a further assumption, as happened with property *B-trusts-NS3* in the previous section. On the contrary, this is unnecessary in *asset-pwd-genuine* as the message proven to be genuine is sent *before* the other one, whose occurrence in past traffic then entails that of the former message.

The proof of the *unicity property* here below, stating that any session key may be used in at most one genuine asset's authentication token, is a good example of how a highly nontrivial protocol property can be proven by applying rules *rtrancl-induct* and *rtrancl-start* in an appropriate combination, so it will be detailed in what follows.

theorem *asset-seskey-unique*:
assumes
A: $s_0 \models s$ **and**
B: $(Asset\ m,\ Token\ m\ (Auth\text{-}PriK\ m)\ B'\ C'\ SK) \in s$ **and**
C: $(Asset\ n,\ Token\ n\ (Auth\text{-}PriK\ n)\ B\ C\ SK) \in s$
shows $m = n \wedge B' = B \wedge C' = C$
⟨*proof*⟩

The starting step is proving the following basic properties, which can easily be done by induction. Particularly, the first two ones are *unicity lemmas*, stating that a given message (in this case, an ephemeral public key) may occur only once in a given set of messages issued by legitimate agents (namely, those exchanged in the second and the third protocol step). The statements of these lemmas, and then their proofs as well, are simpler than the inductive method's unicity lemmas, such as lemmas *no-nonce-NS1-NS2* and *unique-NB* in the previous section, in that they are not expressed in terms of the inductively defined function *parts*. In other words, they deal with genuine stand-alone messages only, rather than with any component of whatever, possibly fake, exchanged message. As explained above, this is yet another benefit of having overcome the inductive method's limitations in modeling spy's fresh message generation.

proposition *asset-ii-unique*:
$[\![s_0 \models s;\ (Asset\ m,\ \{\!|Num\ 2,\ PubKey\ A|\!\}) \in s;$
$(Asset\ n,\ \{\!|Num\ 2,\ PubKey\ A|\!\}) \in s]\!] \Longrightarrow$
$m = n$
⟨*proof*⟩

proposition *asset-ii-owner-ii*:
$[\![s_0 \models s;\ (Asset\ m,\ \{\!|Num\ 2,\ PubKey\ A|\!\}) \in s;$
$(Owner\ n,\ \{\!|Num\ 1,\ PubKey\ A|\!\}) \in s]\!] \Longrightarrow$
False
⟨*proof*⟩

proposition *asset-iv-state*:
$[\![s_0 \models s;\ (Asset\ n,\ Token\ n\ (Auth\text{-}PriK\ n)\ B\ C\ SK) \in s]\!] \Longrightarrow$
$\exists A\ D.\ fst\ (snd\ SK) = \{A, B\} \wedge snd\ (snd\ SK) = \{C, D\} \wedge$
$(Asset\ n,\ \{\!|Num\ 2,\ PubKey\ B|\!\}) \in s \wedge (Asset\ n,\ \{\!|Num\ 4,\ PubKey\ D|\!\}) \in s \wedge$
$Crypt\ (SesK\ SK)\ (PubKey\ D) \in used\ s \wedge (Asset\ n,\ PubKey\ B) \in s$
⟨*proof*⟩

proposition *owner-seskey-other*:
$[\![s_0 \models s;\ (Owner\ n,\ SesKey\ SK) \in s]\!] \Longrightarrow$
$\exists A\ B\ C\ D.\ fst\ (snd\ SK) = \{A, B\} \wedge snd\ (snd\ SK) = \{C, D\} \wedge$

$(Owner\ n,\ \{\!|Num\ 1,\ PubKey\ A|\!\})\in s\ \wedge$
$(Owner\ n,\ \{\!|Num\ 3,\ PubKey\ C|\!\})\in s\ \wedge$
$(Owner\ n,\ Crypt\ (SesK\ SK)\ (PubKey\ D))\in s$
$\langle proof\rangle$

proposition *seskey-spied*:
$[\![s_0 \models s;\ SesKey\ SK \in spied\ s]\!] \Longrightarrow$
$\exists S\ A\ C.\ fst\ SK = Some\ S \wedge A \in fst\ (snd\ SK) \wedge C \in snd\ (snd\ SK)\ \wedge$
$\{PriKey\ S,\ PriKey\ A,\ PriKey\ C\} \subseteq spied\ s$
$\langle proof\rangle$

The following *transition lemmas* and *regularity lemmas* are also required. Lemmas of the former kind state something about any two states s, s' such that $s \vdash s'$, $P\ X\ s'$, $\neg\ P\ X\ s$, and optionally $s_0 \models s$, $Q\ s$, $R\ s'$, where P, Q (if any), and R (if any) are some properties and X is a given message. Lemmas of the latter kind apply to any state s such that $s_0 \models s$, $P\ X\ s$, and optionally $Q\ s$, $R\ s$. Particularly, the special case where $P = (\lambda X\ s.\ X \in parts\ (used\ s))$ gives rise to regularity lemmas resembling the inductive method's ones, where function *parts* is introduced via the same inductive definition as in the previous section, so that set $parts\ (used\ s)$ is the relational method's counterpart of $parts\ (spies\ evs)$.

If $\neg\ P\ X\ s_0$, possibly by virtue of the further assumptions $Q\ s$ or $R\ s$ (if any), a strategy to prove a regularity lemma for P, Q (if any), and R (if any) is to derive it from the corresponding transition lemma, using rule *rtrancl-start* as a *destruction rule*, as explained above. The regularity lemmas *parts-crypt-pubkey* and *parts-pubkey-false* here below are easily derived in this way from the related transition lemmas *parts-crypt-pubkey-start* and *parts-pubkey-false-start*.

Property *parts-crypt-pubkey-start* simply follows from protocol rules, whereas the proofs of properties *parts-pubkey-false-start* and *asset-ii-spied-start* are nontrivial.

The former proof is as follows. By lemma *parts-crypt-pubkey-start* and the assumptions, there exists an event $(Asset\ n,\ Token\ n\ (Auth\text{-}PriK\ n)\ B\ C\ SK)$ contained in s' but not in s, so there is a cryptogram $Crypt\ (SesK\ SK)\ (PubKey\ D) \in used\ s$ by protocol rule *rel-asset-iv*. By lemma *parts-crypt-pubkey* and the assumptions, there is an event $(Asset\ n',\ Token\ n'\ (Auth\text{-}PriK\ n')\ B'\ D\ SK) \in s$, so that $snd\ (snd\ SK) = \{D,\ D'\}$ and $Crypt\ (SesK\ SK)\ (PubKey\ D') \in used\ s$ by lemma *asset-iv-state*. Therefore, each A in set $snd\ (snd\ SK)$ is such that $Crypt\ (SesK\ SK)\ (PubKey\ A) \in used\ s$. But $C \in snd\ (snd\ SK)$ by the earlier application of lemma *parts-crypt-pubkey-start*, so that $Crypt\ (SesK\ SK)\ (PubKey\ C) \in used\ s$, which contradicts assumption D, QED.

Property *asset-ii-spied-start* is proven in a similar way, using these and other basic lemmas (including unicity ones), transition lemmas, and regularity lemmas.

proposition *parts-crypt-pubkey-start*:
$[\![s \vdash s'; Crypt\ (SesK\ SK)\ (PubKey\ C) \in parts\ (used\ s');$
$Crypt\ (SesK\ SK)\ (PubKey\ C) \notin parts\ (used\ s)]\!] \Longrightarrow$
$C \in snd\ (snd\ SK) \wedge ((\exists n.\ (Owner\ n,\ SesKey\ SK) \in s') \vee$
$(\exists n\ B.\ (Asset\ n,\ Token\ n\ (Auth\text{-}PriK\ n)\ B\ C\ SK) \in s')) \vee$
$SesKey\ SK \in spied\ s'$
⟨*proof*⟩

proposition *parts-crypt-pubkey*:
$[\![s_0 \models s; Crypt\ (SesK\ SK)\ (PubKey\ C) \in parts\ (used\ s)]\!] \Longrightarrow$
$C \in snd\ (snd\ SK) \wedge ((\exists n.\ (Owner\ n,\ SesKey\ SK) \in s) \vee$
$(\exists n\ B.\ (Asset\ n,\ Token\ n\ (Auth\text{-}PriK\ n)\ B\ C\ SK) \in s)) \vee$
$SesKey\ SK \in spied\ s$
⟨*proof*⟩

proposition *parts-pubkey-false-start*:
assumes
A: $s_0 \models s$ **and**
B: $s \vdash s'$ **and**
C: $Crypt\ (SesK\ SK)\ (PubKey\ C) \in parts\ (used\ s')$ **and**
D: $Crypt\ (SesK\ SK)\ (PubKey\ C) \notin parts\ (used\ s)$ **and**
E: $\forall n.\ (Owner\ n,\ SesKey\ SK) \notin s'$ **and**
F: $SesKey\ SK \notin spied\ s'$
shows *False*
⟨*proof*⟩

proposition *parts-pubkey-false*:
$[\![s_0 \models s; Crypt\ (SesK\ SK)\ (PubKey\ C) \in parts\ (used\ s);$
$\forall n.\ (Owner\ n,\ SesKey\ SK) \notin s; SesKey\ SK \notin spied\ s]\!] \Longrightarrow$
False
⟨*proof*⟩

proposition *asset-ii-spied-start*:
assumes
A: $s_0 \models s$ **and**
B: $s \vdash s'$ **and**
C: $PriKey\ B \in spied\ s'$ **and**
D: $PriKey\ B \notin spied\ s$ **and**
E: $(Asset\ n,\ \{\!|Num\ 2,\ PubKey\ B|\!\}) \in s$

shows *Auth-PriKey n* $\in$ *spied s* $\wedge$
($\exists$*C SK*. (*Asset n*, *Token n* (*Auth-PriK n*) *B C SK*) $\in$ *s*)
⟨*proof*⟩

An additional propaedeutic step is proving lemma *asset-iv-unique* here below, which reduces the proof of theorem *asset-seskey-unique* to proving that $B' = B$.

This lemma can be proven as follows. The assumptions and lemma *asset-iv-state* entail that both events (*Asset m*, ⦃*Num 2*, *PubKey B*⦄) and (*Asset n*, ⦃*Num 2*, *PubKey B*⦄) are contained in *s*, so $m = n$ by lemma *asset-ii-unique*. Furthermore, by rule *rtrancl-start*, there are two consecutive states *u* and *v* between s_0 and *s* such that both authentication token sendings are contained in *v*, but not in *u*. Thus, assuming that the tokens do not match, one has already been sent in *u*, so that (*Asset n*, *PubKey B*) $\in u$ by lemma *asset-iv-state*, whereas the other one is sent in the very transition from *u* to *v*, so that (*Asset n*, *PubKey B*) $\notin u$ by protocol rule *rel-asset-iv*. This proves by contradiction that the tokens match, QED.

proposition *asset-iv-unique*:
assumes
A: $s_0 \models s$ **and**
B: (*Asset m*, *Token m* (*Auth-PriK m*) *B C' SK'*) $\in$ *s* **and**
C: (*Asset n*, *Token n* (*Auth-PriK n*) *B C SK*) $\in$ *s*
shows $m = n \wedge C' = C \wedge SK' = SK$
⟨*proof*⟩

The only remaining task for the proof of theorem *asset-seskey-unique*, namely proving that $B' = B$, can be accomplished by contradiction, as follows. The assumptions, along with the additional one $B' \neq B$ and lemma *asset-iv-state*, entail that both events (*Asset m*, ⦃*Num 2*, *PubKey B'*⦄) and (*Asset n*, ⦃*Num 2*, *PubKey B*⦄) are contained in *s*, where *fst* (*snd SK*) = {*B*, *B'*} (*proposition D*). Moreover, *Crypt* (*SesK SK*) (*PubKey C*) $\in$ *parts* (*used s*) by assumption *C*, owing to the structure of authentication tokens. A contradiction can then arise by applying lemma *parts-pubkey-false* and showing that its last two assumptions also hold.

In fact, by lemma *owner-seskey-other*, assuming the existence of an event (*Owner i*, *SesKey SK*) $\in$ *s* implies that of an $A \in$ *fst* (*snd SK*) such that (*Owner i*, ⦃*Num 1*, *PubKey A*⦄) $\in$ *s*, which contradicts proposition *D* by lemma *asset-ii-owner-ii*.

On the other hand, by lemma *seskey-spied*, assumption *SesKey SK* $\in$ *spied s* entails that there exists an $A \in$ *fst* (*snd SK*) such that *PriKey A*

$\in$ *spied s*. By rule *rtrancl-start*, there are two consecutive states u and v between s_0 and s such that this holds for v, but not for u. Hence, by proposition D, there exists an event (*Asset i*, $\{\!|$*Num 2*, *PubKey A*$|\!\}$) $\in$ u, and then also an event (*Asset i*, *Token i* (*Auth-PriK i*) A C'' SK') $\in u$ by lemma *asset-ii-spied-start*. But $SK' = SK$ by the assumptions, proposition D, and lemma *asset-iv-unique*, so *Crypt* (*SesK SK*) (*PubKey* C'') $\in$ *parts* (*used u*). A contradiction then arises by applying lemma *parts-pubkey-false* and showing that its last two assumptions also hold, using lemmas *owner-seskey-other*, *asset-ii-owner-ii*, and *seskey-spied*, QED.

Anonymity properties state that under proper assumptions, some message X_n pertaining to *Asset n*/*Owner n* is anonymous, so that they have a conclusion of the form $\langle n, X_n \rangle \notin$ *spied s*. An example is the following property, stating that if the anonymity of *Owner n*'s unique password is not compromised by means other than attacking the protocol, it is preserved unless the *n*th PACE key, either the *n*th password or the *n*th authentication private key, and the anonymity of either the *n*th authentication public key or the *n*th PACE key, are compromised in the same way.

theorem *pwd-anonymous*:
  assumes
    A: $s_0 \models s$ **and**
    B: $n \notin$ *bad-id-password* **and**
    C: $n \notin$ *bad-shakey* $\cap$ (*bad-pwd* $\cup$ *bad-prikey*) $\cap$ (*bad-id-pubkey* $\cup$ *bad-id-shak*)
  shows $\langle n$, *Pwd n*$\rangle \notin$ *spied s*
$\langle$*proof*$\rangle$

Besides the general-purpose tools available for any protocol property, namely rules *rtrancl-induct* and *rtrancl-start*, one more powerful tool for the proof of logging-dependent message anonymity is the constraints put by protocol rules on the messages observable within logged ones, as they limit the scope of what the spy can use to map messages to agents. These constraints can be formalized as statements on the content of set *key-sets* X (*crypts* (*Log* –‘ *spied s*)) for a given message X, and can be proven by induction after proving rewrite rules for the addition of new items to the message sets input to functions *crypts* and *key-sets*. Since the protocol rules modeling protocol steps add new messages to set *Log* –‘ *spied s*, such rewrite rules are required to simplify the respective subgoals arising from induction, so as to allow for the use of the induction hypothesis. An example of this approach is provided by the proof of the above anonymity property contained in [18].

In the case of logging-independent message anonymity, no such constraint is put by protocol rules, so that anonymity properties have to be proven in much the same way as any other protocol property. An example of such a proof is provided in [20].

Since both the initial state s_0 and protocol rules comply with the principle that the spy can map known messages alone to agents, the secrecy of a message is a sufficient condition for its anonymity. Therefore, every confidentiality property could be turned into an anonymity property by just replacing its conclusion $X \notin spied\ s$ with one of the form $\langle n, X \rangle \notin spied\ s$, while leaving its assumptions unchanged. For instance, the following, alternative anonymity property *pwd-anonymous-2* would be a mere corollary of confidentiality property *pwd-secret*, also reported here below.

theorem *pwd-secret*:
  assumes
    A: $s_0 \models s$ **and**
    B: $n \notin bad\text{-}pwd \cup bad\text{-}shakey \cap bad\text{-}prikey$
  shows $Pwd\ n \notin spied\ s$
$\langle proof \rangle$

corollary *pwd-anonymous-2*:
  assumes
    A: $s_0 \models s$ **and**
    B: $n \notin bad\text{-}pwd \cup bad\text{-}shakey \cap bad\text{-}prikey$
  shows $\langle n, Pwd\ n \rangle \notin spied\ s$
$\langle proof \rangle$

So, why should one embark on the enterprise of proving an anonymity property rather endowed with its own assumptions, such as *pwd-anonymous*? The simple answer is that, though being sufficient, the secrecy of a message is usually not necessary for its anonymity, so that the latter is preserved under weaker assumptions than the former; namely, a message may generally remain anonymous even if it becomes public knowledge, as shown by the common experience of newspaper readers. As a result, an anonymity corollary of a confidentiality property would typically contain unnecessarily strong assumptions. For example, the assumptions of property *pwd-anonymous-2* are stronger than those of property *pwd-anonymous*, since:

- $bad\text{-}id\text{-}password \subseteq bad\text{-}pwd \subseteq bad\text{-}pwd \cup (bad\text{-}shakey \cap bad\text{-}prikey)$.
- $bad\text{-}shakey \cap (bad\text{-}pwd \cup bad\text{-}prikey) \cap (bad\text{-}id\text{-}pubkey \cup bad\text{-}id\text{-}shak)$

$$\subseteq \textit{bad-shakey} \cap (\textit{bad-pwd} \cup \textit{bad-prikey})$$
$$= (\textit{bad-shakey} \cap \textit{bad-pwd}) \cup (\textit{bad-shakey} \cap \textit{bad-prikey})$$
$$\subseteq \textit{bad-pwd} \cup (\textit{bad-shakey} \cap \textit{bad-prikey}).$$

4.4 CONCLUSIONS

This discussion clearly suggests that the formal verification of cryptographic protocols with proof assistants is still far from being an exhausted subject. In fact, existing verification methods are doomed to face ever new challenges arising from the continuous evolution of protocols, as well as of cryptography itself, as shown by the reasons leading from the inductive to the relational method. Further directions of investigation include on the one hand the application of the relational method to more real-world protocols, which might result in extensions of the method enabling to formalize additional cryptographic techniques or to optimize formal definitions and proofs, on the other hand the exploration of entirely new approaches, such as the use of formal reasoning about knowledge in multi-agent systems (cf. [9], section 5.5).

At the same time, the preceding discussion also enables to come to two positive conclusions. First, the inductive and the relational method on the whole allow for the formal verification of a wide range of protocols, in the latter case including some making use of public key cryptography features other than mere encryption and decryption, without any counterexample found at the time of writing. Second, formal protocol verification with proof assistants reveals still another connection between logic and computer science, besides all the ones that already contribute to render both subjects so fascinating.

REFERENCES

1. G. Bella and L. C. Paulson. Kerberos Version IV: Inductive Analysis of the Secrecy Goals. In *Computer Security – ESORICS 98*, 1998.
2. A. D. Brucker and S. A. Mödersheim. Integrating Automated and Interactive Protocol Verification. In *FAST 2009: Formal Aspects in Security and Trust*, 2010.
3. D. F. Butin. *Inductive Analysis of Security Protocols in Isabelle/HOL with Applications to Electronic Voting*. PhD thesis, Dublin City University, Sept. 2012.

4. Common Criteria Maintenance Board. *Common Criteria for Information Technology Security Evaluation – Part 3: Security assurance components, Version 3.1, Revision 5*, Apr. 2017. CCMB-2017-04-003.
5. J. Goubault-Larrecq. Towards Producing Formally Checkable Security Proofs, Automatically. In *CSF 2008: 21st IEEE Computer Security Foundations Symposium*, 2008.
6. T. Hales et al. A formal proof of the Kepler conjecture. *arXiv*, Jan. 2015. eprint: 1501.02155, primaryClass: math.MG.
7. A. V. Hess, S. A. Mödersheim, A. D. Brucker, and A. Schlichtkrull. Performing Security Proofs of Stateful Protocols. In *CSF 2021: 34th IEEE Computer Security Foundations Symposium*, 2021.
8. J. Hoffstein, J. Pipher, and J. H. Silverman. *An Introduction to Mathematical Cryptography*. Springer, 2nd edition, 2014.
9. M. Huth and M. Ryan. *Logic in Computer Science – Modelling and Reasoning about Systems*. Cambridge University Press, 2nd edition, 2004.
10. International Civil Aviation Organization (ICAO). *Doc 9303 – Machine Readable Travel Documents – Part 11: Security Mechanisms for MRTDs*, 8th edition, 2021.
11. G. Lowe. An Attack on the Needham-Schroeder Public-Key Authentication Protocol. *Information Processing Letters*, Nov. 1995.
12. S. Meier, C. J. F. Cremers, and D. A. Basin. Efficient Construction of Machine-Checked Symbolic Protocol Security Proofs. *Journal of Computer Security*, 2013.
13. S. Meier, B. Schmidt, C. J. F. Cremers, and D. A. Basin. The TAMARIN Prover for the Symbolic Analysis of Security Protocols. In *CAV 2013: Computer Aided Verification*, 2013.
14. M. S. Nawaz et al. A Survey on Theorem Provers in Formal Methods. *arXiv*, Dec. 2019. eprint: 1912.03028, primaryClass: cs.SE.
15. R. M. Needham and M. D. Schroeder. Using Encryption for Authentication in Large Networks of Computers. *Communications of the ACM*, Dec. 1978.
16. T. Nipkow. *Programming and Proving in Isabelle/HOL*, Dec. 2021. https://isabelle.in.tum.de/website-Isabelle2021-1/dist/Isabelle2021-1/doc/prog-prove.pdf.
17. T. Nipkow, L. C. Paulson, and M. Wenzel. *Isabelle/HOL – A Proof Assistant for Higher-Order Logic*, Dec. 2021. https://isabelle.in.tum.de/website-Isabelle2021-1/dist/Isabelle2021-1/doc/tutorial.pdf.
18. P. Noce. The Relational Method with Message Anonymity for the Verification of Cryptographic Protocols. *Archive of Formal Proofs*, Dec. 2020. https://isa-afp.org/entries/Relational˙Method.html, Formal proof development.

19. P. Noce. A Shorter Compiler Correctness Proof for Language IMP. *Archive of Formal Proofs*, June 2021. https://isa-afp.org/entries/IMP˙Compiler.html, Formal proof development.

20. P. Noce. Logging-independent Message Anonymity in the Relational Method. *Archive of Formal Proofs*, Aug. 2021. https://isa-afp.org/entries/Logging˙Independent˙Anonymity.html, Formal proof development.

21. L. C. Paulson. Theory HOL-Auth.NS˙Public (included in the Isabelle2021-1 distribution), 1996. https://isabelle.in.tum.de/website-Isabelle2021-1/dist/library/HOL/HOL-Auth/NS˙Public.html.

22. L. C. Paulson. The Inductive Approach to Verifying Cryptographic Protocols. *Journal of Computer Security*, 1998.

23. L. C. Paulson. Inductive Analysis of the Internet Protocol TLS. *ACM Transactions on Information and System Security*, 1999.

24. L. C. Paulson. Proving Security Protocols Correct. In *LICS '99: Proceedings of the 14th Annual IEEE Symposium on Logic in Computer Science*, 1999.

25. L. C. Paulson. Relations Between Secrets: Two Formal Analyses of the Yahalom Protocol. *Journal of Computer Security*, 2001.

26. L. C. Paulson. Gödel's Incompleteness Theorems. *Archive of Formal Proofs*, Nov. 2013. https://isa-afp.org/entries/Incompleteness.html, Formal proof development.

27. M. Sipser. *Introduction to the Theory of Computation.* Cengage Learning, 3rd edition, 2013.

28. M. Wenzel. *Isabelle/jEdit*, Dec. 2021. https://isabelle.in.tum.de/website-Isabelle2021-1/dist/Isabelle2021-1/doc/jedit.pdf.

5 Formal Modeling and Security Analysis of Security Protocols

Paolo Modesti
Teesside University, Middlesbrough, United Kingdom

Rémi Garcia
Teesside University, Middlesbrough, United Kingdom

CONTENTS

DOI: 10.1201/9781003090052-5

Security protocols are critical components for the construction of secure Internet services but their design and implementation are difficult and error prone. Formal modeling and verification can be extremely beneficial to support the development of secure software. In this chapter, we introduce tools and techniques for the formal modeling and security analysis of cryptographic protocols. First, we consider the process of constructing a formal model from a given set of requirements, then we discuss the theoretical and practical challenges in automated verification. Different specification languages and verification tools are considered to cater to different levels of user expertise and complexity of the protocols under analysis. The chapter includes case studies, demonstrating the practical applicability of these tools in different application fields: e-commerce, e-payments, and blockchain.

5.1 INTRODUCTION

The Covid-19 pandemic has offered an incredible opportunity for cybercrime to thrive with millions of people working from home using digital devices secured by accidental circumstances rather than by design. Many organizations were notoriously under-prepared. Cyberattacks are more than just technical events, they are attacks on the very foundations of the modern information society. The Center for Strategic and International Studies estimated the worldwide monetary loss from cybercrime to be around $945 billion in 2020 [102]. This has clearly exposed the gap between the practices adopted by the software industry and the stark reality. Crucially, designing secure and dependable systems is fundamental to protect digital resources and reduce the attack surface. Therefore, experts [110, 51] recommend investing more resources on enhancing the design of secure systems rather than just fixing vulnerabilities in existing systems when they are discovered.

In particular, security protocols are critical components for the construction of secure Internet services and distributed applications, but their design and implementation are difficult and error-prone. Some vulnerabilities in very popular protocols like TLS and SSH [97, 55], which are both expertly built and rigorously tested, have been undetected for years. The core reasons of this situation need to be investigated and novel approaches explored. In this context, techniques for formal modeling and analysis of security protocols could support software engineers, but there is reluctance to embrace these tools outside the research settings because of their complexity [59]. Most practitioners find such methodologies complex and incompatible with their work requirements (e.g. difficulty to write the formal specification, steep learning curve, need to understand the underlying advanced theoretical computer science notions [108, 99]). Therefore, vulnerable software is still deployed, with tens of thousands of new vulnerabilities being discovered in major products each year. Such defective software has a serious negative impact on both individuals and organizations utilizing vulnerable systems every day.

In this chapter, we focus on an approach to formal modeling and verification that can be adequate to the practitioner's skills and needs. In fact, we can bridge the gap between formal representation and actual implementation with a framework adopting a conceptual model aligned with the level of abstraction used for the symbolic (high-level) representation of cryptographic and communication primitives. This would make formal methods and tools more accessible to students and practitioners [90].

OUTLINE OF THE CHAPTER

In Section 5.2, we consider modeling and verification in the context of security protocol development, and in Section 5.3 we introduce the fundamentals like the attacker model and security properties. Then, in Section 5.4, we discuss the theoretical and practical challenges in automated verification. Different specification languages and verification tools are considered in Section 5.5 to address different levels of user expertise and complexity of the protocols under analysis. Section 5.6 includes case studies, demonstrating the practical applicability of these tools in two different application fields: e-payments and blockchain.

5.2 MODELING AND VERIFICATION IN THE CONTEXT OF SECURITY PROTOCOLS DEVELOPMENT

This section details the foundational concepts used in security protocol verification. We present how protocols are modeled in a formal framework, and what a dishonest entity represents when searching for vulnerabilities. The different properties that are commonly checked appear in this section, as well as how the verification tools can capture the behavior of cryptographic primitives. We also discuss the problem of abstraction and its pedagogic implications.

A typical workflow for the development of security protocols is shown in Figure 5.1. Similar workflows have been proposed for example in [58, 90]. The process usually starts by identifying a set of software requirements specifying the expected behavior of the protocol, including the security goals that the application aims to achieve. There may be two different scenarios: 1) *New protocols*: the definition of the requirements is entirely under the control of the protocol designer; 2) *Existing protocols*: the requirements, and sometimes a reference implementation, already exist; the developer needs to consult the documentation to extract the relevant information.

The specification of a protocol typically includes the roles of *agents* involved in the protocol, the information they possess prior to the protocol run (*initial knowledge*), the specification of the messages exchanged between agents (*actions*) during the protocol execution and the *security goals* (e.g. secrecy, authentication) that should hold at the end of the protocol run.

While in software engineering having both explicit and implicit requirements is generally common, in formal modeling implicit requirements are particularly problematic. In fact, even something that may be obvious and unambiguous for a human needs to be defined precisely and explicitly to be interpreted by a machine. Therefore, implicit requirements can be a source of ambiguity or incompleteness, that can lead to the development of a formalized model that does not reflect the reality of the system under consideration.

Requirements can be expressed in different ways: informally, formally or semi-formally. Informal requirements are usually described in a

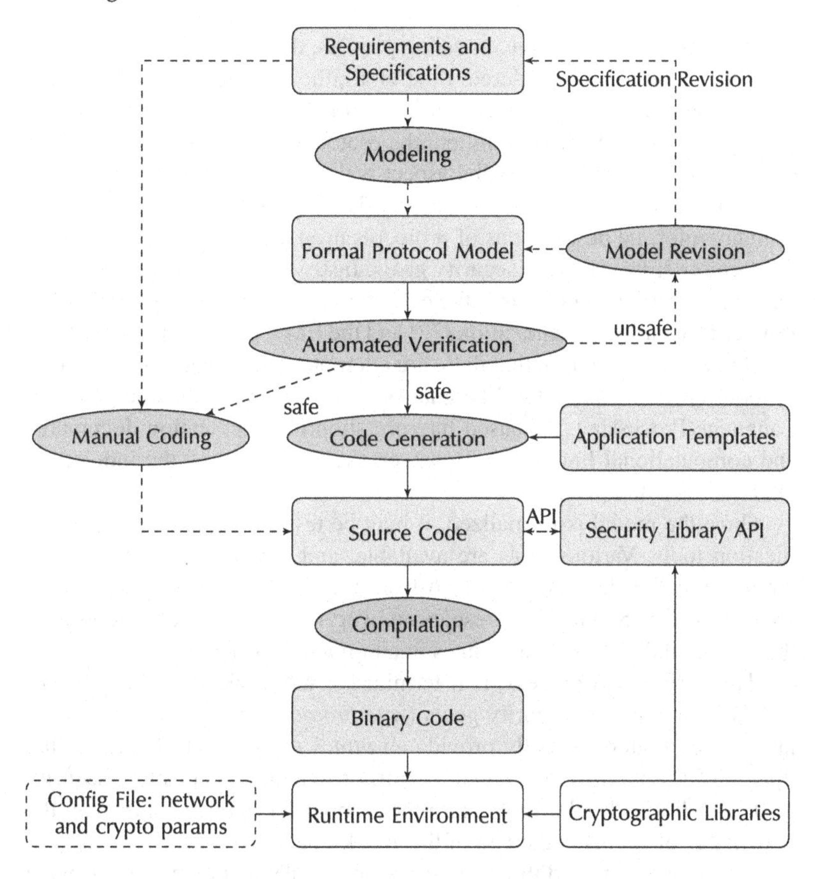

Figure 5.1 Development of security protocols workflow.

natural language, and though common in human communication, may be ambiguous or refer to implicit concepts that may lead to different interpretations by different people. This is quite usual in software engineering as end-users and developers need not only to build a common understanding of the application domain, but also a mutually intelligible communication language. Instead, formal requirements use mathematics-based specification languages to describe in rigorous terms the desired properties of the system under development. Unfortunately, the required formalism is often too complex, not only for the end-users but also for the software developers. Therefore, the adoption of specification languages that can be used by practitioners is extremely important as we will see in Section 5.5.

In the initial phase of the modeling process, there are a few essential aspects that need to be considered. First of all, the specification of the model requires defining not only the actions performed by the agents, but also their initial knowledge. If this is omitted, the protocol may be ambiguous or not even executable. Another crucial aspect is the specification of the security goals that protocols are meant to achieve. This is an often neglected activity as many informal descriptions of protocols used in the industry fail to give a rigorous definition of the security goals. Instead, they use the natural language to describe them. This is the case, for example, of the ISO/IEC 9798 standard for entity authentication [72] and the EMV payment protocol [58].

It is worth to mention that there are attempts at automatic formalization of requirements from natural languages. Such approaches include Natural Language Processing, AI-based interpretation of specification documents and computational linguistics. However, their adoption by the industry is extremely limited [59].

Once the model is formalized, it may be tested with one or more verification tools. Various tools are available, and they differ not only from the specification language but also from the different security properties they can verify. Section 5.4 presents an overview of such tools, along with theoretical and practical challenges in automated verification.

The verification process, if it terminates, will indicate if the protocol satisfies the expected security goals (*safe protocol*) or not (*unsafe*). In the latter case, the tools usually provide an *attack trace*: a list of actions that the *attacker* (aka *intruder*) has to perform to violate one or more security goals. Additionally, the tools can report about the *reached state*, *i.e.* the knowledge of agents at the end of the attack trace.

This output can lead the next design and implementation steps. Unsafe protocols may require an iterative process of revision and verification of the model until the protocol is successfully verified. In some cases, even the requirements may need a revision if they are ambiguous or ill-formed. Although users can employ a verification tool as a black-box security oracle, the attack trace provides elements that help to understand why the protocol fails in satisfying the security goals. Such iterative process is quite standard in formal design of security protocols, and it is aimed at capturing design errors in the very early phases of software development.

After modeling and verification, manual coding or automatic code generation can be used to build an implementation of the protocol. Code generation can be very effective, as this is a phase where implementation errors typically occur [55]. Additionally, code generation often relies on application templates to define the common structure of a program, and on an existing security library API that allows the generated source code to be compiled. Application execution support is provided by the runtime environment of the target platform, including relevant cryptographic libraries and configuration files.

5.3 MODELING FUNDAMENTALS

5.3.1 THE DOLEV-YAO MODEL

To model communications in an adversarial environment, the Dolev-Yao model [53] has been proposed to provide a formal framework of the intruder's capabilities over the network and the honest agents, also called honest principals. In this model, the intruder has complete control over the network. It can overhear and intercept any message, as well as generate its own, following specific rules allowed in a case-by-case basis.

Figure 5.2 shows the intruder rules, and the fact $\mathsf{ik}(m)$ denotes that the intruder knows the term m. Since every communication between honest agents is assumed to be mediated by the intruder, it happens through $\mathsf{ik}(\cdot)$ facts. The model assumes the existence of a set of function symbols (with an associated arity) partitioned into two subsets of *public* and *private* symbols. The first rule describes both asymmetric encryption and signing $(\cdot)$, while the second one models that a ciphertext can be decrypted if the corresponding decryption key is known. $\mathsf{inv}(\cdot)$ is a *private* function symbol representing the secret component of a given key-pair. The third rule allows the attacker to learn the payload of any known signed message. Symmetric encryption $(\{|\cdot|\})$ can be modeled similarly to the first two rules, but in this case the same key is employed for both encryption and decryption. Additionally, there are rules for tupling and projecting tuple elements, and a rule for the application of *public* function symbols to known messages. Constants, including agent identities, are modeled as public functions with zero-arity.

5.3.2 SECURITY PROPERTIES

We introduce here the most common security properties considered in protocol verification: *secrecy* and *authentication*. We also discuss *equational theories* that allow to model important functions and their algebraic properties, like exponentiation used in the Diffie-Hellman key agreement or the *xor* operator used in various cryptographic protocols.

5.3.2.1 Secrecy

When honest agents communicate, the confidentiality of the core payloads is essential to avoid eavesdropping on sensitive information. A term is said

$\mathsf{ik}(M).\mathsf{ik}(K)$	$\Rightarrow$	$\mathsf{ik}(KM)$	*Asymmetric Encryption*
$\mathsf{ik}(KM).\mathsf{ik}(\mathsf{inv}(K))$	$\Rightarrow$	$\mathsf{ik}(M)$	
$\mathsf{ik}(\mathsf{inv}(K)M)$	$\Rightarrow$	$\mathsf{ik}(M)$	
$\mathsf{ik}(M).\mathsf{ik}(K)$	$\Rightarrow$	$\mathsf{ik}(\{\mid M \mid\}_K)$	*Symmetric Encryption*
$\mathsf{ik}(\{\mid M \mid\}_K).\mathsf{ik}(K)$	$\Rightarrow$	$\mathsf{ik}(M)$	
$\mathsf{ik}(M).\mathsf{ik}(N)$	$\Rightarrow$	$\mathsf{ik}(M,N)$	*Tupling*
$\mathsf{ik}(M,N)$	$\Rightarrow$	$\mathsf{ik}(M).\mathsf{ik}(N)$	*Projection*
$\mathsf{ik}(M_1).\cdots.\mathsf{ik}(M_n)$	$\Rightarrow$	$\mathsf{ik}(f(M_1,\ldots,M_n))$	*Function Application*

Figure 5.2 Dolev-Yao intruder rules.

to be secret for a given agent when it cannot be derived from the intruder's knowledge at any point in the protocol execution. It should be noted that secrecy is subjective: an agent receiving a message which has been produced by the intruder may accept it as a legitimate one. This can be, for example, if a simple message is sent encrypted with a public key. The intruder just needs to encrypt its own forged message with the same public key and fool the receiving agent. The term sent at first will still be confidential, but the received one will not, for the honest receiver.

Stronger definitions can include *forward secrecy*. With this property, any message exchanged between non-compromised agents will remain secret after they fall victim to an attack. A common way to implement it is to use one-time keys, preventing information leakage of message history when long-term keys are compromised.

5.3.2.2 Authentication

Authentication properties follow Lowe's definition [77], where they correspond to *agreement* observations. Two kinds of authentication can arise: *non-injective* and the stronger *injective*.

With non-injective agreement, we can have two honest agents A and B, with B wanting to agree with A on a certain message *Msg*. A is then supposed to send *Msg* to B and be authenticated by *Msg*. The agreement is successful when there exists two observations in the protocol execution: B completed a protocol run as receiver, getting *Msg* from apparently A, and A completed it as well as initiator, sending *Msg* to apparently B.

Injective agreement introduces a notion of freshness to avoid replay attacks. The above conditions need to be satisfied, but for each run of B as receiver of *Msg* from apparently A, there must exist only one run of A

as initiator sending *Msg* to apparently *B*. With this guarantee, an intruder could not reuse *Msg* in another session in order to impersonate *A*.

5.3.2.3 Equational Theories

In every formal system, some axioms and symbols are defined, as well as the logic of the relations every symbol implies. The set of formulas that can be coherently derived from the axioms form a global theory of the system.

On a local level, every operator or symbol can have its own theory. Equational theories are expressed in quantifier free first-order logic with an equality relation. Formally, an *equation* is an atomic formula of the form $x = y$. Taking the example of the classical addition, a purely additive equation admits commutativity and associativity for its terms, *i.e.* $x+y = y+x$ and $(x+y)+z = x+(y+z)$.

There are however limits to what can be admissible with this kind of theories in a verification tool. For example, an equational theory must have the *Finite Variant Property* [44], stating that given any term t, t must be rewritable into a finite number of most general *variants* $t_1, \dots, t_n$. A variant can be considered as a pattern describing the canonical form of instances of a term. With this property, an infinite number of possible rewritings for a given term can be expressed using finitely many terms and substitutions.

5.3.3 ABSTRACTIONS AND PEDAGOGY

When applying formal methods, particularly in a practitioner context, we should consider how such techniques can be learned and used correctly. This is a general problem in Computer Science Education (CSE), and learning theories can be useful to understand the challenges newcomers may face. The *constructivist learning theory* [109, 35] claims that knowledge is acquired by combining sensory data (gathered from experience) with existing knowledge to create new cognitive structures. This process is applied recursively to generate new knowledge. Additionally, new knowledge is built reflecting on the existing one. To be effective, learning must be active, and the teacher must guide and support students in their endeavors.

Authors promoting a constructivist perspective in CSE [69, 27] indicate the importance of understanding both abstract and concrete concepts. Ben Ari [27] underlines that the application of constructivism must consider that “a (beginning) computer science student has no effective model of a computer”, and that “the computer forms an accessible ontological reality”. The latter concept manifests itself when students interacting with a computer can immediately realize the effect of the application of their mental models. In other words, the results of misconceptions are discovered instantly.

Rightly, Ben Ari [27] notes that "there is not much point negotiating models of the syntax or semantics of a programming language". Tools used by learners, like compilers used in programming activities, have implications for building mental models.

In formal methods for security, abstractions play a fundamental role. The formalization of a model requires not only a good understanding of the concrete system, but also the ability to identify what elements are relevant, and what can be neglected for the purpose of formal verification. This requires a viable abstract representation of the system, suitable for verification.

There is no general consensus on when abstractions need to be introduced. Object-Oriented Programming is a good example. Adams [5] opposes the idea of postponing teaching OOP until late in the course of studies because at that time it is difficult to have an impact on the learners' low-level model, but he also believes that OOP should not be taught too early when students are not mature enough to assimilate properly the related concepts. Ben Ari [27] remarks that "advocates of an objects-first approach seem to be rejecting Piaget's view that abstraction (or accommodation) follows assimilation". This seems to be confirmed by the fact that practitioners who use abstractions usually have a fairly good knowledge of the underlying concrete model.

An initial crucial question is: what is the appropriate model of the "system"? While in the domain of security protocols there are two main approaches, computational and symbolic, the latter one seems to be more appropriate from a pedagogic point of view. Haberman and Kolikant [68] found that a blackbox-based approach can be used effectively to introduce programming concepts to novices. An important characteristic of the symbolic model is its simplicity, and according to the constructivist approach, the model must be taught explicitly [28, 74].

Along with the adversary model (e.g. Dolev-Yao), it is necessary to learn how to represent cryptographic primitives in the symbolic model and model their properties, to allow the learner to understand the actions that honest agents perform during the protocol execution. At this point, we should recall the recommendation given by [27] regarding the need to explicitly present a viable model one level beneath the one we are teaching. Therefore, learners of tools and techniques for modeling and verification of security protocols cannot ignore the construction of such protocols in a real programming language.

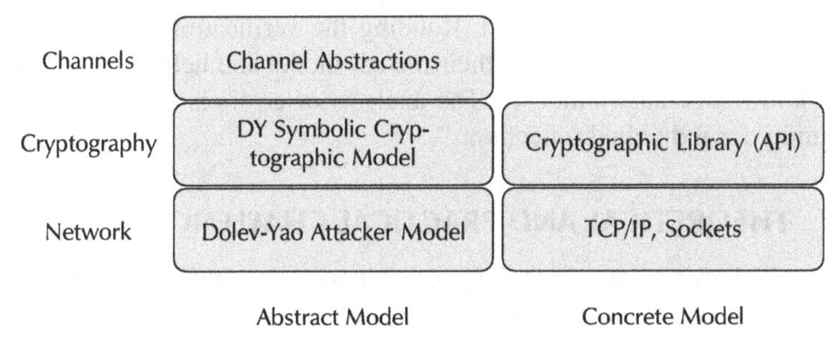

Figure 5.3 Abstract and concrete models.

We can exemplify this with a conceptual framework (Figure 5.3). It consists of both an abstract and concrete models with three different layers: Network, Cryptography and Channels. At the network level, the concrete network protocols run onto, the adversary has full control of the communication medium, and is abstracted by the Dolev-Yao adversary model. The concrete cryptographic primitives are often available through an application programming interface (API), simplifying the access to a set of standard network and cryptographic primitives required to build security protocols implementations. In the abstract model, the symbolic representation of Dolev-Yao intruder rules provides an abstraction of the concrete cryptographic functions.

Abstracting from low-level details where most implementation errors occur [104], the developer can focus on the application design and its security properties. For this reason, the formal methods research community [3, 38, 19] has advocated for the specification of security protocols with high-level programming abstractions, suited for security analysis and automated verification. Alexandron *et al.* [7] suggest that when programmers can work with a less detailed mental model, it becomes easier to work with high-level abstractions. Concretely, we can abstract from cryptographic details at the channel level, using for example a language like *AnBx* [37], an extension of the *AnB* language, where actions are presented in the popular *Alice and Bob* [83] narration style.

Users can model protocols and reason about their security properties using tools for the verification of security protocols in the symbolic model. In a nutshell, the user can specify the security protocol and its security goals and then verify whether the protocol satisfies these goals, or if an attack may occur, with an attack trace being provided in this case.

It should be noted that, with the specification of security goals, the users describe the expectations regarding the security properties of the protocol that reflects their mental model. Running the verification tools provides feedback on the correctness of their mental model and helps to build their own knowledge autonomously. The analysis of errors is therefore an opportunity for individual reflection.

5.4 THEORETICAL AND PRACTICAL CHALLENGES

To precisely formulate a conclusion on whether a security property holds or not, verification tools have to face certain challenges and inherent limitations. Giving a definite and trustworthy answer about non-trivial protocol verification is not always possible, or unreasonably resource-intensive in some cases. The interactions between such processes raise additional concerns in the proof effort, as well as how to reproduce abstract attacks in the real world.

5.4.1 DECIDABILITY

In the protocol verification effort, a major obstacle is the undecidability of security properties. Requirements for a program usually include knowing whether a result will eventually arise, which is unfortunately often equivalent to the undecidable halting problem [105]. In practice, a tool will fail to reach a conclusion or will not terminate in this case. Despite this general limitation, protocol verification can be successfully completed in many cases, under very broad conditions. It has been proven that the security problem is decidable in co-NP time in a Dolev-Yao model of intruders with a bounded number of concurrent sessions [98], including when the intruder is allowed to guess low-entropy messages [6] like passwords instead of random numbers. However, the general case with an infinite number of sessions is undecidable [57], even with a bounded message length [82]. Another undecidability factor with unbounded sessions is the presence of nonces in a protocol. They are fresh values generated by honest agents to prevent replay attacks between concurrent sessions. If the same value appears in two different sessions, it can safely be ignored since we have either an intruder reusing this value, or an honest agent not following the protocol guidelines. Their existence adds complexity to the verification problem, making it undecidable even with a bounded number of nonces [10].

The decidability of a security property also heavily depends on how the cryptographic primitives are considered. In the classic Dolev-Yao model, we assume perfect cryptography and abstract the cryptographic operations as black-box functions. An encrypted message is then akin to random data, and the algebraic properties it exhibits are ignored. While it is true that many attacks exploit design flaws in a protocol more than its underlying cryptographic schemes, a concrete implementation might be vulnerable. Attacks relying on some properties of the encryption function might be missed under the perfect cryptography assumption, and existing results are summarized in a survey [46]. The authors highlight how recent works investigate possible refinings on cryptographic primitives abstraction, in order to allow for a more thorough protocol verification.

5.4.2 VERIFICATION TRUSTWORTHINESS

Whenever a formal result is provided, the trustworthiness of the underlying process needs to be assessed. In fact, a poorly modeled system would lead to an unusable verification result, but the verification in itself should be questioned too. As the effort nowadays is largely automatic, how can we be sure that the verification tools we use are reliable? BAN-logic [40] and its assumptions over the intruder's behavior was found insufficient by Lowe on the Needham-Schroeder protocol [76]. Model checkers then introduced different logics, putting the emphasis on automation.

Doing so, they are complex tools with a large and often complicated codebase, which itself is not necessarily free from bugs. We have then to rely on the programmer's skill and the regular testing of the tool over time to bring us confidence about the verification correctness. Even with a supposedly correct proof system, Gödel second incompleteness theorem [62] states that such a system could not prove its own correctness. *Soundness*, *completeness* and level of certification are key elements in the verification process.

Formally, let Σ be a set of hypotheses and Φ a statement. $\Sigma \models \Phi$, means that Σ logically implies Φ, *i.e.*, in every circumstance in which Σ is true, Φ also holds. Another relation, $\Sigma \vdash \Phi$, states that we can derive Φ starting from Σ, or that Φ is provable from Σ.

A formal deduction system is said *sound* where every conclusion that can be reached in the proof system derives from its premises. Nothing that violates Σ can be proven as true in a sound system. Formally, if $\Sigma \vdash \Phi$, then $\Sigma \models \Phi$.

A *complete* system allows every statement to be provable with our hypotheses. Here, $\Sigma \models \Phi$ implies $\Sigma \vdash \Phi$. If Φ is true given Σ, then we can prove Φ from Σ.

In practice, some tools are sound but not complete, like ProVerif [33]. Here, if the tool states that a security property is true then the property is indeed true in the Dolev-Yao attacker model. However, not all true properties may be provable by the tool.

A way to confidently verify a set of properties is to implement a very small amount of verified mathematics. A *theorem prover* takes logical statements as an input and tests them against an axiomatically accepted set of inference and equivalence rules that can be defined in a trusted kernel. This is the *Logic for Computable Functions* (LCF) [81] paradigm, where *theorems* are an abstract data type, and constructors enforce the underlying logic. With a strong typing enforcement, every user-specified proof step is essentially a composition of the kernel's rules. Modern proof assistants like HOL4 [101] or Isabelle/HOL [93] opt for *Higher-Order Logic* (HOL). HOL allows passing functions as arguments and returning functions, with arbitrary nesting. This extends *First-Order Logic*, where only quantified variables and sets of variables can be considered. Some assistants also apply the *De Bruijn criterion* [23], and generate independently checkable proof objects.

5.4.3 STATE EXPLOSION

Usually, a verification tool explores the realm of possibilities for a given protocol execution. Due to the interleaving between actions and possible messages, the total number of possible traces can be unacceptably high [48] or even infinite. Most systems are represented as state-transition, where the knowledge of an agent is given at each step of a protocol. Great execution complexity can cause the *state explosion* problem, along with potentially enormous time or memory resources needed for protocol verification, which can be limited with various approaches.

Bounded Model Checking is one of the most popular techniques for critical systems verification. Here, the state-transition system is finite and there is a bound on how many transitions the model checker is allowed to explore. Some states are called *error* or *attack* states, and the purpose of the model checker is to try to find a series of transitions which leads to such states. This approach then works by finding a counterexample or bug in the system. Given a bound k, the model checker will typically encode the problem in a propositional Boolean formula in conjunctive normal form $(A \vee B) \wedge (C \vee D \ldots) \wedge \ldots$. With respect to the bound k, the formula is only allowed to contain k conjunctions at most, which denotes a state-space exploration of paths of maximal length k. A SAT solver will in turn test if there is a possible valuation of the literals such as the formula holds true, as does the SATMC [15] tool. If no path of length k in the transition system leads to an unsafe state, then the tested security property is true, given this bound k. Even if it loses precision compared to an exhaustive state-space exploration, a bounded proof can generalize to the unbounded case [78], like when the bound is greater or equal to the *diameter* of the transition system, *i.e.* the smallest number of transitions needed to reach all reachable states.

Binary Decision Diagrams (BDDs) [36] are other representations of interest to explore a state-space efficiently. Instead of listing every possible state individually, this approach allows for state grouping, yielding potentially exponential time and memory savings. A BDD represents a Boolean function with a rooted and directed acyclic graph, where redundant tests of Boolean variables are omitted. Only two leaves exist as Boolean values, and every other node is a Boolean variable. Every non-terminal node represents a variable and has two children, denoting the outcomes of its *true* and *false* valuations. With a brute-force checking process for a Boolean function, every input has to be tested, which equates to 2^n cases with n the number of variables. For example, the expression $A \wedge B$ has four possible valuations. However, having A valued as *false* makes the expression unsatisfiable, and testing B would be useless. In this case, we can omit a B node between A and the *false* Boolean value, so only two variable nodes are required. Tools like OFMC [84] or Tamarin [80] can greatly benefit from this search space reduction.

When the complete model to verify is composed of several asynchronous processes, some of them might be independent of each other. The interleavings between them are then not relevant to the property at hand and could be ignored safely. This simplification is called *Partial Order Reduction* [61]. It makes the precedence relations between events explicit in a dependency graph. If it is found that, from a given state, several paths leading to another state denote equivalent behaviors, then only one *representative* path must be explored. With this technique, paths can be tested for independence on the fly, one local component at the time, saving considerable memory resources. Recent studies show that independence can be tested quasi-optimally in polynomial time [47], avoiding the execution of all possible traces. Similarly to BDDs, this technique can be applied to state-transition-based verification.

Another technique to reduce the state-space size is to work on an abstraction of our model. Distinct levels of abstraction exhibit different levels of detail, and the goal is to find the most abstract model that does not omit relevant details with respect to the tested property. The search of the adequate level of abstraction is then conducted by successive refinements, where a highly abstracted model is verified first. This technique is named *counterexample-guided abstraction refinement* (CEGAR) [43]. A counterexample denotes a bug or vulnerability, which is executed against the concrete system once found. If the execution fails, then the counterexample is called *spurious*, and more detail must be added to the model. The first *non-spurious* counterexample will then embody a genuinely problematic behavior in the concrete model. Hajdu and Micskei introduced the efficient THETA framework [70], along with a survey of other state-of-the-art CEGAR implementations.

5.4.4 COMPOSITIONALITY IN VERIFICATION

As we have seen, in many cases verification tools can automatically provide an answer about the security of a protocol in isolation, even when multiple concurrent sessions are considered. However, in the real-world, protocols do not run in isolation. They are typically components of more complex systems, where they are executed along with different protocols. Several examples have demonstrated that protocol composition is not secure in general (e.g. [11, 65]), even if individual components are secure in isolation. Therefore, researchers have investigated both the verification of complex systems and the identification of sufficient conditions for the safe composition of protocols which are secure in isolation.

While Partial Order Reduction introduces a formalism of asynchronous components, only their execution order is used to devise a simpler proof. In complex protocols, ordered small components can be truly valid in regard to their security properties. Proving the correctness of a multitude of small models would lead to a greatly simplified proof effort, but demonstrating the compositionality of a protocol is non-trivial. A divide and conquer framework was introduced in [11] where large protocols are expressed as multiple smaller ones. In the same spirit, identifying how encryption between protocols can be non-overlapping is presented in [67]. A non-decomposable process is referred to as *atomic* [42]. When protocols are allowed to interact between each other in a stateful way *i.e.* having access to information from previous sessions, secure composition is still possible as outlined in [71]. When components can be isolated and tested separately, we talk about *sequential*, *parallel* or *vertical* compositionality.

Sequential Composition In a sequentially composed security protocol, one component's output is another one's input. To ensure a secure execution, processes are annotated with pre- and post-conditions. When such invariants are satisfied, the protocols' steps are considered securely usable in their composed form. A common application of this philosophy is to have one key-establishment protocol, and another one after it that uses this key for message exchange over a secure channel like in [85]. A sequence of message transmissions and receptions for an agent is called a *strand*, and the relationships between strands for every agent form a graph: the *strand space* [103]. Figure 5.4 shows the strand space for a challenge-response where agents A and B agree on a number N using a private symmetric key *KAB*. The protocol is then considered correct when every subgraph, or *bundle*, contains at least one honest agent behavior, and when injecting an intruder leads to a *strand space* that is isomorphic to the original one.

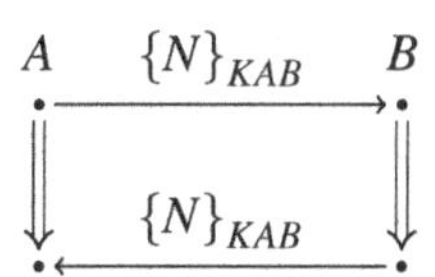

Figure 5.4 Example of strand space in communications.

Parallel Composition Parallel composition is the situation when different protocols run in parallel. The first notion to be considered is protocol independence. According to Guttman and Thayer [67], two protocols are independent if the achievement of goals in one protocol does not depend on the other protocols being in use. They proved a theorem that holds even if the protocols share public key certificates and secret key "tickets" (e.g. Kerberos). In general, interdependence can arise when protocols share a piece of data, like long-term keys. Furthermore, if an honest agent is confused about what protocol is being executed, the intruder might exploit it. To ensure that the received messages are sufficiently distinguishable, Cortier *et al.* opt for a *tagging* approach [45]. A tagged message includes an identifier of the protocol it comes from. For example, adding the protocol's name to an encrypted payload avoids using a ciphertext that was intended for another protocol. This way, even if the two messages are encrypted with the same key, an intruder could not make an honest agent running a protocol accept a message intended for another protocol. Generalizing beyond secrecy goals to the entire geometric fragment proposed by Guttman [66], it is possible to perform secure parallel composition without tagging. This result was presented in [8] where two main syntactic conditions are checked: if a given protocol is type-flaw-resistant and if the protocols in a given set are pairwise parallel-composable. Moreover, requirements are more relaxed: it is sufficient that non-atomic subterms are not unifiable unless they belong to the same protocol and have the same type.

Vertical Composition The last two compositions can be named *horizontal*, as *vertical* composition [65] focuses on protocol encapsulation, like when an application runs on top of a secure TLS channel. While such a composition is reasonable most of the time with disjoint protocols at each layer [85, 42], the self-composition case is more complicated. If the same channel is to be re-used in the protocol stack, stronger conditions must be satisfied. According to Mödersheim and Viganò, message formats disjointness and uniqueness of each payload usage context constitute sufficient conditions [86] for such vertical composition, satisfied in practice by a large class of protocols. A stronger result is presented in [63], where the number of transmissions a channel is allowed to carry is unbounded, and with support for stateful protocols.

5.5 SPECIFICATION LANGUAGES AND TOOLS

To tackle the challenges of security protocols verification, several approaches and tools have been adopted in the past decades. Popular notations like *Alice and Bob* narrations enable high-level specification, while other languages trade simplicity for expressivity. Every concrete tool handles a specific set of behavioral properties and this section presents some of the most widespread or promising frameworks.

5.5.1 ANB LANGUAGE

In the most classical *AnB (Alice and Bob)* notation [83], a protocol is specified by a list of steps (or actions), each representing a message exchange between two honest agents like $A \rightarrow B : Msg$. This notation makes the encoding of security protocols considerably simpler (and more compact) than their equivalents in other formal languages (e.g. process calculi [4, 1]) or real-world programming languages. This intuitive language allows learners to build their own models of security protocols and experiment with them using tools that support such a notation like the model checker OFMC [84] and Tamarin [100].

An example of an *AnB* protocol is displayed in Figure 5.6. The main goal of the protocol is to achieve authentication (precisely the injective agreement defined in [77]) on the message *Msg*. The recipient *B* should have evidence that the message has been endorsed by *A* and is a fresh message. The goal is achieved using asymmetric encryption and a challenge–response technique with a nonce exchange. The abstract functions *pk* and *sk* are used to model asymmetric encryption, mapping agents to their public keys for encryption and signature respectively, while *inv* is a private function modeling the private key of a given public key. It should be noted that private function symbols are used to symbolically represent a notion. They are not concrete functions that can be computed [83].

The succinctness of the *AnB* notation comes with an expressivity drawback. The specification relies on implicit assumptions on how a receiver

```
Protocol: Example_AnBx
Types:
    Agent A,B;
    Certified A,B;
    Number Msg;
    SymmetricKey K;
    Function [Agent,Number -> Number] log
Knowledge:
    A: A,B,log;
    B: B,A,log
Actions:
    A -> B, @(A|B|B): K
    B -> A: {|Msg|}K
    A -> B: {|hash(Msg),log(A,Msg)|}K
Goals:
    K secret between A,B
    Msg secret between A,B
    A authenticates B on Msg
    B authenticates A on K
    B authenticates A on Msg
```

Figure 5.5 *AnBx* protocol example.

processes a message or how agents handle errors with unexpected messages. Therefore, it is potentially ambiguous for a novice unaware of such implicit assumptions. Some research focuses on extending its expressivity with explicit message formats [9] and user-specified equational theories [24].

5.5.2 ANBX LANGUAGE

In this spirit, the *AnBx* language (formally defined in [37]) is built as an extension of *AnB* [83]. The main peculiarity of *AnBx* is to use channels as the main abstraction for communication, providing different authenticity and confidentiality guarantees for message transmission, including a novel notion of *forwarding* channels, enforcing specific security guarantees from the message originator to the final recipient along a chain of intermediate forwarding agents. The translation from *AnBx* to *AnB* can be parametrized using different channel implementations, by means of different cryptographic operations. It also supports private function declarations, useful for example when two honest agents want to share a secret shared key function. It can be explicitly excluded from the intruder's knowledge with a signature as $[ParamTypes \rightarrow\!\!* \ ReturnType]$.

```
Protocol: Example AnB
Types:
    Agent A,B;
    Number Msg,Nonce;
    SymmetricKey K;
    Function pk,sk,hash;
    Function log
Knowledge:
    A: A,B,pk,sk,inv(pk(A)),inv(sk(A)),hash,log;
    B: A,B,pk,sk,inv(pk(B)),inv(sk(B)),hash,log
Actions:
    A -> B: A
    B -> A: {Nonce,B}pk(A)
    A -> B: {{Nonce,B,K}inv(sk(A))}pk(B)
    B -> A: {|Msg|}K
    A -> B: {|hash(Msg),log(A,Msg)|}K
Goals:
    K secret between A,B
    Msg secret between A,B
    A authenticates B on Msg
    B authenticates A on K
    B authenticates A on Msg
    inv(pk(A)) secret between A
    inv(sk(A)) secret between A
```

Figure 5.6 *AnB* protocol example.

Figure 5.5 shows an example protocol in which two agents want to exchange securely a message *Msg*, using a freshly generated symmetric key *K*, *i.e.* a key that is different for each protocol run. If *K* is compromised, neither previous nor subsequent message exchanges will be compromised, only the current one. This is similar to what happens in TLS where a symmetric session key is established (using asymmetric encryption) at the beginning of the exchange. It should also be noted that this setting is more efficient, as symmetric encryption is notoriously faster than the asymmetric kind. Therefore, if the size of the message is significant, using symmetric encryption should be preferable.

The *Types* section includes declarations of identifiers of different types and functions declarations, while the *Knowledge* section denotes the initial knowledge of each agent. An optional section, *Definitions*, can be used to specify macros with parameters. In the *Actions* section, the action $A \rightarrow B, @(A|B|B) : K$ means that the key K is generated by A and sent on a secure channel to B. The notation $@(A|B|B)$ denotes the properties of the channel: the message originates from A, it is freshly generated (@), verifiable by B, and secret for B. How the channel is implemented is delegated to the compiler. The designer can choose between different options, or simply use the default one. This simplifies the life of the designer, who does not need to be in charge of low-level implementation details. A translation to *AnB* is shown in Figure 5.6: in this case the channel is implemented using a challenge-response technique, where *B* freshly generates a nonce (the challenge) encrypted with *pk(A)*, the public key of *A* along with the sender name ($\cdot$ denotes the asymmetric encryption). This guarantees that only *A* would be able to decrypt the incoming message.

The response, along with the challenge, includes the symmetric key *K*. The response is digitally signed with *inv(sk(A))*, the private key of *A* and then encrypted with *pk(B)*, the public key of *B*. This will allow *B* to verify the origin of the message and that *K* is known only by *A* and *B*.

It should be noted that in *AnBx*, we abstract from these cryptographic details, and we simply denote the capacity of *A* and *B* to encrypt and digitally sign using a Public Key Infrastructure (PKI) with the keyword *Certified*. This reflects the customary practice of a Certification Authority to endorse public keys of agents, usually issuing X.509 certificates, allowing every agent to verify the identity associated with a specific public key. Moreover, in *AnBx*, keys for encryption and for signature are distinguished by using two different symbolic functions, *pk* and *sk* respectively.

Once the symmetric key *K* is shared securely between *A* and *B*, then *B* can send the payload *Msg* secretly ($\{| \cdot |\}$ denotes the symmetric encryption). Finally, *A* acknowledges receipt, replying with a digest of the payload computed with the *hash* function (a predefined function available in *AnBx*), and with a value computed with the *log* function.

The section *Goals* denotes the security properties that the protocol is meant to satisfy. They can also be translated into low-level goals suitable for verification with various tools. Supported goals are as follows:

1. *Weak Authentication* goals have the form *B weakly authenticates A on Msg* and are defined in terms of non-injective agreement [77].
2. *Authentication* goals have the form *B authenticates A on Msg* and are defined in terms of injective agreement on the runs of the protocol, assessing the freshness of the exchange.
3. *Secrecy* goals have the form *Msg secret between* $A_1, \ldots A_n$ and are intended to specify which agents are entitled to learn the message *Msg* at the end of a protocol run.

In the example protocol (Figure 5.5), the desirable goals are the secrecy of the symmetric key *K* and of the payload *Msg* that should be known only by *A* and *B*. There are also authentication goals: *B* should be able to verify that *K* originates from *A* and that the key is freshly generated. Finally, two goals express the mutual authentication between A and B regarding Msg, including the freshness of the message. In summary, this protocol allows two agents to securely exchange a message, with guarantees about its origin and freshness.

5.5.3 OFMC MODEL CHECKER

In the cases studies presented in Section 5.6, the OFMC model checker [84] is used for the verification of abstract models. OFMC employs the AVISPA Intermediate Format IF [21] as "native" input language, defining security protocols as an infinite state-transition system using set-rewriting. Notably, OFMC also supports the more intuitive language *AnB* [83], and the *AnB* specifications are automatically translated to IF.

OFMC performs both protocol falsification and bounded session verification by exploring the transition system in a demand-driven way. If a security goal is violated, an attack trace is provided. The two major techniques employed by OFMC are the *lazy intruder*, a symbolic representation of the intruder, and *constraint differentiation*, a general search-reduction technique that combines the lazy intruder with ideas from Partial Order Reduction.

5.5.3.1 Channels as Assumptions

In general, along with the usual insecure channel, the *AnB* language supported by OFMC allows specifying three other types of channels: *authentic*, *confidential*, and *secure*, with variants that allow agents to be identified by a pseudonym rather than by a real identity. The supported standard channels are:

1. $A \rightarrow B : M$, an insecure channel from A to B, under the complete control of a Dolev-Yao intruder [53].
2. $A \bullet\!\!\rightarrow B : M$, an authentic channel from A to B, where B can rely on the facts that A has sent the message M and meant it for B.
3. $A \rightarrow\!\!\bullet B : M$, a confidential channel, where A can rely on the fact that only B can discover the message M.
4. $A \bullet\!\!\rightarrow\!\!\bullet B : M$, a secure channel (both authentic and confidential).

Pseudonymous channels [85] are similar to standard channels, with the exception that one of the secured endpoints is logically tied to a pseudonym instead of a real name. The notation $[A]_\psi$ represents that an agent A is not identified by its real name A, but by the pseudonym ψ. Usually, ψ can be omitted, simplifying the notation to $[A]$, when the role uses only one pseudonym for the entire session, as it is in our case and in many other protocols.

For example, $[A] \bullet\!\!\rightarrow B : M_1$ denotes an authentic channel from A to B, where B can rely on the facts that an agent identified by a pseudonym has sent a message M_1 and that this message was meant for B. If during the same protocol run, another action like $[A] \bullet\!\!\rightarrow B : M_2$ is executed, B can rely on the facts that the same agent (identified by the same pseudonym) has also sent M_2 and again that the message was meant for B.

Assuming that B does not already know the real name of A, the execution of these two actions does not allow B to learn the real identity of A (unless this information is made available during the protocol execution), but B has a guarantee that it was communicating with the same agent during both message exchanges. The term *sender invariance* is used to refer to this property, and the most common example is the TLS protocol without client authentication.

5.5.4 ANBX COMPILER AND CODE GENERATOR

The *AnBx* Compiler and Code Generator [89] is an automatic Java code generator for security protocols specified in *AnBx* or *AnB*. The *AnBx* compiler can be used in the context of Model Driven Development. Provided that a model has been validated with OFMC, the user can automatically generate a Java implementation. Along with this immediate benefit, in a learning context [90] this is useful to familiarize oneself with the software engineering approach of Model Driven Development, but also to compare a manual implementation with a generated one. The main features of the compiler are:

- Automatic computation of the defensive checks that an agent has to perform on incoming messages.
- Optimization of cryptographic operations in order to minimize the number of computational steps and reduce the overall execution time [88].
- Mapping of abstract types and API calls to the concrete ones provided by the AnBxJ library.
- A set of template files is used to generate the code. Template files can be customized, for example, to integrate the generated application in larger systems.

Since the compiler translates the intermediate format to the applied pi-calculus [1], the verification of the protocol logic used for the code emission phase can be performed with the protocol verifier ProVerif [30].

5.5.4.1 AnBx Compiler Architecture

In this section, we introduce the development methodology (Figure 5.7) and provide an overview of the architecture of the *AnBx* compiler (Figure 5.8), a tool built in Haskell, which is one of the key components of this methodology. Other tools, included in the toolkit, are the OFMC symbolic model checker [84] and the cryptographic protocol verifier ProVerif [30]. All components are integrated in the *AnBx* IDE [60], an Eclipse plug-in for the design, verification and implementation of security protocols. Along with the integration of the back-end tools, the IDE includes many features meant to help programmers to increase their productivity, like syntax highlighting, code completion, code navigation and quick fixes.

5.5.4.2 Development Methodology

As seen in Section 5.3, the work of a developer usually begins from gathering the available specification documentation and build a model that can be verified and then used to construct an implementation. Expressing requirements in a simple but formalized language for the specification of security protocols that is amenable for the verification of the model is a key aspect of the approach presented here. Although formal languages like the SPI [4] and applied pi-calculus [2] have been created to model and verify security protocols, their usage among software developers in the industry remains limited due to their complexity. On the contrary, protocol narrations in the *Alice & Bob* style are much closer to the familiar way software engineers use to describe security protocols. Therefore, this methodology adopts the simple *AnB* notation [83] and its extension *AnBx* [37].

One of the main advantages of such languages, which share common traits but are syntactically different, is that they can be quickly learned by developers as they are rather intuitive. Their syntax is similar to the informal or semi-formal languages used in the documentation software engineers are familiar with, but their semantics are formally defined so there are no ambiguities in the way the system interprets them. We have direct experience of students developing a few person-month projects, being able to learn effectively the specification language in a few days or less.

Once the protocol specification has been completed, it is possible to use the *AnBx* compiler to translate the input file to *AnB*, a format which can be verified with the OFMC model checker [84]. It should be underlined that the translation from *AnBx* can be parametrized using different channel implementations that convey the security properties specified at the channel level, by means of different cryptographic operations [37]. The compiler can also directly process protocols in *AnB*. In all cases, the compiler will

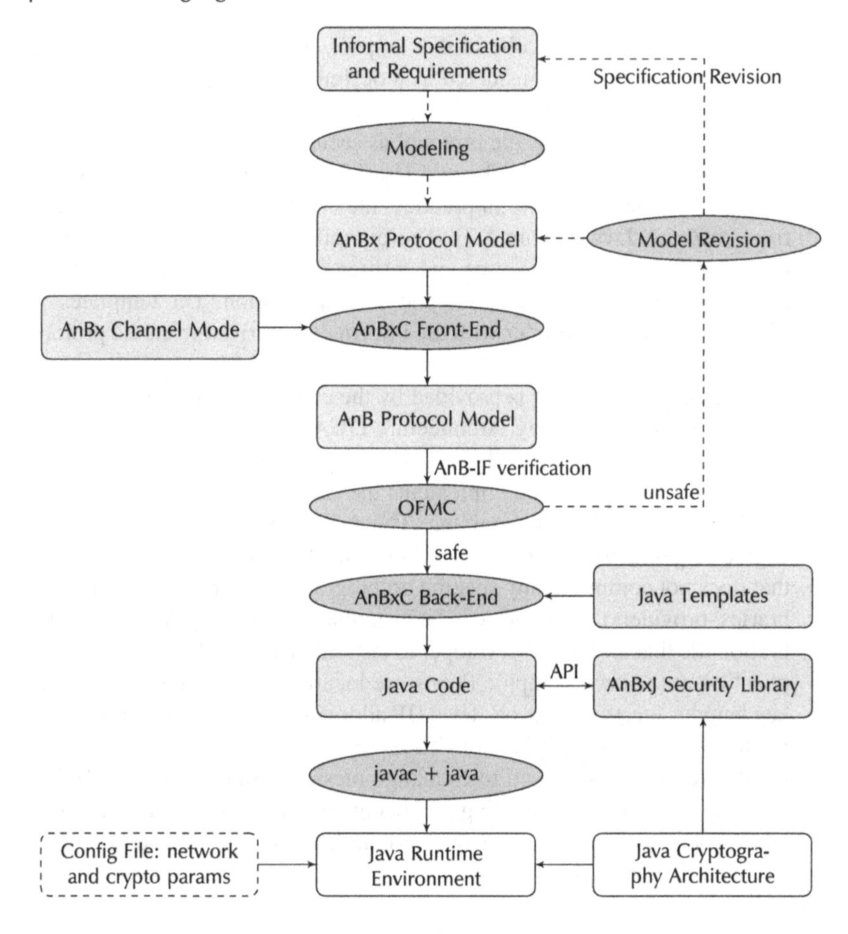

Figure 5.7 Model driven development with *AnBx* (- - - - - manual ——— automatic).

perform a number of sanity and type checks to ensure that the input provided to OFMC is fully sanitized. This is important, because OFMC lacks an *AnB* type checker, and there are situations where the model checker accepts (and might deem secure) protocols which are ill-typed.

If OFMC finds an attack, *i.e.* a security goal can be violated by the intruder, the developer can manually (and iteratively) revise the model until the model checker concludes on a safe model. If necessary, changes can be backported to the original standard. It should be noted that if the user is developing a brand-new protocol, this is an effective way to iteratively prototype security protocols. In that respect, *AnBx*, given the channel abstractions, can also be considered as a design language along to its nature of specification language.

When OFMC verifies the protocol as secure, the *AnB* specification can be passed to the compiler back-end (described later in this section) and generated to Java source code. In practice, the compilation is fully automated from *AnBx* to Java (Figure 5.7). The compiler uses application template files, customizable by the user and written in the target language, to integrate the generated code in the end-user application. The templates are instantiated by the compiler with the information derived from the protocol specification.

The run-time support is provided by the cryptographic services offered by the Java Cryptography Architecture (JCA) [64, 96]. In order to connect to the JCA, a security API (called *AnBxJ* library) wraps, in an abstract way, the JCA interface and implements the custom classes necessary to encode the generated Java programs. The *AnBxJ* library guarantees a high degree of generality and customization, since the API allows to write code that does not commit to any specific cryptographic solution (algorithms, libraries, providers). This code can be instantiated at run-time using a configuration file that allows the developer to customize the deployment of the application at the cryptographic (key store location, aliases, cipher schemes, key lengths, etc.) and network level (IP addresses, ports, etc.) without needing to regenerate the application. The library also gives access to the communication primitives used to exchange messages in the standard TCP/IP network environment, including secure channels like TLS. Communication and cryptographic run-time errors are handled at this level, and exceptions are raised accordingly.

5.5.4.3 *AnBx* Compiler Back-End

The second phase of the compilation process (Figure 5.8) begins with the translation of *AnB* to an *(Optimized) Executable Narration*, a set of actions that operationally encodes how the different agents are expected to execute the protocol. The core of this step is the automatic generation of the consistency checks derived from the static (implicit) information included in the model specification. The checks are defined as consistency formulas; some simplification strategies can be applied to reduce the number of generated formulas and speed up the application. A further step in this direction is the application of other optimization techniques [88], including common sub-expression elimination (CSE), which in general are useful to generate efficient code. In particular, our compiler considers a set of cryptographic operations, which are computationally expensive, to reduce the overall

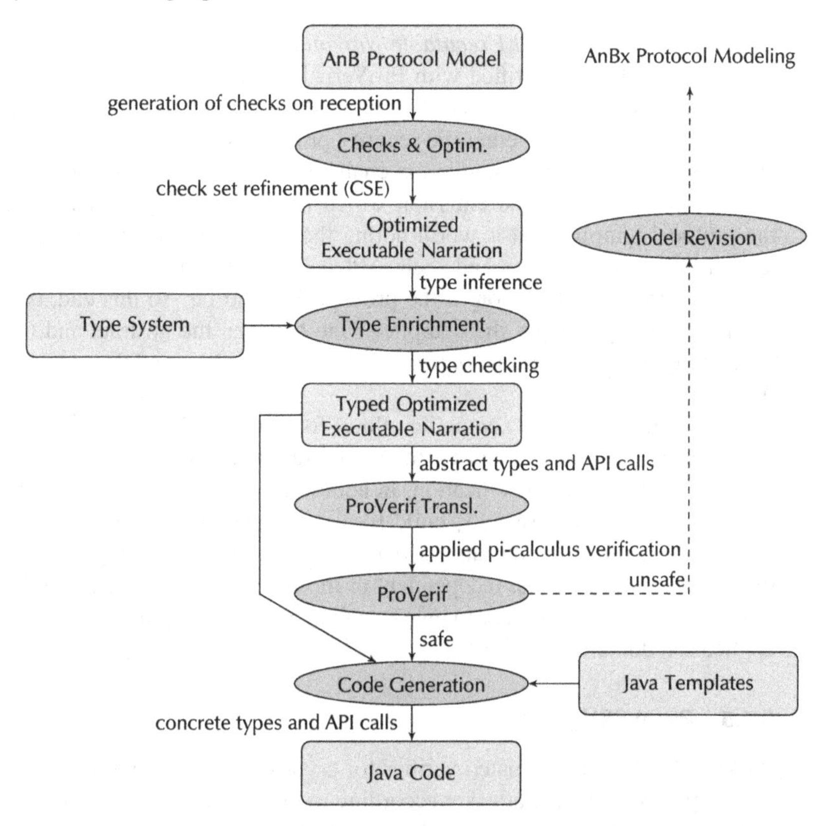

Figure 5.8 Compiler back-end: Type system, code generator, verification (- - - - - manual ——— automatic).

execution time by storing partial results and reordering instructions with the purpose of minimizing the overall number of cryptographic operations.

The next stage is the construction of the *Typed Optimized Executable Narration*, a typed abstract representation of the security-related portion of a generic procedural language supporting a rich set of abstract cryptographic primitives. For that purpose, a type system infers the types of expressions and variables, ensuring that the generated code is well-typed. It has the additional benefit of detecting at run-time whether the structure of the incoming messages is equal to the expected one, according to the protocol specification. This step is necessary, as the type system of *AnB* is too simple, and unsuitable to represent the complexity of a complete implementation of a protocol, in Java for example.

The *Typed Optimized Executable Narration* can be translated into applied pi-calculus and verified with ProVerif [30]. To generate applied pi-calculus, *AnB* security goals need to be analyzed at the initial phase and specific annotations modeling the security properties need to be generated and preserved along the compilation chain. The verification with ProVerif is preliminary to the code emission which is performed by instantiating the protocol templates. It is worth noting that only at this final stage, the language-specific features and their API calls are actually bound to the typed abstract representation of the protocol built so far. To this end, two mappings are applied by the compiler. One between the abstract and the concrete types, the other one between the abstract actions and the concrete API calls.

In summary, the *AnBx* compiler allows for a one-click code generation of a widely configurable and customizable ready-to-run Java application from an *AnBx* or *AnB* specification. In addition to the Java classes and the configuration file, an Ant [13] build file is also generated to easily build and run the application. It is also important to underline that the application templates are generic, *i.e.* independent of the specific protocol, and can be modified by the user in order to integrate the generated code in the required application domain.

5.5.5 PROVERIF

ProVerif [30] is an automated verifier for cryptographic protocols, modeling the protocol and the attacker according to the Dolev-Yao [53] symbolic approach. Differently from model checkers, ProVerif can model and analyze an unbounded number of parallel sessions of the protocol, by using Horn clauses to model a protocol. However, like model checkers, ProVerif can reconstruct a possible attack trace when it detects a violation of the intended security properties. It can check equivalence properties, which consists in determining that two processes are indistinguishable, from a third-party point of view.

ProVerif may report false attacks, but if a security property is reported as satisfied, then this is true in all cases, so it is necessary to analyze the results carefully when attacks are reported. The checks honest agents should perform on received messages must be explicitly stated, increasing the risk of user error. A recent development [32] enables support for axioms, lemmas and restrictions, allowing users to declare intermediate properties helping ProVerif to complete proofs. Those axioms can now be used to handle global mutable states, particularly useful for contract signing protocols [14]. To verify protocols in the computational model, a variant called CryptoVerif [31] exists.

5.5.5.1 Horn Clauses and Applied pi-calculus

Horn clauses were firstly used by Weidenbach [111] in protocol verification as a sound abstraction technique, overestimating the attacker's possible knowledge. Usually, trying to prove a security property is convenient with Horn clauses, as one would have to test its negation and check for a contradiction later in the proof. A Horn clause is a disjunction of terms such as at most one might be true. There are three types of Horn clauses:

- Definite clause with exactly one true term among false ones. The clause would read as $\neg p \vee \neg q \vee r$ which, rewritten with an implication would be $p \wedge q \Rightarrow r$. This means that r holds as long as p *and* q are true. This clause is then used for deduction.
- Fact clause with a true term only. A lone variable p is equivalent to assuming that p is true.
- Goal clause with only false terms. $\neg p \vee \neg q \vee \neg r$ would mean that the properties encoded by all three variables hold when the expression is unsatisfiable.

As an example, we can take a Dolev-Yao attacker, which then knows the network and what transits on it. We obtain an initial knowledge fact *attacker(network)*. A plaintext message exchange is then performed as $A \rightarrow B : Msg$ and Msg is sent over the network, which is another fact we can write as *send(network,Msg)*. The classical eavesdropping attacker rule follows as $send(x,y) \wedge attacker(x) \Rightarrow attacker(y)$. With this rule, we can derive the attacker's knowledge *attacker(Msg)*, proving that *Msg* is not secret in this case.

ProVerif uses the applied pi-calculus [1], suited to specify the behavior of concurrent processes, emphasized around communicating agents. In applied pi-calculus, each participant of the protocol is represented by a process and the messages exchanged by processes are the messages of the protocol. For example, the exchange: "The first process sends a on channel c, the second one inputs this message, puts it in variable x and sends x on channel d" is encoded by *out(c, a) || in(c, x).out(d, x)*. The core feature of ProVerif is to translate it to Horn clauses, but only an approximation of a translation is possible sometimes, which can lead to non-termination.

5.5.6 IDENTIFICATION OF ATTACKS

A protocol can have multiple desired properties that must be verified, as expressed in the *AnB Goals* section. A verification tool or framework should be able to tell the user exactly what failed and in what circumstances. OFMC, while providing one of the simplest attack traces, only outputs one failing goal, the one which failed first. ProVerif prints every one of them, but it is easy to have an unverified goal passing the verification because of some user specification error. To have an intuitive visualization of failing goals, the *AnBx* IDE [60] provides a convenient workflow for individual goal verification.

5.5.7 OTHER TOOLS AND COMPARISON

Tamarin [80] is a state-of-the-art symbolic verifier supporting both automatic and interactive modes. The user then has the option to manually guide the proof process with user-specified lemmas.

The Scyther tool [50] shares a backward reasoning approach based on patterns with Tamarin and has guaranteed termination. Maude-NPA [56] shares a lot of characteristics with Tamarin, with a notable exception of not supporting global mutable states and the fact that Maude-NPA models protocols by strands other than multiset rewriting.

Verifpal [73] is a recent tool which focuses on providing an easy to use and to understand design framework. Its specification language resembles an *AnB* protocol narration, and it uses heuristics to limit the state explosion.

In the AVISPA tool suite [17], OFMC is used in conjunction with two other back-ends: CL-AtSe [106] and SATMC [15]. CL-AtSe takes as input a service specified as a set of rewriting rules, and applies rewriting and constraint solving techniques to model all reachable states. SATMC uses SAT solvers to perform a bounded analysis with propositional formulas. Those three tools are now part of the AVANTSSAR platform [16], extending AVISPA capabilities to *Service-Oriented Architectures*.

A comprehensive survey of recent advances in verification can be found in [22], of which we use the comparison criteria in Table 5.1.

Table 5.1
Comparison between tools for symbolic security analysis.

Tool	Unbounded	Equiv	Eq-thy	State	Link
ProVerif [30]	●	●	◐	●	○
Tamarin [80]	●	●	●	●	○
Verifpal [73]	●	◐	◐	●	◐
OFMC [84]	○	○	◐	●	○
CL-AtSe [106]	○	○	●	●	○
SATMC [15]	○	○	○	●	○
Maude-NPA [56]	●	●	●	○	○
Scyther [50]	●	○	○	○	○

5.5.7.1 Supported Properties

All tools except the back-ends from the AVISPA/AVANTSSAR suite support unbounded verification, accepting undecidability in some cases. ProVerif, Tamarin and Maude-NPA can check diff-equivalence properties [52], testing if protocols have the same structure and differ only by the messages they exchange. Verifpal can test some equivalence queries, but limited to any protocol scenario which can be derived such that some shared secrets are not equivalent to one another.

Tamarin, CL-AtSe and Maude-NPA are the only ones to allow for user-defined equational theories with the Finite Variant Property in general [54], CL-AtSe being however limited to subterm-convergent equational theories [20]. ProVerif and OFMC only permit such theories without associative-commutative axioms, Verifpal is limited to specifying simple equations for signature or public-key establishment like with Diffie-Hellman, and the other ones have fixed or no equational theories.

A global mutable state is useful to model databases like key servers or shared memory in general [75, 106], but Maude-NPA and Scyther lack support for it.

The last criterion is about linking the model to an implementation, providing executable protocols exhibiting symbolic security properties. Verifpal can generate Coq implementations of its models, with Go code generation planned as well, while now being usable as a Go library.

5.6 CASE STUDIES

We present here some practical examples of verification and re-engineering of security protocols. The purpose of these case studies is to give a practical demonstration, within the context of an approach of formal methods for security, of modeling and design techniques that can be applied by practitioners working with real-world protocols. We first consider the formal modeling and verification of the Bitcoin payment protocol BIP70 (originally presented in [91]) and then we present the modeling, verification and re-engineering of the *iKP* e-commerce protocols (originally presented in [39, 87]). A similar approach (originally presented in [37]) can be used for *SET*, an e-commerce protocol that for its complexity is considered as a benchmark for protocol analysis.

The specification languages considered here are *AnB* and its extension *AnBx*. These are intuitive languages that can be learned relatively quickly by practitioners. The verification tools are the model checker OFMC (for iKP, SET and BIP70) and the cryptographic protocol verifier ProVerif (for iKP and SET), that can accept the aforementioned languages directly or by translation using the *AnBx* compiler.

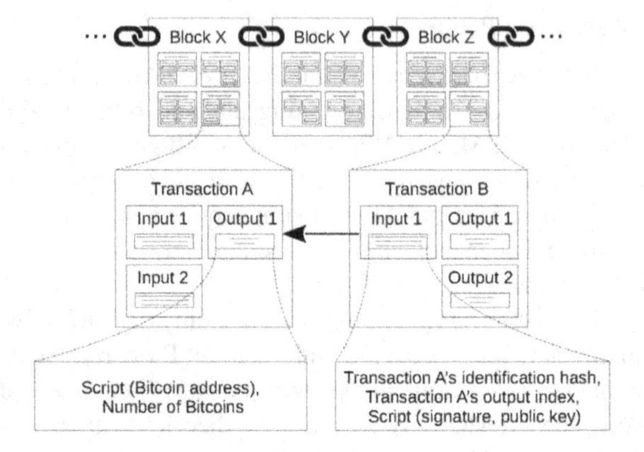

Figure 5.9 Information stored in Bitcoin transactions (courtesy [91]).

5.6.1 BITCOIN PAYMENT PROTOCOL

Before describing the formal modeling technique used in [91] to specify the payment protocol, let us introduce some key concepts of the Bitcoin cryptocurrency.

A *Bitcoin address* in a Bitcoin network is the hash of an Elliptic Curve (EC) public key used as an identifier. The address is a pseudonym associated to the user possessing the corresponding private key. Such private key can be used to claim bitcoins sent to a user and authorize payments to other parties using ECDSA, the Elliptic Curve Digital Signature Algorithm. As the probability of collision is negligible considering the length of the output of the hash function, it is possible to safely assume that these identifiers are unique within the network.

A *transaction* records the transfer of bitcoins. It includes one or more inputs, specifying the origin of bitcoins being spent, and one or more outputs, specifying the new owner's Bitcoin address and the amount of the transaction (see Figure 5.9). To authorize the payment, the sender must specify an input consisting of the previous transaction's identification hash and an index to one of its outputs, and provide the corresponding public key and a valid digital signature. The inputs and outputs are controlled by means of scripts in a Forth-like language specifying the conditions to claim the bitcoins. The *pay-to-pubkey-hash* is the most common script to authorize the payment, requiring a single signature from a Bitcoin address.

The *blockchain* holds the entire transaction history of the network with a secure time-stamp [92] and is organized in blocks of transactions. The *ledger* is an append-only data structure stored in a distributed way by most users of the network. Solving a *proof of work* puzzle allows to append a new transaction to the blockchain. The nodes that solve proofs of work are called *miners,* and they are rewarded in bitcoins for their computational effort. A *proof of work* problem is computationally difficult to solve but easy to verify when a solution is provided.

5.6.1.1 Formal Modeling Approach and Formalization

The formal modeling and security analysis of the BIP70 Payment Protocol presented in [91] involves the symbolic model checker OFMC, and the specification language *AnB*. An important reason for adopting this approach is that this toolkit allows modeling communication channels as abstractions conveying security goals like authenticity and/or secrecy, without the need to specify the concrete implementation used to enforce such goals. It is therefore possible to rely on the assumptions provided by such channels. The result is a simpler model that is tractable by the verifier and can be analyzed more efficiently, mitigating the state explosion problem.

Interestingly in *AnB* channels, agents can be identified by pseudonyms (e.g. ephemeral public keys) rather than by their real identities, as in secure channels like TLS without client authentication. Therefore, the capability of OFMC to verify a range of different channels specified in *AnB* makes the tool suitable for this verification effort. It should be noted that, as discussed in [91], BIP70 runs on top of abstract channels providing security guarantees, and the specific implementation of the underlying protocol is not part of the BIP70 specification. Thus, the analysis should be performed under the assumption that the sufficient conditions for vertical composition (e.g. in [86]) are satisfied.

The BIP70 Payment Protocol [12] was proposed in 2013 by Andresen and Hearn and later adopted by the Bitcoin community as a standard. The goal of the protocol is presented as follows:

> *"This BIP describes a protocol for communication between a merchant and their customer, enabling both a better customer experience and better security against man-in-the-middle attacks on the payment process."*

The communication channel between the customer and the merchant is strongly recommended being over HTTPS with the merchant authentication based on a X.509 certificate issued by a trusted Certificate Authority.

The actions performed and messages exchanged during the protocol run are shown in Figure 5.10. The protocol initiates with the customer clicking on the "Pay Now" button on the merchant's website. This generates a Bitcoin payment URI that enables to open the customer's Bitcoin wallet and

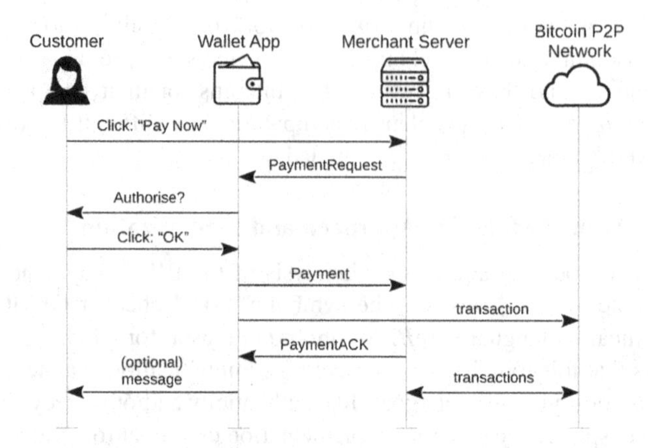

Figure 5.10 BIP70 payment Protocol overview (courtesy [91], adapted from [12]).

download the *Payment Request* from the merchant's website. The digital signature on the *Payment Request* can be verified by the wallet application with the public key of the merchant, checking the validity of the associated certificate. Provided the previous step is completed successfully, the bitcoin amount requested is shown to the customer requesting for authorization of the payment, along with the merchant's name extracted from the X.509 certificate's "common name" field. If the user authorizes the payment, the wallet computes a payment transaction and broadcasts such information to the Bitcoin network. Moreover, transaction and refund addresses within a *Payment* message are sent back to the merchant. The merchant then sends back a *Payment Acknowledgement* to the customer wallet. Finally, the customer receives a confirmation of the payment when the transaction is detected on the blockchain.

5.6.1.1.1 Modeling BIP70

This model considers a BIP70 protocol with n ($n \geq 1$) customers $C_1, \ldots, C_n$ and a single merchant M, as the standard specifies that a transaction may involve more customers. These agents should be able to trade over Bitcoin, and they should know the identity of the merchant. Strictly speaking, since multiple customers can cooperate in the payment of a single merchant, the model requires that at least one of the customers knows the merchant's name at the beginning of the protocol. However, it is not required that the merchant knows the identity of the customers. In fact, there is no provision of a communication mechanism between agents and merchants that explicitly discloses the real identity of the client. Such one-way authentication can be customarily achieved using HTTPS, as recommended in BIP70. In this case, the client has the guarantee that messages are exchanged with the authenticated server, but the server is only guaranteed that the communication channel is shared with the same pseudonymous agent. The pseudonym of the agent C_1 during the protocol run is represented by $[C_1]$.

The model also assumes that C_1 is the only agent that communicates directly with the merchant, while other agents communicate with C_1 to jointly set up the order for the merchant, using a secure channel (or out-of band). This is in accordance with the scenario in which payments may be made from multiple pseudonymous Bitcoin addresses, belonging to one or different entities. It is up to the customer communicating with the merchant to compose the payment transaction, coordinating with all the Bitcoin address holders. The model employs two kinds of channels:

- $[C_1] \bullet\!\!\rightarrow\!\bullet M$ represents a secure (secret and authentic) channel between the client C_1 and the merchant M; M can bind the other end point to a pseudonym $[C_1]$ rather than to the real identity of C.
- $C_i \bullet\!\!\rightarrow\!\bullet C_j$ represents a secure channel between the clients C_i and C_j.

The identifiers used in the messages exchanged during the protocol run are shown in Table 5.2. Additionally, the $\mathscr{H}$ symbol represents the hash function used to generate Bitcoin addresses and the following definitions are used in the message specification:

- $\omega_i = ฿_{C_i}, \mathscr{H}(\mathrm{B}_{C_i})$: the previous transaction outputs for customer C_i.
- $\tau_{C_i} = \mathsf{tr}(\omega_i)$: the previous transaction for customer C_i. Future transactions depend only on unspent/spendable transaction outputs; the function tr that returns a transaction is parameterized on the output used by C_i in the current transaction.
- $\pi_{C_i} = \mathsf{sign}_{\mathsf{inv}(\mathrm{B}_{C_i})}(\mathscr{H}(\tau'_{C_i}), \mathrm{B}_{C_i})$: the transaction input endorsed by C_i.
- $\pi = \pi_{C_1}, \ldots, \pi_{C_n}$: the transaction input, a list of transaction inputs endorsed by the customers.
- $PaymentRequest = \mathsf{sign}_{\mathsf{inv}(\mathsf{sk}(\mathrm{M}))}(\mathscr{H}(\mathrm{B_M}), ฿, t_1, t_2, m_\mathrm{M}, u_\mathrm{M}, z_\mathrm{M})$: the *Payment Request*, a message digitally signed with $\mathsf{inv}(M\mathrm{M})$, the private key of M. The associated public key utilized to verify the digital signature, that we denote as sk(M), is certified by a Certificate Authority and stored in a X.509 certificate.

Table 5.2
Identifiers used to denote the data exchanged (* optional parameters).

Identifier	Description
B_M	Merchant Bitcoin address for the current transaction, a public key freshly generated by M with the corresponding private key denoted by $\mathsf{inv}(B_M)$
B_{C_i}	Customer C_i Bitcoin address for the current transaction, a public key freshly generated by C_i, with the corresponding private key denoted by $\mathsf{inv}(B_{C_i})$
R_{C_i}	Refund address of customer C_i
฿	Number of Bitcoins for the current transaction
$฿_{C_i}$	Number of Bitcoins to be refunded to R_{C_i} in case of a refund
t_1, t_2*	Timestamps indicating *Payment Request* creation and expiry times, resp.
m_M* m_C*, m'_M*	Memo messages included in the *Payment Request* (by M), *Payment* (by C), and *Payment Acknowledgement* (by M) messages
u_M*	Payment URL
z_M*	Payment id provided by the merchant

- $RA_{C_i} = (R_{C_i}, ฿_{Ci})$: the refund address and amount for customer C_i.
- $\tau_C = \pi, (\mathscr{H}(B_M), ฿)$: one or more valid transactions, where π represents the inputs, and $(\mathscr{H}(B_M), ฿)$ represents the output.

5.6.1.1.2 *Agents' Initial Knowledge*

To keep the model simple, the initial knowledge of a single merchant M and two customers C_1, C_2 can be represented as:

- $C_1 : C_1, C_2, M, \mathscr{H}, \mathsf{tr}, \mathsf{sk}, \text{paynow};$
- $C_2 : C_1, C_2, \mathscr{H}, \mathsf{tr}, \mathsf{sk};$
- $M : M, \mathscr{H}, \mathsf{tr}, \mathsf{sk}, \mathsf{inv}(\mathsf{sk}(M)), \text{paynow}, t_1, t_2$

Each agent has an identity and access to the hash function $\mathscr{H}$, the symbolic function tr and a symbolic function sk for modeling digital signatures.

In particular, the sk function allows customers C_i to retrieve MM the public key of agent M from a repository, and verify the corresponding X.509 certificate, provided that they know the name of M.

inv(sk(M)) denotes the private key of M and is known only by M. We should note that in the *AnB* language, inv is a private function. Therefore, neither other agents nor the intruder can use inv to retrieve any agent's private key.

Initially, M does not know the identities of C_1 and C_2, while C_1 and C_2 know each other as they need to collaborate to build the transaction. However, only C_1 knows C_2 since C_1 will be the only customer interacting with the merchant. Finally, various constants (t_1, t_2, paynow) are available to agents.

The initial knowledge can be easily generalized for n customers; it should be noted that a customer does not need to know all other customers prior to the protocol run, but at least one. As customers can coordinate as they wish (including out-of-band communication), only one customer needs to interact with the merchant.

5.6.1.1.3 Security Goals

The following security goals are expected to hold after the protocol completion:

- **Goal 1:** *Refund Addresses Authentication.* M has a guarantee that all refund addresses R_{C_i}, for all $i = 1..n$ are provided by and linked to the customers involved in the transaction. In *AnB*, we denote the goal as:
  M *weakly authenticates* C_i *on* R_{C_i}, ฿$_{C_i}$ (for all $i = 1..n$).
- **Goal 2:** *Refund Address Agreement and Secrecy.* All refund addresses R_{C_i} are secret and known only by the merchant and the customers involved in the transaction. In *AnB*, we denote the goal as:
  $(R_{C_1}, \ldots, R_{C_n})$ *secret between* $M, C_1, \ldots, C_n$.

As the Payment Protocol is built on top of the core Bitcoin protocol and blockchain, a question that should be considered is whether the Payment Protocol is secure assuming the core Bitcoin protocol is secure. In this exercise, the security goals that are expected to be guaranteed by the core Bitcoin protocol and blockchain, such as the double-spending prevention, are assumed to be satisfied. This is a reasonable assumption as the security properties of the core Bitcoin protocol and blockchain have been formally proven in previous works [18, 41]. Similarly, we do not explicitly consider the security issues at the lower layers of the networking stack since the formalized model only encompasses the application layer and assumes that protocols such as TLS are secure. The approach of considering the security properties of different layers in isolation is sound, provided that the conditions of the vertical composition theorem [86] (see 5.4.4) are satisfied. It should be noted that the secrecy goal (2) prevents eavesdropping, and that known prediction and fixation vulnerabilities have been addressed by more recent versions of TLS [29, 49].

5.6.1.1.4 Protocol Actions

Agents are involved in a sequence of message exchanges over the designated channel. On the sender's side, agents should have enough information to compose the message, based on their initial knowledge and the new knowledge acquired during the protocol execution. On the recipient's side, every agent must decompose the incoming messages (e.g, decrypting the message or verifying a digital signature) according to their current knowledge. For the sake of simplicity, it is assumed that all public keys are available, at a certain point of the protocol execution, to the intruder and protocol participants.

$[C_1] \bullet\!\!\rightarrow\!\!\bullet M$	: paynow	C_1 *clicks* "Pay Now"
$M \bullet\!\!\rightarrow\!\!\bullet [C_1]$	: *PaymentRequest*	*Payment Request*
$C_1 \bullet\!\!\rightarrow\!\!\bullet C_2$	: $R_{C_1} M, \mathit{PaymentRequest}, B_{C_1}$	C_1, C_2 *cooperate -*
$C_2 \bullet\!\!\rightarrow\!\!\bullet C_1$	: R_{C_2}, π_{C_2}	*- to build a transaction*
$[C_1] \bullet\!\!\rightarrow\!\!\bullet M$	: $z_M, \tau_C, RA_{C_1}, RA_{C_2}, m_C$	*Payment*
$M \bullet\!\!\rightarrow\!\!\bullet [C_1]$	: $z_M, \tau_C, RA_{C_1}, RA_{C_2}, m_C, m'_M$	*PaymentACK*

5.6.1.1.5 Protocol Message Details

The format of the *Payment Request*, *Payment*, and *Payment Acknowledgement* messages is specified by the BIP70 standard. While the documentation only *recommends* running the protocol over HTTPS, this verification exercise assumes HTTPS is used. Moreover, although the standard supports payment via multiple transactions, the details of the messages considered here are for the case where the customer pays through a single transaction. The formalization and verification results can be easily extended to the case where a payment is made through multiple transactions. The protocol messages are modeled as follows:

- The *Payment Request* consists of the recipient's Bitcoin address $\mathscr{H}(B_M)$, requested Bitcoin amount ฿, timestamps t_1, t_2 corresponding to the request creation and expiry times, a "memo" field m_M for a note showed to the customer, the payment URL u_M where the payment message should be sent, and an identifier z_M for the merchant to link subsequent payment messages with

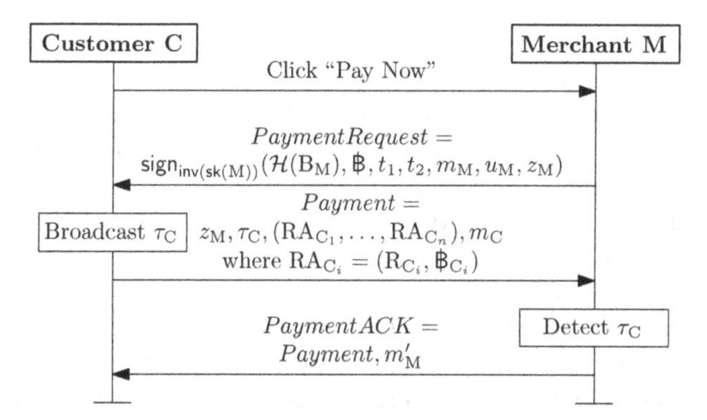

Figure 5.11 Expanded message contents for the Payment Protocol for C and M. Messages are sent over an HTTPS communication channel. $\mathsf{sign}_{\mathsf{inv}(\mathsf{sk}(\mathrm{M}))}(X)$ denotes both the message X and the digital signature on the message by the private key $\mathsf{inv}(\mathsf{sk}(\mathrm{M}))$ (courtesy [91]).

this request. All the fields are collectively signed by the merchant using their private key denoted by $\mathsf{inv}(\mathsf{sk}(\mathrm{M}))$, corresponding to their X.509 certificate public key.

- The *Payment* message consists of the merchant's identifier z_{M}, the payment transaction τ_{C}, a list of pairs of the form $\mathrm{RA}_{\mathrm{C}_i} = (\mathrm{R}_{\mathrm{C}_i}, ฿_{\mathrm{C}i})$ each containing the refund address $\mathrm{R}_{\mathrm{C}_i}$, the amount to be paid to that address $฿_{\mathrm{C}i}$ in case of refund, and an optional customer "memo" field m_{C}.
- The *Payment Acknowledgement* consists of a copy of the *Payment* message sent by the customer and an optional "memo" m'_{M} to be shown to the customer.

The Payment Protocol messages are shown Figure 5.11. Note that the *Payment* message, and especially the refund addresses provided therein, are not signed by the customer, and although protected by HTTPS, they can be subsequently repudiated by the customer. This is an underlying weakness that allows the Silkroad Trader attack discovered by McCorry *et al.* [79].

5.6.1.2 Main Results of BIP70 Security Verification

The formal model presented in [91] is encoded in the *AnBx* language [37], an extension of the *AnB* language supported by OFMC, allowing for macro definitions, functions type signatures and stricter type checking. The model was not tested with ProVerif because this tool does not support pseudonymous channels.

The tests were run on a Windows 10 PC, with Intel Core i7 4700HQ 2.40 GHz CPU with 16 GB RAM and the model was verified for a single session in OFMC in the classic and typed mode. As a result, the model demonstrated that both authentication and secrecy goals were violated. The attack was found in 2.34 seconds.

The authentication goal M *weakly authenticates* C_i *on* $R_{C_i}, ฿_{C_i}$ states that for all customers ($i = 1..n$), the merchant can have a guarantee of endorsement of the refund addresses and amounts.

In particular, the goal is violated because it is not possible to verify the non-injective agreement [77] between the construction of $RA_{C_i} = (R_{C_i}, ฿_{C_i})$ done by C_i and the corresponding values received by M. This is possible because the customers are not required to endorse the value $(R_{C_i}, ฿_{C_i})$ using digital signatures. Therefore, a compromised or dishonest client can easily manipulate the refund address and perform the Silkroad Trader attack [79], where a customer can route Bitcoin payments through an honest merchant to an illicit trader and later deny their involvement.

The secrecy goal $(R_{C_1}, \ldots, R_{C_n})$ *secret between* $M, C_1, \ldots, C_n$ is also violated. The definition of secrecy used in the model implies that all members of the secrecy set know the secret values and agree on them. But in this case, due to a lack of authentication, the customer who is communicating with the merchant can convince other customers that the refund address it is using is different from the one sent to the merchant. For example, R_{C_2}, the refund address of the second customer can be easily replaced with a different address by C_1 before being communicated to M.

It should be noted that, in general, with the automated verification, it is not possible to validate a specific attack trace known a priori, and the analysis usually aims at assessing the absence or presence of at least an attack trace that leads to a violation of a security goal. In particular, in order to verify the protocol, the model checker OFMC builds a state-transition system, and given the initial configuration, analyzes the possible transitions in order to see if any attack state is reachable in the presence of an active attacker. Therefore, the presence of a specific attack trace is not automatically confirmed, rather such automated verification helps to decide whether any attack trace is present or absent, where an attack trace is defined as a sequence of steps leading to a violation of a given security goal. In order to verify a protocol's security, the most important thing is to guarantee the absence of any attack trace.

5.6.2 E-COMMERCE PROTOCOLS

We consider here the specification of a wide and interesting class of protocols, namely e-payment protocols, showing how *AnBx* naturally provides all the necessary primitives to reason about the required high-level security. The case studies we consider are the *iKP* and *SET* e-payment protocols, showing how *AnBx* lends itself to a robust and modular design that captures the increasing level of security enforced by the protocols in the family, depending on the number of agents possessing a certified signing key. Interestingly, as a by-product of these design and verification efforts, a new flaw was identified in the original *iKP* specification and a fix proposed.

5.6.2.1 A Basic e-Payment Scheme

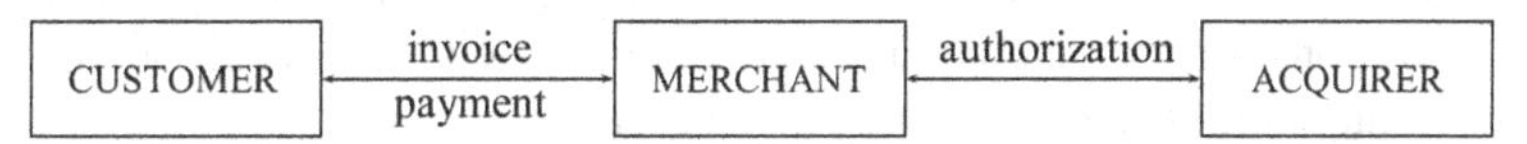

The figure above shows an outline of a bare-bones specification of an e-payment protocol: this abstraction allows us to introduce here many concepts which are common to both of the examples considered.

We assume three agents: a Customer C, a Merchant M and an Acquirer A, *i.e.* a financial institution entitled to process the payment. In the model, each agent starts with an initial knowledge, which may be partially shared with other participants. Indeed, since most e-payment protocols describe only the payment transaction and do not consider any preliminary phase, we assume that the Customer and the Merchant have already agreed on the details of the transaction, including an order description (*desc*) and a *price*. We also assume that the Acquirer shares with the Customer a customer's account number (*can*) comprised of a credit card number and the related PIN. The initial knowledge of the three parties can thus be summarized as follows: C knows *price, desc, can*; M knows *price, desc*; and A knows *can*.

The transaction can be decomposed into the following steps:

1. $C \rightarrow M$: *Initiate*
2. $C \leftarrow M$: *Invoice*
   (In steps 1 and 2 the Customer and the Merchant exchange all the information which is necessary to compose the next payment messages.)
3. $C \rightarrow M$: *Payment Request*
4. $M \rightarrow A$: *Authorization Request*
   (In steps 3 and 4 the Customer sends a payment request to the Merchant. The Merchant uses this information to compose an authorization request for the Acquirer and try to collect the payment.)
5. $M \leftarrow A$: *Authorization Response*
6. $C \leftarrow M$: *Confirm*
   (In steps 5 and 6 the Acquirer processes the transaction information, and then relays the purchase data directly to the issuing bank,

which actually authorizes the sale in accordance with the Customer's account. This interaction is not part of the narration. The Acquirer returns a response to the Merchant, indicating success or failure of the transaction. The Merchant then informs the Customer about the outcome.)

Interestingly, steps (4) and (6) involve forwarding operations, since the Customer never communicates directly with the Acquirer. Still, some credit-card information from the Customer must flow to the Acquirer through the Merchant to compose a reasonable payment request, while the final response from the Acquirer must flow to the Customer through the Merchant to provide evidence of the transaction.

Besides some elements of the initial knowledge, other information needs to be exchanged in the previous protocol template. First, to make transactions unequivocally identifiable, the Merchant generates a fresh transaction ID (tid) for each different transaction. Second, the Merchant associates a $date$ to the transaction or any appropriate timestamp. Both pieces of information must be communicated to the other parties. The transaction is then identified by a *contract*, which comprises most of the previous information: if Customer and Merchant reach an agreement on it, and they can prove this to the Acquirer, then the transaction can be completed successfully. The details on the structure of the contract vary among different protocols. At the end of the transaction, the authorization $auth$ is then returned by the Acquirer, and communicated to the two other participants.

We note that two main confidentiality concerns arise in the previous process: on the one hand, the Customer typically wishes to avoid leaking credit-card information to the Merchant; on the other hand, the Customer and the Merchant would not let the Acquirer know the details of the order or the services involved in the transaction. Specific protocols address such an issue in different ways, as we detail below.

5.6.2.1.1 Message Formats

In the case studies, a message M may be a name m, a tuple of messages $(\tilde{M})$, or a *message digest*. Much in the spirit of the *AnBx* channel abstractions, in fact, it turns out that we can abstract from most explicit cryptographic operations in the examples. Namely, the only transformation on data which we need to consider is the creation of a digest $[M]$ to prove the knowledge of a message M without leaking it.

We also consider digests which are resistant to dictionary attacks, hence presuppose an implementation based on a hashing scheme that combines the message M with a key shared only with the agent which must verify the digest. We note with $[M:A]$ a digest of a message M which is intended to be verified only by A; *hash* and *hmac* functions can be used to implement this in a standard way.

5.6.2.1.2 Protocol goals

A first goal we would like to satisfy for an e-payment system is that all the agents agree on the contract they sign. In terms of security goals, this corresponds to requiring that each participant can authenticate the other two parties on the *contract*. Moreover, the Acquirer should be able to prove to the other two parties that the payment has been authorized and the associated transaction performed: in OFMC this can be represented by requiring that M and C authenticate A on the authorization *auth*.

A stronger variant of the goals described above requires that, after completion of a transaction, each participant is able to provide a non-repudiable proof of the effective agreement by the other two parties on the terms of the transaction. In principle, each agent may wish to have sufficient proofs to convince an external verifier that the transaction was actually carried out as it claims. The lack of some of these provable authorizations does not necessarily make the protocol insecure, but it makes disputes between the parties difficult to settle, requiring to rely on evidence provided by other parties or to collect off-line information.

In summary, the authentication goals we would like to achieve are the following:

1. *M authenticates C on [contract]*, to give evidence to M that C has authorized the payment to M.
2. *C authenticates M on [contract]*, to give evidence to C on the terms of the purchase that M has settled with C.
3. *A authenticates C on [contract]*, to give evidence to A that C authorized A to transfer the money from A's account to M.
4. *A authenticates M on [contract]*, to give evidence to A that M has requested the transfer of the money to M's account.
5. *C authenticates A on [contract],auth*, to give evidence to C that A authorized the payment and performed the transaction.
6. *M authenticates A on [contract],auth*, to give evidence to M that A authorized the payment and performed the transaction.

Finally, we are also interested in some secrecy goals, like verifying that the Customer's credit card information *can* is kept confidential, and transmitted only to the Acquirer. In general, we would like to keep the data exchanged secret among the expected parties. All validated protocol goals are reported for each case study.

Table 5.3
Exchange modes for the revised *iKP* e-payment protocol.

mode/step	→	1KP	2KP	3KP
η_1	$C \to M$	$(-\|-\|-)$	$(-\|-\|M)$	$@(C\|M\|M)$
η_2	$C \leftarrow M$	$(-\|-\|-)$	$@(M\|C\|-)$	$@(M\|C\|C)$
η_3	$C \to M$	$(-\|-\|A)$	$(-\|-\|A)$	$(C\|A\|A)$
η_{4a}	$M \to A$	$(-\|-\|A)$	$(-\|-\|A)$	$(C\|A\|A)$
η_{4b}	$M \to A$	$(-\|-\|A)$	$@(M\|A\|A)$	$@(M\|A\|A)$
η_5	$M \leftarrow A$	$@(A\|C,M\|-)$	$@(A\|C,M\|M)$	$@(A\|C,M\|M)$
η_6	$C \leftarrow M$	$(A\|C,M\|-)$	$(A\|C,M\|-)$	$(A\|C,M\|C)$
certified agents		A	M,A	C,M,A

Table 5.4
Security goals satisfied by Original and Revised *iKP*.

	1KP		2KP		3KP	
Goal	**O**	**R**	**O**	**R**	**O**	**R**
can secret between *C*,*A*	+	+	+	+	+	+
A weakly authenticates *C* on *can*	−	−	−	−	+	+
desc secret between *C*,*M*	+	+	+	+	+	+
auth secret between *C*,*M*,*A*	−	−	−	−	−	+
price secret between *C*,*M*,*A*	−	−	−	+	−	+
M authenticates *A* on *auth*	+*	+	+*	+	+*	+
C authenticates *A* on *auth*	+	+	+	+	+	+
A authenticates *C* on [*contract*]	−	−	−	−	w	+
M authenticates *C* on [*contract*]	−	−	−	−	+	+
A authenticates *M* on [*contract*]	−	−	+	+	w	+
C authenticates *M* on [*contract*]	−	−	+*	+	+*	+
C authenticates *A* on [*contract*],*auth*	+	+	+	+	+	+
M authenticates *A* on [*contract*],*auth*	+*	+	+*	+	+*	+

* goal satisfied only after fixing the definition of Sig_A as in [39]
w = only weak authentication

5.6.3 IKP PROTOCOL FAMILY

The *iKP* protocol family was developed at IBM Research [26, 95] to support credit card-based transactions between customers and merchants (under the assumption that payment clearing and authorization may be handled securely off-line). All protocols in the family are based on public-key cryptography. The idea is that, depending on the number of parties that own certified public key-pairs, we can achieve increasing levels of security, as reflected by the name of the different protocols (1*KP*, 2*KP*, and 3*KP*).

5.6.3.1 Protocol Narration

Despite the complexity of *iKP*, by abstracting from cryptographic details, we can isolate a common communication pattern underlying all the protocols of the family. Namely, a common template can be specified as follows:

1. $C \rightarrow M, \eta_1 : [can{:}A], [desc{:}M]$
2. $C \leftarrow M, \eta_2 : price, tid, date, [contract]$
3. $C \rightarrow M, \eta_3 : price, tid, can, [can{:}A], [contract]$
4. $M \rightarrow A$ (decomposed into two steps to specify different communication modes)
   a. $M \rightarrow A, \eta_{4a} : price, tid, can, [can{:}A], [contract]$
   b. $M \rightarrow A, \eta_{4b} : price, tid, date, [desc{:}M], [contract]$
5. $M \leftarrow A, \eta_5 : auth, tid, [contract]$
6. $C \leftarrow M, \eta_6 : auth, tid, [contract]$

with $contract = (price, tid, date, [can{:}A], [desc{:}M])$.

By instantiating the exchange modes η_j in the previous scheme, one may generate the *AnBx* variants of the different protocols in the *iKP* family, achieving different security guarantees: this is exactly what is shown in Table 5.3. Notice that all the considered protocols rely on blind forwarding at step 4 to communicate sensitive payment information from the Customer to the Acquirer, without disclosing them to the Merchant. Moreover, a forwarding operation is employed at step 6 to preserve the authenticity of the response by the Acquirer.

5.6.3.2 Main Results of *i*KP Security Verification

The *AnBx* protocols described above were verified in [39, 37] and a corresponding analysis of the original specifications of $\{1,2,3\}$KP, as amended in [94], was carried out. Below, we refer to this amended version as the "original" *iKP*, to be contrasted with the "revised" *AnBx* version in Table 5.3. In both cases, the tests were run assuming that the Acquirer is trusted, *i.e.* encoded as a concrete agent *a* rather than as a role *A*; this is often a reasonable assumption in e-payment applications. As mentioned earlier, the *AnBx* specifications are not just more scalable and remarkably simpler, but they also provide stronger security guarantees, which are detailed in Table 5.4 and commented further below.

The *AnBx* specifications of *iKP* (and *SET*, see 5.6.4) were compiled into their respective cryptographic implementations with the *AnBx* compiler. We then verified the generated *CCM* translation with OFMC [84] against the described security goals. The authors also encoded and verified the original versions of *iKP*, and compared the results with those of the revised versions.

The tests with OFMC were done with one and two symbolic sessions. This bounds how many protocol executions the honest agents can engage in, while the intruder is left unbounded thanks to the symbolic lazy intruder technique in OFMC, but with two sessions it was not possible to complete the full verification due to search space explosion. However, by translating the *AnBx* specification with the *AnBx* compiler to ProVerif (see [89]), it is possible to verify the protocol for an unbounded number of sessions.

During the analysis of the original 2*KP* and 3*KP*, a new flaw was found [39]. It is related to the authenticity of the Authorization response *auth* that is generated by the Acquirer and then sent to the other agents at steps 5 and 6. In particular, the starred goals in Table 5.4 are met only after changing the protocol by adding the identities of Merchant and Customer inside the signature of the Acquirer in the original specification. In 2*KP*, since the Customer is not certified, this can be done with an ephemeral identity derived from the credit card number. It is worth noting that, after the completion of the revised and the amended original 3*KP*, each party has evidence of transaction authorization by the other two parties, since the protocol achieves all the authentication goals that can ideally be satisfied, according to the number of certified agents. Moreover, the revised 3*KP*, with respect to the original version, provides the additional guarantee of preserving the secrecy of the authorization response *auth*.

In contrast, the original 3*KP* protocol, the strongest proposed version, fails in two authentication goals: *A* can only weakly authenticate *M* and *C* on [*contract*]. Luckily, if the transaction ID *tid* is unique, this is only a minor problem, since [*contract*] should also be unique, *i.e.* two different contracts cannot be confused.

5.6.4 SET PURCHASE PROTOCOL

Secure Electronic Transaction (*SET*) is a family of protocols for securing credit card transactions over insecure networks. This standard was proposed by a consortium of credit card and software companies led by Visa and MasterCard and involving organizations like IBM, Microsoft, Netscape, RSA and Verisign. The *SET* purchase protocol specification considered here is the one described in [25], where *signed* and *unsigned* variants of *SET* are presented: in the former all the parties possess certified key-pairs, while in the latter the Customer does not. We describe here the *AnBx* models of both variants of the original *SET* protocol, and show how, using the notion of *AnBx* channels, it is possible to re-engineer the protocol and obtain stricter security guarantees than in the original specification.

5.6.4.1 Protocol Narration

Given the complexity of *SET*, to ease the comparison with other works on such a protocol, in this presentation the information exchanged by the agents is denoted with the names commonly used in *SET* specifications. We introduce some basic concepts of the protocol by simply providing a mapping of the exchanged data to the corresponding information in the bare-bones specification presented in Section 5.6.2: this should clarify the role of most of the elements.

We can identify *PurchAmt* with *price*, *OrderDesc* with *desc*, *pan* with *can* and *AuthCode* with *auth*. The initial knowledge of the three parties can then be summarized as follows: *C* knows *PurchAmt*, *OrderDesc* and *pan*; *M* knows *PurchAmt* and *OrderDesc*; *A* knows *pan*.

During the protocol run, the agents generate some identifiers: *LIDM* is a local transaction identifier that the Customer sends to the Merchant, while the Merchant generates another session identifier *XID*; we denote the pair (*LIDM*,*XID*) with *TID*. Finally, we complete the abstraction by stipulating $OIdata = OrderDesc$ and $PIdata = pan$; we let $HOD = ([OIdata{:}M], [PIdata{:}A])$. The latter contains the evidence (digest) of the credit card that the Customer intends to use, and the evidence of the order description that will later be forwarded to the Acquirer.

In the model, *HOD* plays the role of the *dual signature*, a key cryptographic mechanism applied in *SET*, used to let the Merchant and the Acquirer agree on the transaction without needing to disclose all the details. More precisely, the Merchant does not need the customer's credit card number to process an order, but only needs to know that the payment has been approved by the Acquirer. Conversely, the Acquirer does not need to be aware of the details of the order, but just needs evidence that a particular payment must be processed.

Table 5.5
Exchange modes for the revised *SET* e-payment protocol.

mode/step	$\rightarrow$	unsigned SET	signed SET
η_1	$C \rightarrow M$	$(-\mid-\mid M)$	$@(C\mid M\mid M)$
η_2	$C \leftarrow M$	$@(M\mid C\mid -)$	$@(M\mid C\mid C)$
η_{3a}	$C \rightarrow M$	$(-\mid-\mid M)$	$@(C\mid M\mid M)$
η_{3b}	$C \rightarrow M$	$(-\mid-\mid A)$	$(C\mid A\mid A)$
η_{4a}	$M \rightarrow A$	$(-\mid-\mid A)$	$(C\mid A\mid A)$
η_{4b}	$M \rightarrow A$	$@(M\mid A\mid A)$	$@(M\mid A\mid A)$
η_5	$M \leftarrow A$	$@(A\mid C,M\mid M)$	$@(A\mid C,M\mid M)$
η_6	$C \leftarrow M$	$@(A\mid C,M\mid -)$	$@(A\mid C,M\mid C)$
certified agents		M,A	C,M,A

Although many papers on *SET* [25, 34, 107] focus on the signed version of the protocol, we note that both versions expose a common pattern which allows for an easy specification in *AnBx*. The following narration allows to expose the common structure of the protocols:

1. $C \rightarrow M, \eta_1 : LIDM$
2. $M \rightarrow C, \eta_2 : XID$
3. $C \rightarrow M$ (decomposed in two steps to specify different communication modes)
   a. $C \rightarrow M, \eta_{3a} : TID,HOD$
   b. $C \rightarrow M, \eta_{3b} :TID,PurchAmt,HOD,PIdata$
4. $M \rightarrow A$ (decomposed in two steps to specify different communication modes)
   a. $M \rightarrow A, \eta_{4a} : TID,PurchAmt,HOD,PIdata$
   b. $M \rightarrow A, \eta_{4b} : TID,PurchAmt,HOD$
5. $A \rightarrow M, \eta_5 : TID,HOD,AuthCode$
6. $M \rightarrow C, \eta_6 : TID,HOD,AuthCode$

Table 5.5 shows the communication modes we specify to instantiate the previous protocol template with the revised unsigned and signed versions of *SET*.

5.6.4.2 Main Results of SET Security Verification

The *AnBx* specifications of the *SET* purchase protocol were verified with OFMC and ProVerif. With OFMC, the models were verified for

Table 5.6
Security goals satisfied by Original and Revised SET purchase protocol.

	unsigned SET		signed SET	
Goal	**O**	**R**	**O**	**R**
pan secret between *C,A*	+	+	+	+
A weakly authenticates *C* on *pan*	−	+	+	+
OrderDesc secret between *C,M*	+	+	+	+
PurchAmt secret between *C,M,A*	−	−	+	+
AuthCode secret between *C,M,A*	−	−	−	+
M authenticates *A* on *AuthCode*	+	+	+	+
C authenticates *A* on *AuthCode*	−	+	−	+
C authenticates *M* on *AuthCode*	+*	+	+*	+
A authenticates *C* on *contract*	w	+	w	+
M authenticates *C* on *contract*	−	+	+	+
A authenticates *M* on *contract*	−	+	−	+
C authenticates *M* on *contract*	+	+	+	+
C authenticates *A* on *contract,AuthCode*	−	+	−	+
M authenticates *A* on *contract,AuthCode*	+	+	+	+

* goal satisfied only after fixing step 5 as in [25]
w = only weak authentication
Revised *SET*: $cont. = PriceAmt, TID, [PIData{:}A], [OIData{:}M]$
Original *SET*: $cont. = PriceAmt, TID, \mathsf{hash}(PIData), \mathsf{hash}(OIData)$

2 sessions, by incrementing the depth of the search space, up to the available RAM (16Gb). ProVerif allows to verify an unbounded number of

sessions, but in some cases the goals cannot be proved, due to the internal mechanisms of ProVerif and the fact that, in general, verifying protocols for an unbounded number of sessions is undecidable.

The results show that the revised versions of the protocols satisfy stronger security guarantees than the original ones [25], as reported in Table 5.6. It is worth noting, in particular, that the revised versions do not suffer from two known flaws affecting the original *SET* specification. The first flaw [25] involves the fifth step of the protocol, where it is not possible to unequivocally link the identities of the Acquirer and the Merchant with the ongoing transaction and the authorization code. Namely, the original message should be amended to include the identity of the merchant M, otherwise the goal "C authenticates M on *AuthCode*" cannot be satisfied. In the revised version, the exchange at step 5 is automatically compiled into a message including the identities of both the Merchant and the Customer, so the problem is solved.

The same implementation also prevents the second flaw, presented in [34]. In that paper, the specification of the protocol is more detailed than in [25], as it introduces an additional field *AuthRRTags*, which includes the identity of the Merchant. We tested the version of *SET* presented in [34] with OFMC and verified the presence of the flaw, namely an attack against the purchase phase, which exploits a lack of verification in the payment authorization process. It may allow a dishonest Customer to fool an honest Merchant when collaborating with another dishonest Merchant. The attack is based on the fact that neither *LIDM* nor *XID* can be considered unique, so they cannot be used to identify a specific Merchant. Therefore, the customer can start a parallel purchase with an accomplice playing the role of another merchant, and make the Acquirer authorize the payment in favor of the accomplice. Here again, the goal "C authenticates M on *AuthCode*" fails.

During the analysis, we also verified that both the original specifications [25, 34] fail to verify the goals "C authenticates A on *AuthCode*" and "C authenticates M on *contract*,*AuthCode*". To overcome this problem, the protocol must be fixed in the sixth (and final) step, as already outlined in [107]. This issue arises from the fact that the Customer does not have any evidence of the origin of *AuthCode* by the Acquirer and instead has to rely only on information provided by the Merchant. For example, giving to the Customer a proof that the Acquirer authorized the payment requires a substantial modification of the sixth step of the protocol. In fact, instead of letting the Merchant sign a message for the Customer, we exploit the *AnBx* forward mode to bring to the Customer the authorization of the payment signed directly by the Acquirer. It is worth noticing that, employing a *fresh* forward mode in the sixth step, we can achieve the desired strong authenticity goal on the pair, even though the transaction identifier is not unique.

The results confirm what is outlined in [107], in showing that, while *iKP* meets all the non-repudiation goals, the original specification of *SET* does not. It is important to notice that, to achieve non-repudiation, each participant must have sufficient proofs to convince an external verifier that the transaction was actually carried out as it claims. A way to obtain this is to assume that the authentication is obtained by means of digital signatures computed with keys which are valid within a Public Key Infrastructure and are issued by a trusted third party (Certification Authority). Although this limits the way authentic channels in *AnBx* could be implemented, in practice it does not represent a significant restriction, since in the considered protocols, digital signatures are the standard means of authentication.

5.7 CONCLUSION

In this chapter, we consider the formalization of security protocols with the purpose of automatically verifying if they satisfy their expected security goals. We also discussed theoretical and practical challenges in automated verification, along with different specification languages and verification tools available. The case studies demonstrated the practical applicability of some of these tools and techniques to real-world security protocols.

REFERENCES

1. Martín Abadi and Cédric Fournet. Mobile values, new names, and secure communication. In Chris Hankin and Dave Schmidt, editors, *Conference Record of POPL 2001: The 28th ACM SIGPLAN-SIGACT Symposium on Principles of Programming Languages, London, UK, January 17-19, 2001*, pages 104–115. ACM, 2001.

2. Martín Abadi and Cédric Fournet. Mobile values, new names, and secure communication. *ACM SIGPLAN Notices*, 36(3):104–115, 2001.

3. Martín Abadi, Cédric Fournet, and Georges Gonthier. Secure implementation of channel abstractions. *Information and Computation(Print)*, 174(1):37–83, 2002.

4. Martín Abadi and Andrew D. Gordon. A calculus for cryptographic protocols: The Spi calculus. In *CCS '97, Proceedings of the 4th ACM Conference on Computer and Communications Security, Zurich, Switzerland, April 1-4, 1997*, pages 36–47. ACM, 1997.

5. Joel C Adams. Object-centered design: a five-phase introduction to object-oriented programming in cs1–2. In *ACM SIGCSE Bulletin*, volume 28, pages 78–82. ACM, 1996.

6. Pedro Adão, Paulo Mateus, and Luca Viganò. Protocol insecurity with a finite number of sessions and a cost-sensitive guessing intruder is np-complete. *Theoretical Computer Science*, 538:2–15, 2014.

7. Giora Alexandron, Michal Armoni, Michal Gordon, and David Harel. Scenario-based programming: reducing the cognitive load, fostering abstract thinking. In *Companion Proceedings of the 36th International Conference on Software Engineering*, pages 311–320. ACM, 2014.

8. Omar Almousa, Sebastian Mödersheim, Paolo Modesti, and Luca Viganò. Typing and compositionality for security protocols: A generalization to the geometric fragment. In *Computer Security - ESORICS 2015 - 20th European Symposium on Research in Computer Security, Vienna, Austria, September 21-25, 2015, Proceedings, Part II*, pages 209–229, 2015.

9. Omar Almousa, Sebastian Mödersheim, and Luca Viganò. Alice and Bob: Reconciling formal models and implementation. In Chiara Bodei, Gian-Luigi Ferrari, and Corrado Priami, editors, *Programming Languages with Applications to Biology and Security: Essays Dedicated to Pierpaolo Degano on the Occasion of His 65th Birthday*, volume 9465 of *Lecture Notes in Computer Science*, pages 66–85. Springer International Publishing, 2015.

10. Roberto M. Amadio and Witold Charatonik. On name generation and set-based analysis in the dolev-yao model. In Lubos Brim, Petr Jancar, Mojmír Kretínský, and Antonín Kucera, editors, *CONCUR 2002 - Concurrency Theory, 13th International Conference, Brno, Czech Republic, August 20-23, 2002, Proceedings*, volume 2421 of *Lecture Notes in Computer Science*, pages 499–514. Springer, 2002.

11. Suzana Andova, Cas Cremers, Kristian Gjøsteen, Sjouke Mauw, Stig Fr. Mjølsnes, and Sasa Radomirovic. A framework for compositional verification of security protocols. *Inf. Comput.*, 206(2-4):425–459, 2008.

12. G. Andresen and M. Hearn. BIP 70: Payment Protocol. *Bitcoin Improvement Process*, July 2013. https://github.com/bitcoin/bips/blob/master/bip-0070.mediawiki.

13. Apache Foundation. The Apache Ant Project, 2019. http://ant.apache.org.

14. Myrto Arapinis, Joshua Phillips, Eike Ritter, and Mark Dermot Ryan. Statverif: Verification of stateful processes. *Journal of Computer Security*, 22(5):743–821, 2014.

15. A. Armando and L. Compagna. SATMC: A SAT-based model checker for security protocols. *Lecture Notes in Computer Science*, pages 730–733, 2004.

16. Alessandro Armando, Wihem Arsac, Tigran Avanesov, Michele Barletta, Alberto Calvi, Alessandro Cappai, Roberto Carbone, Yannick Chevalier, Luca Compagna, Jorge Cuéllar, et al. The AVANTSSAR platform for the automated validation of trust and security of service-oriented architectures. In *Tools and Algorithms for the Construction and Analysis of Systems*, pages 267–282. Springer, 2012.

17. Alessandro Armando, David Basin, Yohan Boichut, Yannick Chevalier, Luca Compagna, Jorge Cuéllar, P Hankes Drielsma, Pierre-Cyrille Héam, Olga Kouchnarenko, Jacopo Mantovani, et al. The AVISPA tool for the automated validation of internet security protocols and applications. In *Computer Aided Verification*, pages 281–285. Springer, 2005.

18. Nicola Atzei, Massimo Bartoletti, Stefano Lande, and Roberto Zunino. A formal model of bitcoin transactions. In Meiklejohn and Sako, editors, *Financial Cryptography and Data Security (FC 2018)*, volume 10957 of *LNCS*, pages 541–560. Springer, 2018.

19. Matteo Avalle, Alfredo Pironti, and Riccardo Sisto. Formal verification of security protocol implementations: a survey. *Formal Aspects of Computing*, 26(1):99–123, 2014.

20. Tigran Avanesov, Yannick Chevalier, Michaël Rusinowitch, and Mathieu Turuani. Intruder deducibility constraints with negation. decidability and application to secured service compositions. *J. Symb. Comput.*, 80:4–26, 2017.

21. AVISPA. Deliverable 2.3: The Intermediate Format. Available at www.avispa-project.org, 2003.

22. Manuel Barbosa, Gilles Barthe, Karthik Bhargavan, Bruno Blanchet, Cas Cremers, Kevin Liao, and Bryan Parno. Sok: Computer-aided cryptography. In *42nd IEEE Symposium on Security and Privacy, SP 2021, San Francisco, CA, USA, 24-27 May 2021*, pages 777–795. IEEE, 2021.

23. Henk Barendregt and Herman Geuvers. Proof-assistants using dependent type systems. In John Alan Robinson and Andrei Voronkov, editors, *Handbook of Automated Reasoning (in 2 volumes)*, pages 1149–1238. Elsevier and MIT Press, 2001.

24. David A. Basin, Michel Keller, Sasa Radomirovic, and Ralf Sasse. Alice and bob meet equational theories. In Narciso Martí-Oliet, Peter Csaba Ölveczky, and Carolyn L. Talcott, editors, *Logic, Rewriting, and Concurrency - Essays dedicated to José Meseguer on the Occasion of His 65th Birthday*, volume 9200 of *Lecture Notes in Computer Science*, pages 160–180. Springer, 2015.

25. G. Bella, F. Massacci, and L.C. Paulson. Verifying the SET purchase protocols. *Journal of Automated Reasoning*, 36(1):5–37, 2006.

26. M. Bellare, JA Garay, R. Hauser, A. Herzberg, H. Krawczyk, M. Steiner, G. Tsudik, E. Van Herreweghen, and M. Waidner. Design, implementation, and deployment of the iKP secure electronic payment system. *IEEE Journal on Selected Areas in Communications*, 18(4):611–627, 2000.

27. Mordechai Ben-Ari. Constructivism in computer science education. *Journal of Computers in Mathematics and Science Teaching*, 20(1):45–74, 2001.

28. Mordechai Ben-Ari and Tzippora Yeshno. Conceptual models of software artifacts. *Interacting with Computers*, 18(6):1336–1350, 2006.

29. Karthikeyan Bhargavan, Bruno Blanchet, and Nadim Kobeissi. Verified models and reference implementations for the TLS 1.3 standard candidate. In *2017 IEEE Symposium on Security and Privacy, SP 2017, San Jose, CA, USA, May 22-26, 2017*, pages 483–502. IEEE Computer Society, 2017.

30. Bruno Blanchet. An efficient cryptographic protocol verifier based on Prolog rules. In *Computer Security Foundations Workshop, IEEE*, pages 0082–0082. IEEE Computer Society, 2001.

31. Bruno Blanchet. A computationally sound mechanized prover for security protocols. *IEEE Transactions on Dependable and Secure Computing*, 5(4):193–207, 2008.

32. Bruno Blanchet, Vincent Cheval, and Véronique Cortier. ProVerif with Lemmas, Induction, Fast Subsumption, and Much More. In *43RD IEEE Symposium on Security and Privacy (S&P'22)*, San Francisco, United States, May 2022.

33. Bruno Blanchet et al. Modeling and verifying security protocols with the applied pi calculus and proverif. *Foundations and Trends in Privacy and Security*, 1(1-2):1–135, 2016.

34. S. Brlek, S. Hamadou, and J. Mullins. A flaw in the electronic commerce protocol SET. *Information Processing Letters*, 97(3):104–108, 2006.

35. Jerome S Bruner. *The Process of Education.* Harvard University Press, 2009.

36. Randal E. Bryant. Graph-based algorithms for boolean function manipulation. *IEEE Transactions on Computers*, 35(8):677–691, 1986.

37. Michele Bugliesi, Stefano Calzavara, Sebastian Mödersheim, and Paolo Modesti. Security protocol specification and verification with anbx. *J. Inf. Secur. Appl.*, 30:46–63, 2016.

38. Michele Bugliesi and Riccardo Focardi. Language based secure communication. In *Computer Security Foundations Symposium, 2008. CSF'08. IEEE 21st*, pages 3–16, 2008.

39. Michele Bugliesi and Paolo Modesti. AnBx-Security protocols design and verification. In *Automated Reasoning for Security Protocol Analysis and Issues in the Theory of Security: Joint Workshop, ARSPA-WITS 2010*, pages 164–184. Springer-Verlag, 2010.

40. Michael Burrows, Martín Abadi, and Roger M. Needham. A logic of authentication. *ACM Transactions on Computer Systems*, 8(1):18–36, 1990.

41. Kaylash Chaudhary, Ansgar Fehnker, Jaco van de Pol, and Mariëlle Stoelinga. Modeling and verification of the bitcoin protocol. In van Glabbeek, Groote, and Höfner, editors, *Proceedings Workshop on Models for Formal Analysis of Real Systems, MARS 2015*, volume 196 of *EPTCS*, pages 46–60, 2015.

42. Ştefan Ciobâcă and Véronique Cortier. Protocol composition for arbitrary primitives. In *Proceedings of the 23rd IEEE Computer Security Foundations Symposium, CSF 2010, Edinburgh, United Kingdom, July 17-19, 2010*, pages 322–336. IEEE Computer Society, 2010.

43. Edmund M. Clarke, Orna Grumberg, and David E. Long. Model checking and abstraction. *ACM Transactions on Programming Languages and Systems*, 16(5):1512–1542, 1994.

44. Hubert Comon-Lundh and Stéphanie Delaune. The finite variant property: How to get rid of some algebraic properties. In Jürgen Giesl, editor, *Term Rewriting and Applications, 16th International Conference, RTA 2005, Nara, Japan, April 19-21, 2005, Proceedings*, volume 3467 of *Lecture Notes in Computer Science*, pages 294–307. Springer, 2005.

45. Véronique Cortier and Stéphanie Delaune. Safely composing security protocols. *Formal Methods in System Design*, 34(1):1–36, 2009.

46. Véronique Cortier, Stéphanie Delaune, and Pascal Lafourcade. A survey of algebraic properties used in cryptographic protocols. *Journal of Computer Security*, 14(1):1–43, 2006.

47. Camille Coti, Laure Petrucci, César Rodríguez, and Marcelo Sousa. Quasi-optimal partial order reduction. *Formal Methods in System Design*, 57(1):3–33, 2021.

48. C. Cremers and P. Lafourcade. Comparing state spaces in automatic security protocol verification. In *Proceedings of the 7th International Workshop on Automated Verification of Critical Systems (AVoCS'07), Oxford, UK, September*, pages 49–63. Citeseer, 2007.

49. Cas Cremers, Marko Horvat, Jonathan Hoyland, Sam Scott, and Thyla van der Merwe. A comprehensive symbolic analysis of TLS 1.3. In Bhavani M. Thuraisingham, David Evans, Tal Malkin, and Dongyan Xu, editors, *Proceedings of the 2017 ACM SIGSAC Conference on Computer and Communications Security (CCS 2017)*, pages 1773–1788. ACM, 2017.

50. Cas JF Cremers. The scyther tool: Verification, falsification, and analysis of security protocols. In *Computer Aided Verification*, pages 414–418. Springer, 2008.

51. Melissa Dark, Steven Belcher, Matt Bishop, and Ida Ngambeki. Practice, practice, practice... secure programmer! In *Proceeding of the 19th Colloquium for Information System Security Education*, 2015.

52. Stéphanie Delaune and Lucca Hirschi. A survey of symbolic methods for establishing equivalence-based properties in cryptographic protocols. *J. Log. Algebraic Methods Program.*, 87:127–144, 2017.

53. D. Dolev and A. Yao. On the security of public-key protocols. *IEEE Transactions on Information Theory*, 2(29), 1983.

54. Jannik Dreier, Charles Duménil, Steve Kremer, and Ralf Sasse. Beyond subterm-convergent equational theories in automated verification of stateful protocols. In Matteo Maffei and Mark Ryan, editors, *Principles of Security and Trust - 6th International Conference, POST 2017, Held as Part of the European Joint Conferences on Theory and Practice of Software, ETAPS 2017, Uppsala, Sweden, April 22-29, 2017, Proceedings*, volume 10204 of *Lecture Notes in Computer Science*, pages 117–140. Springer, 2017.

55. Zakir Durumeric, James Kasten, David Adrian, J Alex Halderman, Michael Bailey, Frank Li, Nicolas Weaver, Johanna Amann, Jethro Beekman, Mathias Payer, et al. The matter of heartbleed. In *Proceedings of the 2014 Conference on Internet Measurement Conference*, pages 475–488. ACM, 2014.

56. Santiago Escobar, Catherine A. Meadows, and José Meseguer. Maude-npa: Cryptographic protocol analysis modulo equational properties. In Alessandro Aldini, Gilles Barthe, and Roberto Gorrieri, editors, *Foundations of Security Analysis and Design V, FOSAD 2007/2008/2009 Tutorial Lectures*, volume 5705 of *Lecture Notes in Computer Science*, pages 1–50. Springer, 2007.

57. Shimon Even and Oded Goldreich. On the security of multi-party ping-pong protocols. In *24th Annual Symposium on Foundations of Computer Science, Tucson, Arizona, USA, 7-9 November 1983*, pages 34–39. IEEE Computer Society, 1983.

58. Leo Freitas, Paolo Modesti, and Martin Emms. A Methodology for Protocol Verification applied to EMV. In *Formal Methods: Foundations and Applications - 21th Brazilian Symposium, SBMF 2018, Salvador, Brazil, November*

28-30, 2018, Proceedings, volume 11254 of *Lecture Notes in Computer Science*. Springer, 2018.

59. Hubert Garavel, Maurice H. ter Beek, and Jaco van de Pol. The 2020 expert survey on formal methods. In Maurice H. ter Beek and Dejan Nickovic, editors, *Formal Methods for Industrial Critical Systems - 25th International Conference, FMICS 2020, Vienna, Austria, September 2-3, 2020, Proceedings*, volume 12327 of *Lecture Notes in Computer Science*, pages 3–69. Springer, 2020.

60. Rémi Garcia and Paolo Modesti. An IDE for the design, verification and implementation of security protocols. In *2017 IEEE International Symposium on Software Reliability Engineering Workshops, ISSRE Workshops 2017, Toulouse, France, October 23-26, 2017*, pages 157–163, 2017.

61. Patrice Godefroid. *Partial-Order Methods for the Verification of Concurrent Systems - An Approach to the State-Explosion Problem*, volume 1032 of *Lecture Notes in Computer Science*. Springer, 1996.

62. Kurt Gödel. On formally undecidable propositions of principia mathematica and related systems. *The undecidable. Hew*, 1964.

63. Sébastien Gondron and Sebastian Mödersheim. Vertical composition and sound payload abstraction for stateful protocols. In *34th IEEE Computer Security Foundations Symposium, CSF 2021, Dubrovnik, Croatia, June 21-25, 2021*, pages 1–16. IEEE, 2021.

64. L. Gong, G. Ellison, and M. Dageforde. *Inside Java 2 Platform Security: Architecture, Api Design, and Implementation*. Addison-Wesley, 2003.

65. Thomas Groß and Sebastian Mödersheim. Vertical protocol composition. In *Proceedings of the 24th IEEE Computer Security Foundations Symposium, CSF 2011, Cernay-la-Ville, France, 27-29 June, 2011*, pages 235–250. IEEE Computer Society, 2011.

66. Joshua D. Guttman. Establishing and preserving protocol security goals. *Journal of Computer Security*, 22(2):203–267, 2014.

67. Joshua D. Guttman and F. Javier Thayer. Protocol independence through disjoint encryption. In *Proceedings of the 13th IEEE Computer Security Foundations Workshop, CSFW '00, Cambridge, England, UK, July 3-5, 2000*, pages 24–34. IEEE Computer Society, 2000.

68. Bruria Haberman and Yifat Ben-David Kolikant. Activating "black boxes" instead of opening "zipper"- a method of teaching novices basic cs concepts. In *Proceedings of the 6th Annual Conference on Innovation and Technology in Computer Science Education*, ITiCSE '01, pages 41–44, New York, NY, USA, 2001. ACM.

69. Said Hadjerrouit. A constructivist framework for integrating the java paradigm into the undergraduate curriculum. In *ACM SIGCSE Bulletin*, volume 30, pages 105–107. ACM, 1998.

70. Ákos Hajdu and Zoltán Micskei. Efficient strategies for cegar-based model checking. *J. Autom. Reason.*, 64(6):1051–1091, 2020.

71. Andreas V. Hess, Sebastian Alexander Mödersheim, and Achim D. Brucker. Stateful protocol composition. In Javier López, Jianying Zhou, and Miguel Soriano, editors, *Computer Security - 23rd European Symposium on Research in Computer Security, ESORICS 2018, Barcelona, Spain, September 3-7, 2018, Proceedings, Part I*, volume 11098 of *Lecture Notes in Computer Science*, pages 427–446. Springer, 2018.

72. ISO/IEC. ISO/IEC 9798-1:2010. Information technology – Security techniques – Entity authentication – Part 1: General, 2010.

73. Nadim Kobeissi, Georgio Nicolas, and Mukesh Tiwari. Verifpal: Cryptographic protocol analysis for the real world. In Karthikeyan Bhargavan, Elisabeth Oswald, and Manoj Prabhakaran, editors, *Progress in Cryptology - INDOCRYPT 2020 - 21st International Conference on Cryptology in India, Bangalore, India, December 13-16, 2020, Proceedings*, volume 12578 of *Lecture Notes in Computer Science*, pages 151–202. Springer, 2020.

74. Herman Koppelman and Betsy van Dijk. Teaching abstraction in introductory courses. In *Proceedings of the Fifteenth Annual Conference on Innovation and Technology in Computer Science Education*, ITiCSE '10, pages 174–178, New York, NY, USA, 2010. ACM.

75. Steve Kremer and Robert Künnemann. Automated analysis of security protocols with global state. *Journal of Computer Security*, 24(5):583–616, 2016.

76. G. Lowe. An attack on the needham-schroeder public-key authentication protocol. *Information Processing Letters*, 56(3):131–133, 1995.

77. Gavin Lowe. A hierarchy of authentication specifications. In *CSFW'97*, pages 31–43. IEEE Computer Society Press, 1997.

78. Gavin Lowe. Towards a completeness result for model checking of security protocols. *Journal of Computer Security*, 7(1):89–146, 1999.

79. Patrick McCorry, Siamak F. Shahandashti, and Feng Hao. Refund attacks on bitcoin's payment protocol. In *20th Financial Cryptography and Data Security Conference*, 2016.

80. Simon Meier, Benedikt Schmidt, Cas Cremers, and David A. Basin. The TAMARIN prover for the symbolic analysis of security protocols. In

Natasha Sharygina and Helmut Veith, editors, *Computer Aided Verification - 25th International Conference, CAV 2013, Saint Petersburg, Russia, July 13-19, 2013. Proceedings*, volume 8044 of *Lecture Notes in Computer Science*, pages 696–701. Springer, 2013.

81. Robin Milner. Logic for computable functions: description of a machine implementation. 1972.

82. J Mitchell, A Scedrov, N Durgin, and P Lincoln. Undecidability of bounded security protocols. In *Workshop on Formal Methods and Security Protocols*. Citeseer, 1999.

83. Sebastian Mödersheim. Algebraic properties in Alice and Bob notation. In *International Conference on Availability, Reliability and Security (ARES 2009)*, pages 433–440, 2009.

84. Sebastian Mödersheim and Luca Viganò. The open-source fixed-point model checker for symbolic analysis of security protocols. In Alessandro Aldini, Gilles Barthe, and Roberto Gorrieri, editors, *Foundations of Security Analysis and Design V, FOSAD 2007/2008/2009 Tutorial Lectures*, volume 5705 of *Lecture Notes in Computer Science*, pages 166–194. Springer, 2009.

85. Sebastian Mödersheim and Luca Viganò. Secure pseudonymous channels. In Michael Backes and Peng Ning, editors, *Computer Security - ESORICS 2009, 14th European Symposium on Research in Computer Security, Saint-Malo, France, September 21-23, 2009. Proceedings*, volume 5789 of *Lecture Notes in Computer Science*, pages 337–354. Springer, 2009.

86. Sebastian Mödersheim and Luca Viganò. Sufficient conditions for vertical composition of security protocols. In *Proceedings of the 9th ACM Symposium on Information, Computer and Communications Security*, ASIA CCS '14, pages 435–446, New York, NY, USA, 2014. ACM.

87. Paolo Modesti. *Verified Security Protocol Modeling and Implementation with AnBx*. PhD thesis, Università Ca' Foscari Venezia (Italy), 2012.

88. Paolo Modesti. Efficient Java code generation of security protocols specified in AnB/AnBx. In *Security and Trust Management - 10th International Workshop, STM 2014, Proceedings*, pages 204–208, 2014.

89. Paolo Modesti. AnBx: Automatic generation and verification of security protocols implementations. In *8th International Symposium on Foundations & Practice of Security*, volume 9482 of *LNCS*. Springer, 2015.

90. Paolo Modesti. Integrating formal methods for security in software security education. *Informatics in Education*, 19(3):425–454, 2020.

91. Paolo Modesti, Siamak F. Shahandashti, Patrick McCorry, and Feng Hao. Formal modeling and security analysis of bitcoin's payment protocol. *Comput. Secur.*, 107:102279, 2021.

92. S. Nakamoto. Bitcoin: A Peer-to-Peer Electronic Cash System, November 2008. https://bitcoin.org/bitcoin.pdf.

93. Tobias Nipkow, Lawrence C. Paulson, and Markus Wenzel. *Isabelle/HOL — A Proof Assistant for Higher-Order Logic*, volume 2283 of *LNCS*. Springer, 2002.

94. K. Ogata and K. Futatsugi. Formal analysis of the iKP electronic payment protocols. *Lecture Notes in Computer Science*, pages 441–460, 2003.

95. D. O'Mahony, M. Peirce, and H. Tewari. *Electronic payment systems for e-commerce*. Artech House Publishers, 2001.

96. M. Pistoia, N. Nagaratnam, L. Koved, and A. Nadalin. *Enterprise Java 2 Security: Building Secure and Robust J2EE Applications*. Addison Wesley, 2004.

97. Erik Poll and Aleksy Schubert. Verifying an implementation of SSH. In *WITS*, volume 7, pages 164–177, 2007.

98. Michaël Rusinowitch and Mathieu Turuani. Protocol insecurity with a finite number of sessions, composed keys is np-complete. *Theor. Comput. Sci.*, 299(1-3):451–475, 2003.

99. Rupert Schlick, Michael Felderer, Istvan Majzik, Roberto Nardone, Alexander Raschke, Colin Snook, and Valeria Vittorini. A proposal of an example and experiments repository to foster industrial adoption of formal methods. pages 249–272, 2018.

100. Benedikt Schmidt, Sebastian Meier, Cas Cremers, and David Basin. Automated analysis of Diffie-Hellman protocols and advanced security properties. In *Computer Security Foundations Symposium (CSF), 2012 IEEE 25th*, pages 78–94. IEEE, 2012.

101. Konrad Slind and Michael Norrish. A brief overview of HOL4. In Otmane Aït Mohamed, César A. Muñoz, and Sofiène Tahar, editors, *Theorem Proving in Higher Order Logics, 21st International Conference, TPHOLs 2008, Montreal, Canada, August 18-21, 2008. Proceedings*, volume 5170 of *Lecture Notes in Computer Science*, pages 28–32. Springer, 2008.

102. Zhanna Malekos Smith and Eugenia Lostri. The hidden costs of cybercrime. Technical report, 2020.

103. F. Javier Thayer, Jonathan C. Herzog, and Joshua D. Guttman. Strand spaces: Proving security protocols correct. *Journal of Computer Security*, 7(1):191–230, 1999.

104. K. Tsipenyuk, B. Chess, and G. McGraw. Seven pernicious kingdoms: a taxonomy of software security errors. *IEEE Security Privacy*, 3(6):81–84, 2005.

105. A. Turing. On computable numbers, with an application to the entscheidungsproblem. *Proceedings of The London Mathematical Society*, 41:230–265, 1937.

106. M. Turuani. The cl-atse protocol analyzer. *Lecture Notes in Computer Science*, 4098:277, 2006.

107. E. Van Herreweghen. Non-repudiation in SET: Open issues. *Lecture Notes in Computer Science*, pages 140–156, 2001.

108. J Voas and K Schaffer. Whatever happened to formal methods for security? *Computer*, 49(8):70, 2016.

109. Barry J Wadsworth. *Piaget's Theory of Cognitive and Affective Development: Foundations of Constructivism.* Longman Publishing, 1996.

110. James Walden and Charles E. Frank. Secure software engineering teaching modules. In *Proceedings of the 3rd Annual Conference on Information Security Curriculum Development*, InfoSecCD '06, pages 19–23, New York, NY, USA, 2006. ACM.

111. Christoph Weidenbach. Towards an automatic analysis of security protocols in first-order logic. In Harald Ganzinger, editor, *Automated Deduction - CADE-16, 16th International Conference on Automated Deduction, Trento, Italy, July 7-10, 1999, Proceedings*, volume 1632 of *Lecture Notes in Computer Science*, pages 314–328. Springer, 1999.

6 Cryptographic Web Applications: from Security Engineering to Formal Analysis

Michele Bugliesi
Universitá Ca' Foscari Venezia, Italy

Stefano Calzavara
Universitá Ca' Foscari Venezia, Italy

Alvise Rabitti
Universitá Ca' Foscari Venezia, Italy

CONTENTS

DOI: 10.1201/9781003090052-6

In this chapter we study the problem of securely deploying cryptography within web applications. We overview the most significant security issues in cryptographic web applications and review the solutions proposed in the literature. We focus in particular on principled techniques amenable to formal verification, given the strong security requirements of applications like encrypted web storage and the complexity of the web platform. Our discussion covers both traditional network attackers in the Dolev-Yao style and web attackers, abusing web technologies like HTML and JavaScript. Moreover, we distinguish between attacks performed in the symbolic model and attacks performed in the computational model of cryptography.

6.1 INTRODUCTION

The Web is the key access point to a plethora of security-sensitive data sources and services that we use on a daily basis, yet it was not designed with security and privacy in mind [92]. As a result, it is largely left to web developers to understand security best practices and to take actions to implement appropriate defenses for their applications. Over the last few years, browser vendors have been very active in fostering security on the Web, by designing new client-side defenses [61, 90] and by strongly promoting security by default [1, 2]. Yet, as shown by a consistently large number of papers at major computer security conferences, we are still far from having achieved a safe and secure web platform design.

In this chapter we focus on a specific topic in web security: the secure deployment of *cryptography* within web applications. Web application security and cryptography are normally considered two separate research areas, and indeed they are studied by different communities and often independently. Nevertheless, for an increasingly large class of real-world web applications, cryptography constitutes one of the core tools to implement the desired functionality; and given the open nature and complexity of the Web, deploying cryptography securely in web applications is far from obvious. We refer to such class of web applications as *cryptographic web applications* and we introduce *encrypted web storage* [10] as a paradigmatic example.

The typical use-case scenario for an encrypted web storage application is one in which a user wants to securely store encrypted data on a server by operating a standard web browser, e.g., using a high-security cloud storage service like SpiderOak (https://spideroak.com/) or a password manager like LastPass (https://www.lastpass.com/). In such applications, encryption is performed at the client side with a key known only to the user, e.g., derived from her password, so that the user's secrets are safeguarded even against malicious or compromised servers. At a high-level, encrypted web storage operates as follows:

1. The user authenticates at the web application by providing the required access credentials, e.g., username and password. After authentication, the web application responds with an HTML page providing a user interface to store encrypted data on the server.
2. The user encrypts the secret data s using her cryptographic key k and sends the ciphertext $\{s\}_k$ to the remote server through the user interface. Of course, concrete details about the encryption scheme may significantly vary across different web applications, but the high-level description given here suffices for now.
3. A user willing to access her data s again can access the web application to download the ciphertext $\{s\}_k$ from the server and decrypt it at the client using her cryptographic key k.

Though the above scheme looks straightforward, the complexity of the web platform hides a number of challenges which may undermine security. A first problem comes from *securing the network communication*: an attacker gaining access to the user's password when it is communicated during the authentication process would be able to access the web application under the user's identity. This would not leak the cryptographic key k, as the key is managed at the client, but it may nevertheless reveal the nature of the stored files based on public metadata and likely enable offline bruteforce attacks against encrypted content. Using the HTTPS protocol to protect the transport layer is, therefore, a necessary ingredient for security in this case, yet it is insufficient without additional safeguards. Indeed, the activation of HTTPS may be blocked by attackers having full control of the network traffic [72], and even when HTTPS is active, *cryptographic flaws* and other *low-level implementation mistakes* may void its intended confidentiality and integrity guarantees [31]. A further potential problem is related to *client-side web security*, as user keys entered as input to the browser are exposed to a range of attacks, e.g., from scripts running in the same web page where the keys are input. Preventing such key management attacks in real browsers is surprisingly difficult, in particular due to the undisciplined nature and weak isolation properties of JavaScript [98].

6.1.1 SCOPE OF THE CHAPTER

We overview the most significant security issues in cryptographic web applications and review the solutions proposed in the literature. We focus in particular on principled techniques amenable to *formal verification*, for two reasons. First, cryptographic web applications are expected to satisfy strong security and privacy requirements, hence provably sound (or at the very least rigorous) guarantees are desirable. Second, the complexity of the web platform and the huge extension of the attack surface against cryptographic web applications call for a systematic, formal assessment of the entire implementation stack from specification to production code.

6.1.2 STRUCTURE OF THE CHAPTER

After a brief review of the basic technical background (Section 6.2), we introduce encrypted web storage as a motivating example (Section 6.3). Then we review existing research relevant to the development of secure cryptographic web applications. In Section 6.4 we focus on *web application security*, with a specific emphasis on JavaScript – the de-facto standard for developing the client-side logic of web applications. First, we discuss the techniques and approaches to build the basic isolation and information-flow control safeguards required for any secure deployment of Javascript code, including cryptographic web applications. Then we review the main proposals specifically designed for the correct use of cryptography within JavaScript. In Section 6.5 we discuss *protocol security*. First we focus on the analysis of web protocols, i.e., protocols designed to run in the web setting, and discuss their specific challenges. Then we review a rich research body on the verification of the cryptographic guarantees offered by the HTTPS protocol, which is the cornerstone of secure communication on the Web. In Section 6.6 we conclude by providing our perspective on the current state of the art for the development of secure cryptographic web applications, discussing its key strengths and its most important limitations.

6.2 BACKGROUND

6.2.1 ENCRYPTION AND DECRYPTION

We present a brief overview of basic cryptographic concepts required to understand the chapter.

- **Symmetric key encryption:** In symmetric key encryption, the same cryptographic key is used both to encrypt the plaintext and to recover the plaintext from the ciphertext. The communicating parties are thus required to have a shared secret, the *symmetric key*, whose confidentiality is critical to the security of encryption. How to securely share this secret is one of the biggest problems with symmetric key encryption. Moreover, any of the two parties sharing the symmetric key may use it to produce a valid ciphertext, meaning that it is not possible to attribute the construction of an encrypted message to any of them. Some examples of symmetric key encryption algorithms are AES, DES and Blowfish.
- **Asymmetric key encryption:** Asymmetric key encryption is based on a pair of jointly generated keys, known as *private key* and *public key*. Every message encrypted with the public key can be decrypted only with the corresponding private key: this makes it easy to securely communicate with the owner of the private key, because she can openly advertise the public key and everyone who has access to it can construct encrypted messages. Dually, every message encrypted (signed) with the private key can be decrypted (verified) only with the corresponding public key: this means that every message which decrypts correctly can be attributed to the owner of the private key. Prominent examples of asymmetric key encryption algorithms include RSA and ElGamal.
- **Secure uses of encryption:** Using appropriate encryption algorithms is a necessary, yet not sufficient condition for security, because cryptographic code can go wrong in many different ways.

> Attacks against cryptography normally leverage *side-channels* (e.g., timing) or other weaknesses of the encryption/decryption process to leak information about the plaintext or the cryptographic key used to protect it.

To provide a concrete example of a potential issue, encryption algorithms normally operate on *blocks* of fixed size (*block ciphers*). A *block cipher mode* is an algorithm that uses a standard fixed-length block cipher to protect an arbitrarily long message: first, the message is split into a list of blocks, possibly padding the data therein, and then blocks are individually encrypted. Different techniques have been devised to chain together the different blocks, with different degrees of security. The Electronic Codebook Mode (ECB) just encrypts every block independently from the others, hence any two blocks that are the same in plaintext will be the same also in the ciphertext: this is insecure because it may reveal patterns in the plaintext even after encryption. Luckily, stronger modes offering improved security guarantees are available, such as Cipher Block Chaining (CBC) and Counter (CTR).

6.2.2 THE WEB PLATFORM

Documents on the Web are shipped in the form of *web pages*, hypertext files connected to other documents via hyperlinks. The structure of a web page and all the elements included therein are defined using the HyperText Markup Language (HTML), which is parsed and rendered by a *web browser*. The page content can be dynamically updated at the browser by using *JavaScript*, a full-fledged dynamically typed scripting language. JavaScript code included in a web page can manipulate its content by operating on the *Document Object Model* (DOM), a tree-like representation of the web page structure. The ability to change the DOM, possibly as a reaction to user inputs, is one of the keys to develop rich, interactive web applications providing a user experience similar to traditional desktop applications.

Web pages are requested and served over the *Hyper Text Transfer Protocol* (HTTP), a request-response protocol based on the client-server paradigm. The browser acts as the client and sends HTTP requests for resources hosted at remote servers; the servers, in turn, provide HTTP responses containing the requested resources if available. All the HTTP traffic flows in clear, hence the HTTP protocol does not guarantee the confidentiality and the integrity of the communication, nor does it provide any mechanism for server authentication, thus leaving users vulnerable to server impersonation attacks. To protect the exchanged data and

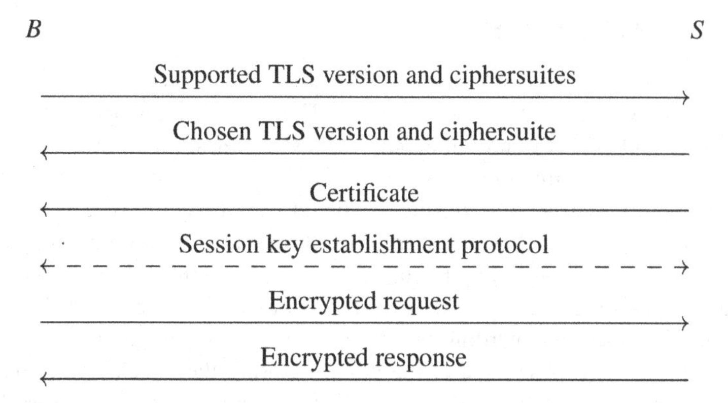

Figure 6.1 High-level description of the TLS protocol.

enable server authentication, the *HTTP Secure* (HTTPS) protocol wraps plain HTTP traffic within an encrypted channel established through the *Transport Layer Security* (TLS) protocol. TLS leverages public key cryptography to establish a fresh symmetric key, called the *session key*, between the client and the server; the session key is then used to ensure both the confidentiality and the integrity of the exchanged messages of the session. The public key of the server is stored within a *certificate*, which binds the public key to the server's identity and is signed by a trusted Certification Authority (CA). Figure 6.1 shows a high-level conceptual description of TLS. The session key is established via the *Handshake* protocol of TLS, while the protection of encrypted session messages is enforced by the *Record* protocol.

Both HTTP and its secure variant HTTPS are *stateless*, hence each browser request is treated by the server as independent from all the other ones. Most web applications, however, need to keep track of information about previous requests, for instance to implement user authentication or to verify whether a user has performed the expected payment steps prior to concluding the purchase process. HTTP *cookies* are the most common mechanism to maintain state information about the requesting client and implement *stateful sessions* on the Web. Roughly, a cookie is just a key-value pair, set by the server for the client and automatically attached by the client to all subsequent requests to the server. Cookies may either directly store state information or, more commonly, just include a unique random session identifier, which allows the server to identify the requesting client and restore the corresponding session state when processing multiple requests by the same client.

6.2.3 THREAT MODEL

Following the traditional approach in web security literature [4], we focus on two different types of attackers. The *web attacker* operates a web browser and uses it to mount attacks against target users and web applications. We assume that this attacker also has control of a malicious website, possibly shipping a valid certificate signed by a trusted CA; this website may be leveraged to host arbitrary content that are accessed by the target user or may be imported by the target web application, e.g., in the form of malicious scripts. We only focus on attacks based on the use of HTML and JavaScript, thus disregarding attacks mounted by malware or other malicious software downloaded at the client granting the attacker full control of the client's host. Examples of classic web attacks include Cross Site Scripting (XSS), where the attacker exploits insufficient input sanitization to inject malicious JavaScript code in target web pages [68], and Cross Site Request Forgery (CSRF), where the attacker forces the victim's browser to fire authenticated security-sensitive requests to vulnerable servers [12]. The *network attacker* is more powerful than the web attacker and extends her capabilities with complete control of the network traffic, e.g., because they operate a malicious WiFi access point. The network attacker thus operates from a man-in-the-middle position, which grants her the ability to compromise the confidentiality and the integrity of the unencrypted HTTP traffic.

Given our focus on cryptographic web applications, we refine the standard threat model above by further categorizing attackers based on the model of cryptography, as done in the literature on protocol verification [22]. The *symbolic attacker*, first introduced in the traditional Dolev-Yao model [45], operates in a world of perfect cryptography, e.g., where the plaintext can only be recovered from the ciphertext by knowing the corresponding decryption key. Accordingly, the symbolic attacker can break cryptography by gaining access to the right cryptographic keys or exploiting logical bugs which block the use of cryptography. The *computational attacker*, instead, can break cryptography by exploiting low-level weaknesses of existing ciphersuites, e.g., legacy encryption algorithms, insecure encryption modes like ECB, or side-channel attacks leaking the plaintext bit by bit.

6.3 MOTIVATING EXAMPLE

6.3.1 ENCRYPTED WEB STORAGE PROTOCOL

Figure 6.2 provides a high-level conceptual description of an encrypted web storage protocol. Clearly, real-world implementations of similar

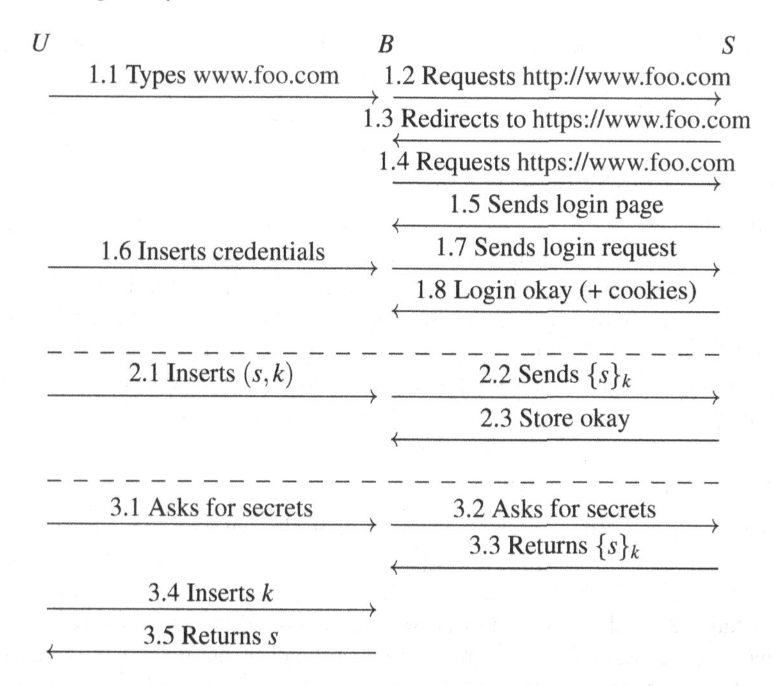

Figure 6.2 High-level description of an encrypted web storage protocol.

protocols may be quite more complex than this simple conceptual description, with concrete details varying significantly across implementations. However, this picture suffices to convey the key messages of the security problems of interest for our present purposes.

The protocol involves three parties – a user U, a browser B and a server S, hosted at www.foo.com – and can be conceptually split in three stages, marked by the dashed lines in the figure and discussed below.

6.3.1.1 Authentication

The user types www.foo.com in the address bar of her browser (1.1), which sends a request for that host over HTTP (1.2). The server upgrades the security of the session by enforcing a redirection to HTTPS (1.3), which is automatically performed by the browser (1.4). The server uses HTTPS to provide a login page (1.5), where the user inserts her access credentials (1.6). When the user submits the login form, her browser sends a login request including the access credentials (1.7); if the credentials are valid, the server authenticates the user and sets authentication cookies in her browser

(1.8). From now on, the authentication cookies are automatically attached by the browser in every request to the server, thus connecting all the subsequent requests to the current session.

6.3.1.2 Data Storage

Authenticated users can store data on the server without trusting it, by means of client-side encryption. A secret s is encrypted at the browser side with a key k only known to the user (2.1), e.g., using a JavaScript library, to produce a ciphertext $\{s\}_k$ sent to the server for storage (2.2). The server can notify the correct receipt of the stored data (2.3), though it has no way to inspect its content. In most practical scenarios, encrypted data are enriched with public *metadata*, e.g., a textual label describing what they are.

6.3.1.3 Data Retrieval

When the user wants to access her data again, she can request it via the user interface of the web application (3.1), which composes an HTTP request to the server (3.2); the data selection process may involve access to the metadata, which are stored in clear. The encrypted data $\{s\}_k$ are returned back to the client (3.3), which can request the decryption key k to the user (3.4) to retrieve the plaintext s (3.5). Again, the client-side logic of the decryption stage is normally implemented by means of JavaScript.

6.3.1.4 Security Considerations

Writing secure web applications is well known to be hard, and encrypted web storage makes no exception in spite of its conceptual simplicity. While there are several ways in which security can break, here we focus on the problems of specific interest to the cryptography community, providing concrete examples of attacks to help understand the security issues arising from the interaction between web application security and cryptography. We refer to the numbered steps of Figure 6.2 to clarify how the attacks work.

6.3.1.5 Symbolic Web Attacks

Symbolic web attacks may compromise the confidentiality or the integrity of sensitive data by exploiting logical flaws and quirks of JavaScript code to prevent the use of client-side cryptography, or otherwise make it entirely useless. An example of symbolic web attack is *key theft via a malicious script*. When the user's key is input for client-side encryption at step 2.1, it is necessarily made available to JavaScript. A malicious script running on the same web page, e.g., injected through an XSS vulnerability, may then get access to the key and exfiltrate it to the attacker's website. Of course, this largely voids the benefits of client-side encryption, because an attacker with access to the encrypted content will then be able to decrypt it.

6.3.1.6 Symbolic Network Attacks

Symbolic network attacks exploit the weak security guarantees of HTTP to compromise the confidentiality or the integrity of sensitive data. An example of symbolic network attack is *SSL stripping* [72]: in our example, the attacker intercepts the first request (at step 1.2) sent over HTTP, thus preventing the redirect over HTTPS at step 1.3. With the session downgraded to HTTP, the network attacker can now access the victim's password at step 1.7 and mount an impersonation attack. This, in turn, might reveal the nature of the stored files, e.g., through their unencrypted metadata, and enable offline bruteforce attacks against local copies of the encrypted files which can be downloaded by the attacker.

6.3.1.7 Computational Web Attacks

Computational web attacks may exploit incorrect cryptographic practices in JavaScript code to harm the confidentiality or the integrity of sensitive data. An example is a side-channel attack enabled by the use of client-side encryption based on the ECB mode. Recall that ECB is the simplest block cipher mode of operation, which splits the plaintext in independent blocks of the same length and independently encrypts them, thus allowing the attacker to identify reoccurring patterns in encrypted data. This might reveal partial information about the plaintext through the ciphertext.

6.3.1.8 Computational Network Attacks

Computational network attacks may exploit cryptographic flaws in the implementation of protected network communication to compromise the confidentiality or the integrity of sensitive data. An example of computational network attack is the *DROWN attack* against TLS [8], which exploits flaws in the SSLv2 protocol, deprecated but still widely supported, and certificate reuse practices. This allows an attacker to use an SSLv2-supporting server as a Bleichenbacher padding oracle to decrypt communications between the victim and all the servers sharing the same private key, even though they are using modern TLS encryption.

Table 6.1
Examples of attacks against encrypted web storage.

	Web	**Network**
Symbolic	Key theft via malicious script	SSL stripping attack
Computational	Side-channel via ECB mode in JS	DROWN attack on TLS

6.3.2 DISCUSSION

Table 6.1 reviews the four attacks against encrypted web storage described in the present section and contextualizes them within our threat model. It is clear that these attacks are quite different in nature and require countermeasures at very different levels. Web attacks normally exploit *web application flaws*, i.e., flaws in the web application logic or its implementation which undermine the use of client-side cryptography. To address these issues, one has to deal with the intricacies of JavaScript and the complexity of modern web browsers. For example, secure key management requires the robust isolation of secrets from malicious scripts running in the same web page (e.g., as the result of a successful XSS exploitation, a class of attacks studied in the literature from multiple perspectives). We focus on this class of problems in Section 6.4.

Network attacks, instead, leverage *protocol flaws*, i.e., flaws in the communication flow between the client and the server. Protocol verification has been the subject of a massive research effort in recent years, and a variety of techniques, including automated formal frameworks such as those discussed in other chapters of the handbook, are now available for security assessment. However, web protocols have peculiarities that make them even harder to verify and require custom solutions to achieve meaningful security guarantees. For example, browsers are unorthodox protocol participants, as they are engaged in a variety of tasks that may affect their reactive behavior, and hence interfere with the protocol flow they are supposed to adhere to [55]. We focus on this class of problems in Section 6.5.

6.4 WEB APPLICATION SECURITY

JavaScript is the de facto standard for implementing the client-side logic of (cryptographic) web applications, and yet it is fundamentally ineffective and weak as a support for a safe and secure script deployment in an open context, such as the one characterizing web applications [16].

6.4.1 PROBLEM OVERVIEW

Web applications are typically structured as compositions (*mashups*) of different active components and contents originating from multiple sources. The interaction among components is governed by the Same-Origin Policy (SOP), the base security mechanism implemented by web browsers to restrict the ways that a document or script loaded by one origin can interact with a resource from another origin[1]. SOP generally blocks access to resources loaded from a different origin, but cross-origin references are still allowed in some cases: in particular, SOP does not prevent cross-origin scripts to be embedded within a web page. Though this is arguably an inevitable choice to not undermine the functionality that scripting is meant to provide in practice, the very nature of JavaScript makes the coexistence of scripts from different origins on the same page a major threat for security.

To exemplify, consider the following code snippet implementing a function where a secret message is first encrypted and then sent to a remote server:

```
key = "Xf13mZYp89rs";
api = function(msg) {
          e = Crypto.encrypt(msg, key);
          xhr = new XMLHttpRequest();
          xhr.open("GET", "https://secure.com/");
          xhr.send(e);
}
```

This code is insecure when executed in a page hosting a malicious script, because the variable key is global and can be accessed by any other active component in the page. Moving key within the api function, as in the modified snippet below:

```
api = function(msg) {
          key = "Xf13mZYp89rs";
          e = Crypto.encrypt(msg, key);
          xhr = new XMLHttpRequest();
          xhr.open("GET", "https://secure.com/");
```

[1]An *origin* is defined as a triple including a protocol, a port and a host; for example, a web page at http://www.foo.com/bar has an origin including the HTTP protocol, its standard port 80 and the host www.foo.com. We refer the interested reader to RFC 6454 for full details (available at https://datatracker.ietf.org/doc/html/rfc6454).

```
            xhr.send(e);
    }
```

does not solve the problem, as key is still global due to the *variable hoisting* mechanism peculiar of the JavaScript scope rules. To prevent this effect, one can add the keywords var or let to variable declarations:

```
    var api = function(msg) {
            var key = "Xf13mZYp89rs";
            var e = Crypto.encrypt(msg, key);
            var xhr = new XMLHttpRequest();
            xhr.open("GET", "https://secure.com/");
            xhr.send(e);
    }
```

While this fixes the problem with key, the code is still insecure due to the dynamic code overriding capabilities of JavaScript. Indeed, a malicious script running in the same page can redefine the behavior of the Crypto.encrypt function so as to make it leak its second argument (the key) to an external server. *Inlining* the implementation of the encryption function within a local variable is a possible workaround, but is clearly sub-optimal from a programming perspective.

Taking a more general perspective, what the example shows is that basic isolation properties that can be taken for granted in traditional programming languages like Java or C++ turn out to be surprisingly hard to ensure in JavaScript. This, in turn, makes it correspondingly hard to deploy secure web applications. We discuss research efforts to counter this weakness in Section 6.4.2 below. A further JavaScript weakness arises from its dynamic type system, whose extremely permissive typing rules pose little-to-no constraint on any syntactically well-formed code fragment, thus leaving room to a variety of insecure uses of cryptographic APIs and sensitive content. We discuss the proposals on how to approach this class of problems in Section 6.4.3.

6.4.2 PROTECTING AGAINST MALICIOUS JAVASCRIPT

Various approaches have been investigated to provide safeguards against the threats of cross-origin script execution. One solution is to selectively identify potentially dangerous content and filter it out, or otherwise block its execution: we look at three different approaches to accomplish that in Section 6.4.2.1. Two further alternatives from the literature are *sandboxing*, i.e., techniques to provide isolation between different scripts running in the same web page (Section 6.4.2.2), and *information flow control*, whereby scripts are not isolated but their interaction is controlled so that secrets are used according to an intended security policy (Section 6.4.2.3).

6.4.2.1 Blocking Dangerous Content

Several techniques are available to prevent the injection of malicious code, or at least make it much less likely in practice.

6.4.2.1.1 Sanitization and encoding

Input sanitization [80] ensures that all the untrusted information to be processed by the web application is first filtered to remove malicious content, e.g., script tags leading to code execution. Output encoding, in turn, ensures that malicious content is never interpreted in a dangerous way by the browser, i.e., it is never turned into executable code.

Appropriate definition and placement of sanitizers and encoders is a challenging task, even for experienced web developers [9]. Formal methods are thus particularly important to support the security of this delicate process. BEK is a domain-specific language for sanitizers, which enables automated formal analysis based on symbolic finite state automata [62]. BEK sanitizers that are deemed safe by the analysis can be compiled into traditional programming languages, while preserving their correctness properties. More recent research devised techniques for the black-box inference of formal models of sanitizers based on symbolic finite state transducers [5]. These models can be turned into BEK code to take advantage of the existing analysis toolchain and prove the correctness of sanitization routines. Even correct sanitizers might fall short of accomplishing their task, however, when they are misplaced in the web application code. This motivated further research on formal techniques for automated sanitizers placement [89, 70]. Modern templating systems for web application development allow one to clearly identify content injection points, which can be automatically encoded to avoid vulnerabilities leading to script execution [88].

6.4.2.1.2 Content Security Policy (CSP)

While input sanitization and output encoding are effective against content injection, their complexity calls for an in-depth defense approach based on multiple layers. *Content Security Policy (CSP)* is a declarative defensive mechanism available in web browsers to control the allowed sources for remote script inclusion and forbid dangerous JavaScript features which may lead to malicious script injection [90, 103, 33]. Remarkably, a large fragment of CSP has been formalized in terms of a simple denotational semantics called CoreCSP, which supports rigorous reasoning. One instance of this line of work is our own proposal of sound techniques to reason about protection against script injection and to compare the permissiveness of two CSPs by means of simple syntactic checks [34].

6.4.2.1.3 *JavaScript standards*

Introducing standards to prevent the use of dangerous JavaScript constructs is a further alternative to block malicious code snippets by purely syntactic means. That is the case of ECMAScript Strict Mode [78], a standardized subset of JavaScript in which execution in strict mode may be activated for entire scripts or for single functions to forbid the use of a few, well identified idiosyncratic JavaScript features. Among these, the `eval()` function used to turn strings into executable code, the weird scoping mechanisms associated with uses of `with`, `this` and object properties like `caller` and `prototype`, all of which open doors to ways to break the intended scope and namespaces one would like to confine.

6.4.2.2 JavaScript Sandboxing

JavaScript sandboxing has been studied for years now. In its original formulation, sandboxing is defined as the problem of confining untrusted code running in an otherwise trusted web page to restrict its functionality. More specifically, its goal is to prevent third party code from unrestrictedly accessing the DOM and/or other resources (data and code) within the host page, storing sensitive information that originate from the client (e.g., user passwords) or from the page origin (e.g., cookies or program variables).

The most recent survey on the topic [98] classifies Javascript sandboxing techniques in three categories: (i) sandboxing through subsets and rewriting, (ii) sandboxing through browser modifications and (iii) sandboxing without browser modifications. Though largely consistent with [98], our presentation below uses different labeling for the three categories and in few cases a different categorization of the existing proposals.

6.4.2.2.1 *Sandboxing by code transformation*

Sandboxing through code transformation relies on mechanisms akin to those adopted in ECMAScript Strict Mode, discussed earlier, to identify a safe language subset, free of any idiosyncratic construct, and then process any third party script to make it compliant with the identified subset. Unlike pure subsetting mechanisms like ECMAScript Strict Mode, which simply reject the occurrences of forbidden code (by, e.g., throwing an exception in case the execution flows through any of them), code transformation techniques provide mechanisms to confine any unwanted construct within special contexts (wrappers) that mediate their behavior and their interaction with the rest of the code and the DOM. Examples of such techniques include pure code-transformation tools such as BrowserShield [85], Self-Protecting JavaScript [82] and, more recently SecureJS [66], as well as mixed subsetting-and-transformation approaches such as those found in Google Caja [76] and ADSafe [40], which simply reject certain constructs (e.g., `with` and `eval()`) and mediate the execution for others (e.g., `this` and direct references to the DOM).

6.4.2.2.2 *Sandboxing by browser modifications*

An alternative approach to provide confinement safeguards is to rely on browser modifications supporting mechanisms to control the runtime execution of JavaScript code, by interposing a filter on each access that third party code may attempt on the page or browser components one wishes to protect. Various such mechanisms have been explored so far, ranging from callbacks, as in BEEP [64], capability-based access control monitors, as in Contego [71], or other special-purpose techniques (e.g., advices as in ConScript [75] and WebJail [97]), all of which are implemented as browser extensions, plugins or browser modifications.

6.4.2.2.3 *Sandboxing by language-based isolation mechanisms*

Though effective for developing proof-of-concept sandboxing implementations, browser instrumentation falls short of enabling viable engineered, production-scale solutions, as it relies on end-users installing the corresponding extensions / plugins / other browser modifications, and it is costly to maintain.

A more robust approach is to leverage the new functionality of the maturing ECMA JavaScript standards (and associated new APIs) implemented by all modern browsers to construct language-based isolation mechanisms and other mediation patterns for the DOM or other sensitive resources accessed by untrusted code. These mechanisms typically rely on metaprogramming language extensions and APIs that enable one to access the program internal representation and/or override language core operations such as object field lookup, function invocation and many others. Based on such capabilities, various proposals have emerged that provide for different implementations of wrappers [79, 105], object views [74], membranes [3], or even whole JavaScript meta interpreters [96] to mediate the behavior of untrusted code and its interaction with the trusted computing base of the host page. Other proposals introduce such capabilities to provide for mediated interaction between scripts from different origins after having encapsulated and isolated them within separate iframes [95] or separate, parallel threads (e.g., web workers) as in [100].

6.4.2.2.4 *Other flavors of sandboxing*

In its original formulation, sandboxing aims at confining untrusted code running in a trusted web page. A complementary problem is to protect the execution of trusted code within untrusted web pages. Here the challenge is even harder, as one cannot assume that trusted code runs before untrusted code, which is a popular assumption for JavaScript sandboxing - and an important one, because the trusted code can implement safeguards before the untrusted code is executed. The most relevant reference for such techniques is the Defensive JavaScript proposal [17], which has also been used to develop a defensive cryptographic library implementation called DJCL.

6.4.2.3 Information Flow Control for JavaScript

JavaScript sandboxing allows one to isolate different scripts running in the same page, or at the very least to mediate their communication interfaces. However, sandboxing may turn out to be insufficient in practice, because once a secret is shared with another script, there is no way to control how it is used. To exemplify, assume that script A is willing to share a cryptographic key k with script B, but not with script C. Sandboxing may appear to be the right tool here but it is not, because B has multiple ways to leak k to C once it gets access to it. For example, the key may be leaked one bit at the time over a communication channel established between B and C, even independently of the DOM or any other shared resources.

Information flow control [87] provides a more robust and flexible tool than sandboxing. In contrast to sandboxing, information flow control does not entirely prevent information sharing, but rather constrains information propagation. The key idea is to label both scripts and data with *security levels* and enforce a security discipline whereby a script at level ℓ can only observe data at level $\ell' \leq \ell$, so that the public (low) observable behavior of a script is independent of the script's secret (high) input data. This intuitive idea is commonly formalized in terms of a semantic security notion, called *non-interference* [53].

Information flow control has been investigated extensively from both a theoretical and a practical perspective in different scenarios, including the web [21]. Although the first works in the area adopted traditional approaches based on static analysis [37, 56], later efforts have increasingly focused on dynamic analysis and enforcement techniques, most notably due to the challenges related to the unconventional semantics of JavaScript. We review such efforts below.

6.4.2.3.1 Dynamic information flow control for JavaScript

Dynamic information flow control is enforced by means of runtime security monitors. The first significant work in the area is by Hedin and Sabelfeld [60]. They formalized a small-step operational semantics for a core of JavaScript and proposed a dynamic type system designed to enforce non-interference. A custom JavaScript interpreter called JSFlow [59] provides a prototype implementation of their system. Similar work in the area was proposed for JavaScript bytecode, based on a modification of the WebKit JavaScript engine [20, 84]. A different approach was taken by Chudnov and Naumann, with their proposal of inline information flow monitoring for JavaScript [36] based on the instrumentation of JavaScript programs, so that their execution can be monitored for information flow violations without browser changes.

6.4.2.3.2 Secure multi-execution

Secure multi-execution is a relatively novel technique for the dynamic enforcement of non-interference [43]. The key idea is to execute programs multiple times, once for each security level, using special rules for input/output operations. Outputs are only produced in the execution corresponding to their security level, while inputs are replaced by default inputs except in executions corresponding to their security level or higher. This policy enforces non-interference by construction. The approach is particularly valuable for a complex language like JavaScript, because it treats the monitored program as a black-box, focusing exclusively on its input/output behavior. Notably, the approach was also implemented in an experimental web browser called FlowFox [54].

6.4.2.3.3 JavaScript and its browser context

JavaScript is a key building block of many web applications, but it is just one of the many components of a standard web browser that may be exploited as attack vectors to access the sensitive data manipulated by cryptographic web applications. To provide holistic security guarantees, Bauer *et al.* extended the Chromium browser with dynamic information flow control [13]. Their implementation was formalized as a state transition system and proved to satisfy a formal notion of non-interference across the full browser surface.

6.4.3 SECURING THE USE OF CLIENT-SIDE CRYPTOGRAPHY

Cryptographic code can be broken in many different ways, even when it runs with adequate isolation safeguards. Dalskov and Orlandi performed an independent security review of the popular SpiderOak cloud service provider, detecting several security flaws in their use of cryptography by reverse engineering of the implementation [41]. Though their analysis is performed on the desktop client, the class of problems they address is just as relevant for web-based JavaScript clients. This research line has received only limited attention by the community so far: we review some of the relevant approaches and references below.

6.4.3.1 JavaScript Cryptography

Unlike programming languages like Java and C++, JavaScript has long suffered from the lack of standardized cryptographic libraries. This led to security incidents that gave JavaScript cryptography a bad name: for example, original version (now discontinued) of the Crypto.cat[2] encrypted chat application not only recreated their own cryptographic routines in JavaScript, but also deployed them insecurely [27]. While other, well-designed cryptographic libraries have been available for a while now [91], no standard cross-browser cryptographic library has existed up to 2017, with the public release of the W3C Web Cryptography API [58]. This API provides a direct interface to the thoroughly vetted cryptographic primitives implemented in the browser, e.g., for TLS. Moreover, it simplifies key management by exposing *handles* to secret keys rather than secret keys themselves, which provides strong guarantees against key exfiltration. Remarkably, the Web Cryptography API has undergone a rigorous formal analysis, which has identified a few flaws in its original design, subsequently fixed before its release [27]. Although the Web Cryptography API is not fully compliant with all the best practices (e.g., it still includes some low-security encryption algorithms for backward compatibility), it can significantly simplify the deployment of secure cryptographic web applications.

6.4.3.2 Verification of Crypto APIs Usage in JavaScript

Identifying a carefully reviewed and provably safe cryptographic APIs is a necessary prerequisite for security. However, one must also ensure that the APIs are used correctly, and consistently with the secrecy and integrity properties intended for the data manipulated within scripts. In their work on API confinement [94], Taly *et al.* propose a static analysis for (a restricted

[2]https://en.wikipedia.org/wiki/Cryptocat

subset of) JavaScript to assess whether the security-critical resources available to the script are only accessed through the intended APIs, i.e., are confined there. Mitchell and Kinder [77] make a further step forward in the same direction, with their proposal of a system of security annotations, i.e., type-like tags used to mark cryptographic material (ciphertexts vs cryptographic keys), to support a run-time monitoring mechanism to assess that scripts comply with the intended uses of the APIs and the cryptographic resources they manipulate. Interestingly, they exemplify their system at work on the W3C Web Cryptography API.

6.4.3.3 Beyond JavaScript

Recent research on cryptographic web applications advocated WebAssembly (Wasm) as a better-suited basis for a principled development of client-side scripting. Wasm is a low-level language intended to serve as the compiler target of traditional languages like C and C++ and is now supported by the major commercial JavaScript engines, including V8 (Chrome) and SpiderMonkey (Firefox). Interestingly, Wasm is based on a formal semantics, amenable to analysis and verification [57, 101]. Recent work proposed a formally verified tool-chain to compile F^* (an advanced dependently-typed programming language) into Wasm, while preserving the original correctness and security guarantees of the source code [83]. The authors used their compiler to build a verified Wasm implementation of the Signal protocol, which is used by several cryptographic web applications, including Whatsapp. Another relevant work on Wasm is CT-Wasm [102], an extension of Wasm that supports static proofs of information flow security and constant time guarantees. In other words, not only CT-Wasm code ensures that secret data cannot be leaked either directly or indirectly, but its computation over secrets cannot introduce timing side-channels. We anticipate more work on Wasm in the upcoming years due to its expected rise in popularity.

6.5 PROTOCOL SECURITY

Protocol verification boasts an extensive research tradition [22], and several state-of-the-art tools can automatically establish security proofs even for complicated protocols [6, 23, 73]. Protocol verification tools have been applied to the web setting as well, with interesting success stories such as the previously unknown attack against the Google variant of the SAML single sign-on protocol discovered by the SATMC model-checker [7]. However, traditional protocol verification tools are inherently unfit for formal security proofs in the web setting, as browsers are rather unconventional protocol participants, which do not simply follow the intended protocol

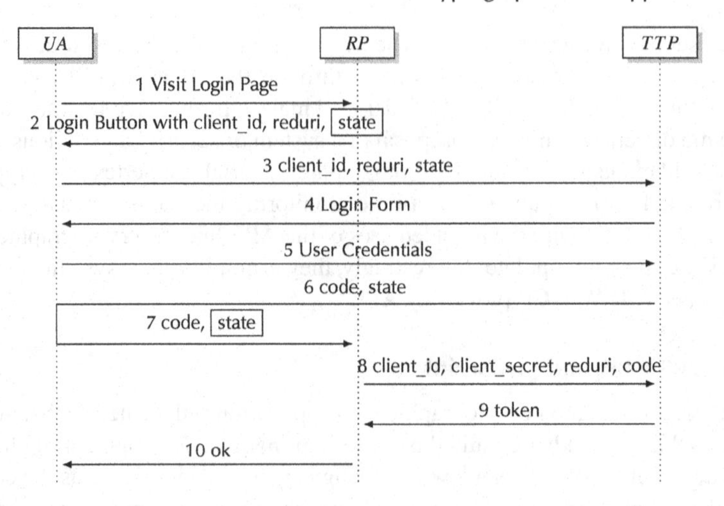

Figure 6.3 The OAuth 2.0 protocol (authorization code grant).

flow, but automatically react to external inputs as part of their functionality [55].

6.5.1 PROBLEM OVERVIEW

To exemplify, consider the *authorization code grant* of the OAuth 2.0 protocol in Figure 6.3. The protocol allows a Relying Party (*RP*) to integrate services from a Trusted Third Party (*TTP*) within the scope of a User Agent (*UA*), which is normally a standard web browser. OAuth 2.0 is often used to implement single sign-on, e.g., to use Facebook or Google as a third party offering an authentication service[3]. The protocol starts (step 1) with the *UA* visiting the *RP*'s login page. A login button is provided back that, when clicked, triggers a request to the *TTP* (steps 2-3). The request comprises the following pieces of information: client_id, the identifier registered for the *RP* at the *TTP*; reduri, the URI at *RP* to which the *TTP* will redirect the *UA* after access has been granted; state, which is normally a freshly generated nonce. The *UA* authenticates with the *TTP* (steps 4-5), which in turn redirects the *UA* to the reduri at *RP* with a freshly generated value code and the state value (steps 6-7). The *RP* verifies the validity of code in a back-channel exchange with the *TTP* (steps 8-9): the *TTP* acknowledges

[3] Strictly speaking, OAuth 2.0 is an authorization protocol used to securely access resources at a third-party server, rather than an authentication protocol. Site operators who are only interested in deploying single sign-on should rather prefer the OpenID Connect protocol to avoid potential security flaws [52].

the validity of code by sending back a freshly generated token, indicating that the *UA* has been authenticated. Finally, the *RP* confirms the successful authentication to the *UA* (step 10) and the token can be later used to access other resources at the *TTP*.

The security of OAuth 2.0 depends on many subtleties, including the invariant that the state parameter freshly generated at step 2 matches the corresponding value received at step 7, as indicated by the boxes in the figure. If this does not happen, a *session swapping* attack becomes possible [93]. The attack starts with the attacker signing in to the *TTP* with her own credentials and obtaining a valid code (step 6). The attacker then blocks the protocol execution and tricks the victim's *UA* to send the attacker's code to the *RP* by composing a message with the format required at step 7, e.g., by exploiting a CSRF vulnerability[4]. This makes the victim's *UA* authenticate at the *RP* with the attacker's identity: from there on, the attacker can track the activity of the victim at the *RP* just by accessing her own account. This attack is out of scope for traditional protocol verification tools, because it makes the victim's *UA* start the protocol at step 7 by means of CSRF, something that a honest protocol participant would never do in traditional protocol models. We discuss approaches to web protocol security in Section 6.5.2.

A final, important remark concerns TLS, one of the cornerstones for web protocol security (note that OAuth 2.0 itself silently relies on TLS, as its behavior depends on several exchanges over HTTPS [51]). Indeed, in many respects TLS is the elephant in the room of secure web communication, because it is assumed to guarantee message confidentiality and integrity, and yet in protocol verification it is typically abstracted away by a simplified scheme in which a fresh symmetric key encrypts each client request and is sent to the server encrypted under the server's public key [11]. This is a simple yet imprecise abstraction of a rather complex protocol like TLS, whose properties critically depend on correct configuration and low-level implementation details [31]. For example, the *special DROWN* attack can reconstruct the session key used to protect the TLS traffic in just a matter of minutes on vulnerable configurations [8]. Ensuring the correctness of the TLS protocol and its implementations is a matter of utmost importance for web security, which motivated many research efforts. We review the state of the art on the topic in Section 6.5.3.

[4]A CSRF vulnerability happens when a web application authenticates a security-sensitive request based on its cookies alone. Indeed, recall that cookies are automatically attached by the browser, hence a web attacker can forge potentially authenticated requests from the victim's browser by using HTML or JavaScript.

6.5.2 WEB PROTOCOL SECURITY

Prior research on web protocol security can be essentially categorized along two lines. The first is *security verification*, i.e., the development of formal techniques designed to assess web protocol security against a precise set of goals (Section 6.5.2.1). The second, instead, is *security enforcement*, which involves the design and the implementation of techniques aimed at strengthening deployed web protocols against attacks (Section 6.5.2.2).

6.5.2.1 Security Verification

Several proposals for the security verification of web protocols appeared in the literature. They are all based on formal models of the web platform, most often mechanized with existing verification tools.

6.5.2.1.1 Toward formal foundations of web security

The first realistic model of the web platform was designed by Akhawe *et al.* in 2010 [4]. Their work presented a mechanized model written in Alloy, a declarative language based on a relational extension of first-order logic [63]. Alloy supports the automated identification of satisfying models or counter-examples for a given assertion, i.e., a property of interest which is expected to hold for the underlying specification. Though the bounded verification performed by Alloy cannot be used to establish security proofs, it is still useful to find attacks within a reasonably detailed model of the Web. In fact, one of the most important result of this research work was the careful definition of a meaningful threat model for the Web, including both the web attackers and network attackers we are considering in the present chapter. Remarkably, the model was applied to analyze five real-world case studies and find violations of at least one security goal in all of them, leading to the identification of a new attack against the popular WebAuth protocol. Other researchers took advantage of Alloy to assess the design of novel web security mechanisms. Chen *et al.* applied the model to validate the effectiveness of App Isolation, an experimental web browser providing stronger isolation guarantees for web applications [35]. De Ryck *et al.*, instead, used the model to formally evaluate the design of CsFire, a browser-side protection mechanism against CSRF attacks [86].

6.5.2.1.2 The WebSpi library for ProVerif

WebSpi [11] is a library designed to provide support for web security analysis within the popular ProVerif protocol verification tool [23]. ProVerif translates protocols expressed in a dialect of the applied pi-calculus into

Horn clauses, which are then used to automatically verify both secrecy and authenticity properties of protocols by resolution. ProVerif can prove security for an unbounded number of protocol executions, though the analysis it implements is not guaranteed to terminate in general. WebSpi complements ProVerif by defining all the basic ingredients needed to model web applications, web protocols and their security properties within the applied pi-calculus. WebSpi also defines a web attacker model in terms of an applied pi-calculus process and provides facilities for fine-tuning the security analysis by enabling or disabling different classes of attacks, e.g., code injection attacks against trusted web applications.

WebSpi was used by its authors to perform a formal security analysis of the OAuth 2.0 authorization protocol, discovering many previously unknown vulnerabilities in major websites such as Yahoo! and WordPress when they connect to social networks such as Twitter and Facebook [11]. The same authors also applied WebSpi to analyze the security of several cloud-based storage services, including popular services like Dropbox, SpiderOak and 1Password [10].

6.5.2.1.3 The Web Infrastructure Model (WIM)

The most expressive model of the web platform to date has been proposed by Fett *et al.* [49]. It is based on a generic process algebra in which processes have addresses and messages are modeled as first-order terms with equational theories defining the behavior of cryptographic primitives. Though abstract enough to allow for formal reasoning, this pen-and-paper model tries to follow as closely as possible the existing web standards and it spans several pages of the technical report accompanying the original paper. The model is not directly amenable for automation, due to several features which are useful, but challenging for automated tools. For instance, the set of first-order terms (messages) is infinite and the treatment of state information in the model is non-monotonic, e.g., cookies can be deleted from the browser cookie jar and not only added.

The model was originally employed by the authors to analyze BrowserID, a complex single sign-on system by Mozilla allowing websites to delegate user authentication to email providers. The security analysis unveiled several attacks, including a critical one which allowed an attacker to hijack the sessions of any user owning a GMail or Yahoo! address [49]. In the same paper, the model was also used to prove the security of a revised variant of the BrowserID system. The analysis of BrowserID was extended to deal with privacy aspects in later work by the same authors [50], revealing additional pitfalls suggesting the need for a major overhaul of the system. Other applications of the model include formal security analyzes of OAuth 2.0 [51], OpenID Connect [52], the OpenID Financial-grade API [48] and the Web Payment API [44].

6.5.2.1.4 Web session security

Web sessions are a specific type of custom web protocols that we use on a daily basis. In a web session, a user authenticates at a website by presenting her access credentials and gets back a set of cookies, which are used by the browser to keep the authenticated session alive. Securing web sessions is tricky and several solutions have been proposed in the past, as reviewed in a relatively recent survey [32]. We focus here on proposals which have been formalized and proved sound.

Bugliesi *et al.* presented CookiExt, a browser extension designed to provide automated protection against session hijacking attacks by ensuring the confidentiality of session cookies [25]. The design of CookiExt was proved formally sound using Featherweight Firefox, a Coq model of the core of a standard web browser [24]. The same research group also introduced a general definition of web session integrity, going beyond protection against session hijacking, and proposed browser-side modifications to enforce it correctly on any website [26]. Desirable session security guarantees have also been formalized in terms of non-interference and enforced by means of programmable browser-side security monitors [28]. Finally, Calzavara *et al.* studied how to enforce web session integrity without browser modifications by means of type-checking of web application code [29]. With the exception of [25], all these works used simple pen-and-paper models of the web platform, which heavily abstract from the complexity of real browsers and web applications. Nevertheless, they are useful because they improve our understanding of the root causes underlying web session insecurity and propose formally sound defensive techniques based on semantic security properties.

6.5.2.2 Security Enforcement

Even when a web protocol design is proved sound in theory, its real-world implementation might hide subtle vulnerabilities. For this reason, several authors have proposed runtime monitors designed to supervise the protocol execution and dynamically prevent the exploitation of possible vulnerabilities [38, 69, 104]. Here we focus on the few attempts pursuing formal security guarantees [30, 99].

6.5.2.2.1 WPSE

The Web Protocol Security Enforcer (WPSE) is a browser extension designed to strengthen web protocols by means of three key ingredients: message confidentiality, message integrity and compliance with the intended protocol flow [30]. WPSE takes as input a protocol specification described in terms of a finite state automaton, extended with confidentiality and integrity annotations, and monitors the HTTP messages at the browser side to ensure compliance with respect to the protocol specification. The soundness theorem of WPSE intuitively states that, if a web protocol is proved secure for a given trace property assuming that the browser follows the protocol specification, then WPSE preserves security in presence of a compromised browser which tries to deviate from the intended protocol. The approach was proved general enough to deal with known attacks against real-world protocols like OAuth 2.0 and SAML.

6.5.2.2.2 Bulwark

Bulwark is the first formally sound solution for the holistic security monitoring of web protocols [99]. The motivating observation is that previous work in the area puts specific restrictions on the placement of security monitors, e.g., on the client rather than on a specific server, leading to a limited coverage of the attack surface and other types of problems for practical real-world deployments. Bulwark proposes a solution whereby security monitors are placed exactly where they are required for security. It requires users to express the web protocol they want to monitor in ProVerif [23], along with its intended security properties, which are verified by ProVerif itself. In a second step, selected protocol participants can be marked as *inattentive*, i.e., compliant with the protocol flow, yet lazy in the implementation of security checks. For each inattentive protocol participant, Bulwark introduces a corresponding security monitor in the ProVerif specification, trying out different monitor placement options to minimize the deployment cost. If the monitored protocol still satisfies the intended security properties, Bulwark generates real-world implementations of the monitors as JavaScript service workers or Python proxies.

6.5.3 TRANSPORT LAYER SECURITY

The literature on the formal analysis of the TLS protocol can be categorized in two main classes. A first line of work focuses on the high-level *protocol analysis* based on formal models designed around a variety of underlying assumptions. Clearly, logical flaws in the protocol design may have an enormous impact of the web ecosystem, given that every HTTPS host makes use of TLS. We discuss key results on the TLS protocol analysis in Section 6.5.3.1.

A second line of work, instead, focuses on the analysis of *TLS implementations*. Since TLS is a large and complex protocol, its secure implementation is far from straightforward. Though implementation bugs may only affect specific hosts and can more easily be fixed through software updates, their impact in the wild should not be underestimated, as shown by the Heartbleed security incident [47]. We discuss the most prominent work on the analysis of TLS implementations in Section 6.5.3.2.

6.5.3.1 Protocol Analysis

TLS is a complex protocol, which includes multiple sub-protocols, among which the Handshake and Record protocols are the most important components. A formal analysis of the entire TLS stack, with all the sub-protocols and options, was performed using higher-order logic and the inductive method in a seminal paper by Paulson [81]. Paulson was able to prove all the expected security guarantees of TLS in the symbolic model, but his analysis covers TLS 1.0 and is thus nowadays outdated. A more recent analysis covering both the Handshake and the Record protocols of TLS 1.2 was performed in the computational model [67]. In the following we just report on analyzes of the most recent TLS specification (1.3). We note that the final version of TLS 1.3 only came after draft 28: the models used in research studies have been designed earlier, but they are still meaningful because the design of the protocol did not change significantly.

6.5.3.1.1 Symbolic model

Existing analyzes of TLS in the symbolic model mostly focus on the Handshake sub-protocol, which handles the mutual authentication between the client and the server, and the selection of various parameters like the ciphersuite and the key to be used during the subsequent encrypted data transfer. The critical role of this protocol and its complexity make it a very interesting target for formal verification.

The state of the art in the area is a model of TLS 1.3 draft 21 implemented in the Tamarin prover [39], an advanced tool for the symbolic analysis of security protocols. The formal model covers all the handshake modes supported in TLS 1.3 and, rather than examining each in isolation, considers all the interactions between them. The model includes six out of the eight TLS 1.3 security properties, and succeeds in verifying them under reasonable assumptions for most handshake modes. The model also highlights an unexpected behavior in which a client does not know whether data sent by the server was provided under the assumption that the client was already authenticated or not.

6.5.3.1.2 *Computational model*

During the development of TLS 1.3 (at draft 5 and the corresponding dh fork) a comprehensive evaluation of the cryptographic security of the handshake protocol has been performed in [46]. This work formally proves the security of the handshake protocol for both the draft and the fork versions, and then builds on previous composition results for multi-stage key exchange protocols to extend the security proof to the Record and Session Resumption protocols under proper assumptions. One of the main limitations is that the analysis of 0-RTT, one of the TLS 1.3 new features, was not formalized, because there was no specification of the 0-RTT handshake in the drafts at the time of writing.

6.5.3.2 Implementation Analysis

Although progress in verification tools has enabled the automated analysis of complex security protocols like TLS, the actual verification of their implementation is still a difficult task. Black-box testing may help at detecting bugs, but it is still limited to the generated test cases, while a formally verified implementation requires to reconcile cryptographic verification with program analysis [15].

We discuss below the major verified implementations of TLS developed by the research community. It may be worth noting that not much formal analysis of existing TLS implementations, such as the widespread OpenSSL library[5], has been done. A notable exception is a formal analysis of the correctness and security of the HMAC construction of OpenSSL for the legacy 0.9.1c version [14].

6.5.3.2.1 *miTLS*

miTLS [18] is a modular implementation of the whole TLS 1.2 protocol in F#, structured to enable its automatic verification. The security specifications of its main components are verified using the F7 typechecker. As an example, F7 catches a side-channel padding oracle attack framing it as a type error. Cryptographic security is achieved by typing, leveraging a feature of the F7 typechecker which allows one to annotate types with logical specifications and then verify their consistency. The main limitations of the analysis are the missing support for some popular algorithms and ciphersuites, and strong assumptions about the Handshake protocol.

[5] https://www.openssl.org/

The analysis of the protocol has later been extended, constructing modular proofs for its implementation including ciphersuite negotiation, key exchange, renegotiation, and resumption [19]. Further research [42] extended the miTLS library with support for the TLS 1.3 Record protocol, which has many differences with respect to previous versions. The high-level security properties of the Record protocol were reduced to cryptographic assumptions on its ciphers and implemented in F*, verifying each reduction step by typing an F* module and proving the security of the implementation. The implementation was plugged in the miTLS library and tested with Chrome and Firefox.

6.5.3.2.2 *nqsb-TLS*

Not Quite So Broken-TLS [65] is a TLS specification built as a collection of pure functions over abstract datatypes and then implemented using the OCaml functional language. This implementation can be used as an oracle to check other implementations, or leveraged in real-world TLS applications. OCaml is a memory safe and statically typed programming language which allows one to encode complex state machines with statically enforced invariants, thus avoiding a large source of vulnerabilities. nsqb-TLS implements the entire set of TLS RFCs but for some rarely used features, does not support legacy options or SSLv3 and, most notably, misses elliptic curve cryptography. The library is built as a stack of modules: the Core that handles data transfer, the Nocrypto module which provides cryptography, the X.509 module takes care of X.509 certificates and the ASN.1 module manages the parsing of the Abstract Syntax Notation One interface description language. The structure of nqsb-TLS and its development process rules out the root causes of many vulnerabilities discovered in different TLS implementations and other known issues have been mitigated; furthermore, the library was tested through fuzzing and by running a public bug bounty program, which went unassigned.

6.6 PERSPECTIVE AND CONCLUSION

We hope our analysis met its goal of raising the reader's awareness of the challenges that developers of cryptographic web applications have to face. Indeed, it should by now be apparent that developing secure cryptographic web applications is a daunting task for many reasons (see Table 6.2 for a summary). First, the attack surface against cryptographic web applications is extremely large. Traditional web applications already face threats from JavaScript (web attackers) and HTTP (network attackers); on top of that, cryptographic web applications must also account for a range

Table 6.2
Key Concerns and Mitigation Techniques Discussed in the Chapter.

Key Concerns	Mitigation	Reference
Taming malicious JavaScript	Blocking dangerous content JavaScript sandboxing Information flow control	Sec. 6.4.2
JavaScript cryptography	Choosing libraries Correct use of libraries	Sec. 6.4.3
Web authentication	Securing web protocols Securing web sessions	Sec. 6.5.2
Securing communication	Verification of TLS Secure implementation of TLS	Sec. 6.5.3

of subtle cryptographic threats, e.g., side channels. Such threats are already difficult to tame when working with strongly-typed programming languages providing appropriate isolation guarantees, let alone when struggling with JavaScript and its peculiar semantics. Today, given the massive codebases available in the wild, JavaScript sandboxing is still a security cornerstone. Yet, one might argue that the huge research corpus on JavaScript sandboxing has largely failed, as none of the proposed solutions has ever seen widespread practical adoption [98]. Indeed, we observe a decreasing interest in the problem by the community and we are not aware of any verified sandboxing solution working in current browsers. Likewise, we acknowledge that industrial engineering efforts have similar troubles at taming JavaScript in practice: for example, Google archived the Caja project[6] in January 2021, due to an increasing number of reported security vulnerabilities.

As a result, the community is now exploring different research directions. For example, recent papers have advocated the use of WebAssembly for the development of secure cryptographic web applications [83, 102]. We agree with this standpoint, and ourselves advocate WebAssembly as a more disciplined language for cryptographic web applications to be adopted in the future. Moving to a new language may also spawn interesting research directions, because many relevant security problems studied in the JavaScript setting remain unsolved in WebAssembly, which just provides more robust foundations to enforce security.

The outlook on web protocol security is more positive. Several papers have leveraged the large research body on protocol verification to analyze

[6] https://developers.google.com/caja/

web protocols, with promising, albeit still partial, results [11, 99]. The key challenge for web protocols arises from the peculiar nature of web browser as protocol participants [55], a complexity which should clearly be accounted for by a realistic security assessment and yet represents a still largely unsolved problem. On the one hand, we have automated approaches based on highly simplified browsers models [10, 11, 99], which are great for attack finding but unfit for providing reliable security proofs. On the other hand, we have manual approaches based on expressive pen-and-paper browser models [48, 49, 50], which can provide meaningful security proofs, but clearly do not scale and, inevitably, might suffer from human errors. A further limitation of most of the current frameworks for web protocol verification is their focus on the symbolic model: generalizing those results to the computational model is certainly a relevant research direction to be considered in future work.

Finally, we acknowledge that the tremendous effort spent on the formal verification of TLS is paying off. In particular, we note that the HACL* formally verified cryptographic library [106] based on F* has entered the Mozilla Firefox codebase since 2017[7]. Cryptographic libraries shipped in web browsers are paramount, because they are used both in TLS and in the Web Cryptography API, hence provide immediate benefits to a large number of web applications. Though the formal techniques used to verify the security of cryptographic libraries are very sophisticated and hardly accessible to average developers, they nevertheless proved effective to provide off-the-shelf secure solutions to an extremely wide audience.

6.6.1 ACKNOWLEDGMENTS

We would like to thank Ralf Kuesters and the anonymous reviewers for their useful comments on a preliminary version of the present chapter.

REFERENCES

1. Autoupgrade Image Mixed Content. https://www.chromestatus.com/feature/4926989725073408. Accessed: 22-06-2021.

2. Cookies default to SameSite=Lax. https://www.chromestatus.com/feature/5088147346030592. Accessed: 22-06-2021.

3. Pieter Agten, Steven Van Acker, Yoran Brondsema, Phu H. Phung, Lieven Desmet, and Frank Piessens. Jsand: Complete client-side sandboxing of third-party javascript without browser modifications. In *Proceedings of the*

[7] https://blog.mozilla.org/security/2017/09/13/verified-cryptography-firefox-57/

28th Annual Computer Security Applications Conference, ACSAC '12, page 1–10, New York, NY, USA, 2012. Association for Computing Machinery.

4. Devdatta Akhawe, Adam Barth, Peifung E. Lam, John C. Mitchell, and Dawn Song. Towards a formal foundation of web security. In *Proceedings of the 23rd IEEE Computer Security Foundations Symposium, CSF 2010, Edinburgh, United Kingdom, July 17-19, 2010*, pages 290–304. IEEE Computer Society, 2010.

5. George Argyros, Ioannis Stais, Aggelos Kiayias, and Angelos D. Keromytis. Back in black: Towards formal, black box analysis of sanitizers and filters. In *IEEE Symposium on Security and Privacy, SP 2016, San Jose, CA, USA, May 22-26, 2016*, pages 91–109. IEEE Computer Society, 2016.

6. Alessandro Armando, et.al. The AVANTSSAR platform for the automated validation of trust and security of service-oriented architectures. In Cormac Flanagan and Barbara König, editors, *Tools and Algorithms for the Construction and Analysis of Systems - 18th International Conference, TACAS 2012, Tallinn, Estonia, March 24 - April 1, 2012. Proceedings*, volume 7214 of *LNCS*, pages 267–282. Springer, 2012.

7. Alessandro Armando, Roberto Carbone, Luca Compagna, Jorge Cuéllar, and Llanos Tobarra. Formal analysis of SAML 2.0 web browser single sign-on: breaking the saml-based single sign-on for google apps. In Vitaly Shmatikov, editor, *Proceedings of the 6th ACM Workshop on Formal Methods in Security Engineering, FMSE 2008, Alexandria, VA, USA, October 27, 2008*, pages 1–10. ACM, 2008.

8. Nimrod Aviram, et. al. DROWN: breaking TLS using sslv2. In Thorsten Holz and Stefan Savage, editors, *25th USENIX Security Symposium, USENIX Security 16, Austin, TX, USA, August 10-12, 2016*, pages 689–706.

9. Davide Balzarotti, Marco Cova, Viktoria Felmetsger, Nenad Jovanovic, Engin Kirda, Christopher Kruegel, and Giovanni Vigna. Saner: Composing static and dynamic analysis to validate sanitization in web applications. In *2008 IEEE Symposium on Security and Privacy (S&P 2008), 18-21 May 2008, Oakland, California, USA*, pages 387–401. IEEE Computer Society, 2008.

10. Chetan Bansal, Karthikeyan Bhargavan, Antoine Delignat-Lavaud, and Sergio Maffeis. Keys to the cloud: Formal analysis and concrete attacks on encrypted web storage. In David A. Basin and John C. Mitchell, editors, *Principles of Security and Trust - Second International Conference, POST 2013, Held as Part of the European Joint Conferences on Theory and Practice of Software, ETAPS 2013, Rome, Italy, March 16-24, 2013. Proceedings*, volume 7796 of *Lecture Notes in Computer Science*, pages 126–146. Springer, 2013.

11. Chetan Bansal, Karthikeyan Bhargavan, Antoine Delignat-Lavaud, and Sergio Maffeis. Discovering concrete attacks on website authorization by formal analysis. *J. Comput. Secur.*, 22(4):601–657, 2014.

12. Adam Barth, Collin Jackson, and John C. Mitchell. Robust defenses for cross-site request forgery. In Peng Ning, Paul F. Syverson, and Somesh Jha, editors, *Proceedings of the 2008 ACM Conference on Computer and Communications Security, CCS 2008, Alexandria, Virginia, USA, October 27-31, 2008*, pages 75–88. ACM, 2008.

13. Lujo Bauer, Shaoying Cai, Limin Jia, Timothy Passaro, Michael Stroucken, and Yuan Tian. Run-time monitoring and formal analysis of information flows in chromium. In *22nd Annual Network and Distributed System Security Symposium, NDSS 2015, San Diego, California, USA, February 8-11, 2015*. The Internet Society, 2015.

14. Lennart Beringer, Adam Petcher, Katherine Q. Ye, and Andrew W. Appel. Verified correctness and security of openssl HMAC. In Jaeyeon Jung and Thorsten Holz, editors, *24th USENIX Security Symposium, USENIX Security 15, Washington, D.C., USA, August 12-14, 2015*, pages 207–221. USENIX Association, 2015.

15. Benjamin Beurdouche, Karthikeyan Bhargavan, Antoine Delignat-Lavaud, Cédric Fournet, Markulf Kohlweiss, Alfredo Pironti, Pierre-Yves Strub, and Jean Karim Zinzindohoue. A messy state of the union: taming the composite state machines of TLS. *Commun. ACM*, 60(2):99–107, 2017.

16. Karthikeyan Bhargavan, Antoine Delignat-Lavaud, and Sergio Maffeis. Defensive javascript - building and verifying secure web components. In Alessandro Aldini, Javier López, and Fabio Martinelli, editors, *Foundations of Security Analysis and Design VII - FOSAD 2012/2013 Tutorial Lectures*, volume 8604 of *Lecture Notes in Computer Science*, pages 88–123. Springer, 2013.

17. Karthikeyan Bhargavan, Antoine Delignat-Lavaud, and Sergio Maffeis. Language-based defenses against untrusted browser origins. In Samuel T. King, editor, *Proceedings of the 22th USENIX Security Symposium, Washington, DC, USA, August 14-16, 2013*, pages 653–670. USENIX Association, 2013.

18. Karthikeyan Bhargavan, Cédric Fournet, Markulf Kohlweiss, Alfredo Pironti, and Pierre-Yves Strub. Implementing TLS with verified cryptographic security. In *2013 IEEE Symposium on Security and Privacy, SP 2013, Berkeley, CA, USA, May 19-22, 2013*, pages 445–459. IEEE Computer Society, 2013.

19. Karthikeyan Bhargavan, Cédric Fournet, Markulf Kohlweiss, Alfredo Pironti, Pierre-Yves Strub, and Santiago Zanella Béguelin. Proving the TLS

handshake secure (as it is). In Juan A. Garay and Rosario Gennaro, editors, *Advances in Cryptology - CRYPTO 2014 - 34th Annual Cryptology Conference, Santa Barbara, CA, USA, August 17-21, 2014, Proceedings, Part II*, volume 8617 of *Lecture Notes in Computer Science*, pages 235–255. Springer, 2014.

20. Abhishek Bichhawat, Vineet Rajani, Deepak Garg, and Christian Hammer. Information flow control in webkit's javascript bytecode. In Martín Abadi and Steve Kremer, editors, *Principles of Security and Trust - Third International Conference, POST 2014, ETAPS 2014, Grenoble, France, April 5-13, 2014, Proceedings*, volume 8414 of *Lecture Notes in Computer Science*, pages 159–178. Springer, 2014.

21. Nataliia Bielova. Survey on javascript security policies and their enforcement mechanisms in a web browser. *J. Log. Algebraic Methods Program.*, 82(8):243–262, 2013.

22. Bruno Blanchet. Security protocol verification: Symbolic and computational models. In Cormac Flanagan and Barbara König, editors, *Tools and Algorithms for the Construction and Analysis of Systems - 18th International Conference, TACAS 2012, Tallinn, Estonia, March 24 - April 1, 2012. Proceedings*, volume 7214 of *LNCS*, pages 3–29. Springer, 2012.

23. Bruno Blanchet. Modeling and verifying security protocols with the applied pi calculus and proverif. *Found. Trends Priv. Secur.*, 1(1-2):1–135, 2016.

24. Aaron Bohannon and Benjamin C. Pierce. Featherweight firefox: Formalizing the core of a web browser. In John K. Ousterhout, editor, *USENIX Conference on Web Application Development, WebApps'10, Boston, Massachusetts, USA, June 23-24, 2010*. USENIX Association, 2010.

25. Michele Bugliesi, Stefano Calzavara, Riccardo Focardi, and Wilayat Khan. Cookiext: Patching the browser against session hijacking attacks. *J. Comput. Secur.*, 23(4):509–537, 2015.

26. Michele Bugliesi, Stefano Calzavara, Riccardo Focardi, Wilayat Khan, and Mauro Tempesta. Provably sound browser-based enforcement of web session integrity. In *IEEE 27th Computer Security Foundations Symposium, CSF 2014, Vienna, Austria, 19-22 July, 2014*, pages 366–380. IEEE Computer Society, 2014.

27. Kelsey Cairns, Harry Halpin, and Graham Steel. Security analysis of the W3C web cryptography API. In Lidong Chen, David A. McGrew, and Chris J. Mitchell, editors, *Security Standardisation Research - Third International Conference, SSR 2016, Gaithersburg, MD, USA, December 5-6, 2016, Proceedings*, volume 10074 of *Lecture Notes in Computer Science*, pages 112–140. Springer, 2016.

28. Stefano Calzavara, Riccardo Focardi, Niklas Grimm, and Matteo Maffei. Micro-policies for web session security. In *IEEE 29th Computer Security Foundations Symposium, CSF 2016, Lisbon, Portugal, June 27 - July 1, 2016*, pages 179–193. IEEE Computer Society, 2016.

29. Stefano Calzavara, Riccardo Focardi, Niklas Grimm, Matteo Maffei, and Mauro Tempesta. Language-based web session integrity. In *33rd IEEE Computer Security Foundations Symposium, CSF 2020, Boston, MA, USA, June 22-26, 2020*, pages 107–122. IEEE, 2020.

30. Stefano Calzavara, Riccardo Focardi, Matteo Maffei, Clara Schneidewind, Marco Squarcina, and Mauro Tempesta. WPSE: fortifying web protocols via browser-side security monitoring. In William Enck and Adrienne Porter Felt, editors, *27th USENIX Security Symposium, USENIX Security 2018, Baltimore, MD, USA, August 15-17, 2018*, pages 1493–1510.

31. Stefano Calzavara, Riccardo Focardi, Matús Nemec, Alvise Rabitti, and Marco Squarcina. Postcards from the post-http world: Amplification of HTTPS vulnerabilities in the web ecosystem. In *2019 IEEE Symposium on Security and Privacy, SP 2019, San Francisco, CA, USA, May 19-23, 2019*, pages 281–298. IEEE, 2019.

32. Stefano Calzavara, Riccardo Focardi, Marco Squarcina, and Mauro Tempesta. Surviving the web: A journey into web session security. *ACM Comput. Surv.*, 50(1):13:1–13:34, 2017.

33. Stefano Calzavara, Alvise Rabitti, and Michele Bugliesi. Content security problems?: Evaluating the effectiveness of content security policy in the wild. In Edgar R. Weippl, Stefan Katzenbeisser, Christopher Kruegel, Andrew C. Myers, and Shai Halevi, editors, *Proceedings of the 2016 ACM SIGSAC Conference on Computer and Communications Security, Vienna, Austria, October 24-28, 2016*, pages 1365–1375.

34. Stefano Calzavara, Alvise Rabitti, and Michele Bugliesi. Semantics-based analysis of content security policy deployment. *ACM Trans. Web*, 12(2):10:1–10:36, 2018.

35. Eric Yawei Chen, Jason Bau, Charles Reis, Adam Barth, and Collin Jackson. App isolation: get the security of multiple browsers with just one. In Yan Chen, George Danezis, and Vitaly Shmatikov, editors, *Proceedings of the 18th ACM Conference on Computer and Communications Security, CCS 2011, Chicago, Illinois, USA, October 17-21, 2011*, pages 227–238.

36. Andrey Chudnov and David A. Naumann. Inlined information flow monitoring for javascript. In Indrajit Ray, Ninghui Li, and Christopher Kruegel, editors, *Proceedings of the 22nd ACM SIGSAC Conference on Computer and Communications Security, Denver, CO, USA, October 12-16, 2015*, pages 629–643.

37. Ravi Chugh, Jeffrey A. Meister, Ranjit Jhala, and Sorin Lerner. Staged information flow for javascript. In Michael Hind and Amer Diwan, editors, *Proceedings of the 2009 ACM SIGPLAN Conference on Programming Language Design and Implementation, PLDI 2009, Dublin, Ireland, June 15-21, 2009*, pages 50–62.

38. Luca Compagna, Daniel Ricardo dos Santos, Serena Elisa Ponta, and Silvio Ranise. Aegis: Automatic enforcement of security policies in workflow-driven web applications. In Gail-Joon Ahn, Alexander Pretschner, and Gabriel Ghinita, editors, *Proceedings of the Seventh ACM Conference on Data and Application Security and Privacy, CODASPY 2017, Scottsdale, AZ, USA, March 22-24, 2017*, pages 321–328.

39. Cas Cremers, Marko Horvat, Jonathan Hoyland, Sam Scott, and Thyla van der Merwe. A comprehensive symbolic analysis of TLS 1.3. In Bhavani M. Thuraisingham, David Evans, Tal Malkin, and Dongyan Xu, editors, *Proceedings of the 2017 ACM SIGSAC Conference on Computer and Communications Security, CCS 2017, Dallas, TX, USA, October 30 - November 03, 2017*, pages 1773–1788.

40. Douglas Crockford. Adsafe - a safe javascript widget framework for advertising and other mashups. https://github.com/douglascrockford/ADsafe. Accessed: 25-10-2021.

41. Anders P. K. Dalskov and Claudio Orlandi. Can you trust your encrypted cloud?: An assessment of spideroakone's security. In Jong Kim, Gail-Joon Ahn, Seungjoo Kim, Yongdae Kim, Javier López, and Taesoo Kim, editors, *Proceedings of the 2018 on Asia Conference on Computer and Communications Security, AsiaCCS 2018, Incheon, Republic of Korea, June 04-08, 2018*, pages 343–355. ACM, 2018.

42. Antoine Delignat-Lavaud, Cédric Fournet, Markulf Kohlweiss, Jonathan Protzenko, Aseem Rastogi, Nikhil Swamy, Santiago Zanella Béguelin, Karthikeyan Bhargavan, Jianyang Pan, and Jean Karim Zinzindohoue. Implementing and proving the TLS 1.3 record layer. In *2017 IEEE Symposium on Security and Privacy, SP 2017, San Jose, CA, USA, May 22-26, 2017*, pages 463–482. IEEE Computer Society, 2017.

43. Dominique Devriese and Frank Piessens. Noninterference through secure multi-execution. In *31st IEEE Symposium on Security and Privacy, S&P 2010, 16-19 May 2010, Berleley/Oakland, California, USA*, pages 109–124. IEEE Computer Society, 2010.

44. Quoc Huy Do, Pedram Hosseyni, Ralf Küsters, Guido Schmitz, Nils Wenzler, and Tim Würtele. A formal security analysis of the W3C web payment apis: Attacks and verification. *IACR Cryptol. ePrint Arch.*, page 1012, 2021.

45. Danny Dolev and Andrew Chi-Chih Yao. On the security of public key protocols. *IEEE Trans. Inf. Theory*, 29(2):198–207, 1983.

46. Benjamin Dowling, Marc Fischlin, Felix Günther, and Douglas Stebila. A cryptographic analysis of the TLS 1.3 handshake protocol candidates. In Indrajit Ray, Ninghui Li, and Christopher Kruegel, editors, *Proceedings of the 22nd ACM SIGSAC Conference on Computer and Communications Security, Denver, CO, USA, October 12-16, 2015*, pages 1197–1210. ACM, 2015.

47. Zakir Durumeric, James Kasten, David Adrian, J. Alex Halderman, Michael Bailey, Frank Li, Nicholas Weaver, Johanna Amann, Jethro Beekman, Mathias Payer, and Vern Paxson. The matter of heartbleed. In Carey Williamson, Aditya Akella, and Nina Taft, editors, *Proceedings of the 2014 Internet Measurement Conference, IMC 2014, Vancouver, BC, Canada, November 5-7, 2014*, pages 475–488. ACM, 2014.

48. Daniel Fett, Pedram Hosseyni, and Ralf Küsters. An extensive formal security analysis of the openid financial-grade API. In *2019 IEEE Symposium on Security and Privacy, SP 2019, San Francisco, CA, USA, May 19-23, 2019*, pages 453–471. IEEE, 2019.

49. Daniel Fett, Ralf Küsters, and Guido Schmitz. An expressive model for the web infrastructure: Definition and application to the browser ID SSO system. In *2014 IEEE Symposium on Security and Privacy, SP 2014, Berkeley, CA, USA, May 18-21, 2014*, pages 673–688. IEEE Computer Society, 2014.

50. Daniel Fett, Ralf Küsters, and Guido Schmitz. SPRESSO: A secure, privacy-respecting single sign-on system for the web. In Indrajit Ray, Ninghui Li, and Christopher Kruegel, editors, *Proceedings of the 22nd ACM SIGSAC Conference on Computer and Communications Security, Denver, CO, USA, October 12-16, 2015*, pages 1358–1369. ACM, 2015.

51. Daniel Fett, Ralf Küsters, and Guido Schmitz. A comprehensive formal security analysis of oauth 2.0. In Edgar R. Weippl, Stefan Katzenbeisser, Christopher Kruegel, Andrew C. Myers, and Shai Halevi, editors, *Proceedings of the 2016 ACM SIGSAC Conference on Computer and Communications Security, Vienna, Austria, October 24-28, 2016*, pages 1204–1215. ACM, 2016.

52. Daniel Fett, Ralf Küsters, and Guido Schmitz. The web SSO standard openid connect: In-depth formal security analysis and security guidelines. In *30th IEEE Computer Security Foundations Symposium, CSF 2017, Santa Barbara, CA, USA, August 21-25, 2017*, pages 189–202. IEEE Computer Society, 2017.

53. Joseph A. Goguen and José Meseguer. Security policies and security models. In *1982 IEEE Symposium on Security and Privacy, Oakland, CA, USA, April 26-28, 1982*, pages 11–20. IEEE Computer Society, 1982.

54. Willem De Groef, Dominique Devriese, Nick Nikiforakis, and Frank Piessens. Flowfox: a web browser with flexible and precise information flow control. In Ting Yu, George Danezis, and Virgil D. Gligor, editors, *the ACM Conference on Computer and Communications Security, CCS'12, Raleigh, NC, USA, October 16-18, 2012*, pages 748–759. ACM, 2012.

55. Thomas Groß, Birgit Pfitzmann, and Ahmad-Reza Sadeghi. Browser model for security analysis of browser-based protocols. In Sabrina De Capitani di Vimercati, Paul F. Syverson, and Dieter Gollmann, editors, *Computer Security - ESORICS 2005, 10th European Symposium on Research in Computer Security, Milan, Italy, September 12-14, 2005, Proceedings*, volume 3679 of *Lecture Notes in Computer Science*, pages 489–508. Springer, 2005.

56. Salvatore Guarnieri and V. Benjamin Livshits. GATEKEEPER: mostly static enforcement of security and reliability policies for javascript code. In Fabian Monrose, editor, *18th USENIX Security Symposium, Montreal, Canada, August 10-14, 2009, Proceedings*, pages 151–168. USENIX Association, 2009.

57. Andreas Haas, Andreas Rossberg, Derek L. Schuff, Ben L. Titzer, Michael Holman, Dan Gohman, Luke Wagner, Alon Zakai, and J. F. Bastien. Bringing the web up to speed with webassembly. In Albert Cohen and Martin T. Vechev, editors, *Proceedings of the 38th ACM SIGPLAN Conference on Programming Language Design and Implementation, PLDI 2017, Barcelona, Spain, June 18-23, 2017*, pages 185–200. ACM, 2017.

58. Harry Halpin. The W3C web cryptography API: motivation and overview. In Chin-Wan Chung, Andrei Z. Broder, Kyuseok Shim, and Torsten Suel, editors, *23rd International World Wide Web Conference, WWW '14, Seoul, Republic of Korea, April 7-11, 2014, Companion Volume*, pages 959–964. ACM, 2014.

59. Daniel Hedin, Arnar Birgisson, Luciano Bello, and Andrei Sabelfeld. Jsflow: tracking information flow in javascript and its apis. In Yookun Cho, Sung Y. Shin, Sang-Wook Kim, Chih-Cheng Hung, and Jiman Hong, editors, *Symposium on Applied Computing, SAC 2014, Gyeongju, Republic of Korea - March 24 - 28, 2014*, pages 1663–1671. ACM, 2014.

60. Daniel Hedin and Andrei Sabelfeld. Information-flow security for a core of javascript. In Stephen Chong, editor, *25th IEEE Computer Security Foundations Symposium, CSF 2012, Cambridge, MA, USA, June 25-27, 2012*, pages 3–18. IEEE Computer Society, 2012.

61. Jeff Hodges, Collin Jackson, and Adam Barth. HTTP strict transport security (HSTS). *RFC*, 6797:1–46, 2012.

62. Pieter Hooimeijer, Benjamin Livshits, David Molnar, Prateek Saxena, and Margus Veanes. Fast and precise sanitizer analysis with BEK. In *20th*

USENIX Security Symposium, San Francisco, CA, USA, August 8-12, 2011, Proceedings. USENIX Association, 2011.

63. Daniel Jackson. Alloy: a lightweight object modelling notation. *ACM Trans. Softw. Eng. Methodol.*, 11(2):256–290, 2002.

64. Trevor Jim, Nikhil Swamy, and Michael Hicks. Defeating script injection attacks with browser-enforced embedded policies. In *Proceedings of the 16th International Conference on World Wide Web*, WWW '07, page 601–610, New York, NY, USA, 2007. Association for Computing Machinery.

65. David Kaloper-Mersinjak, Hannes Mehnert, Anil Madhavapeddy, and Peter Sewell. Not-quite-so-broken TLS: lessons in re-engineering a security protocol specification and implementation. In Jaeyeon Jung and Thorsten Holz, editors, *24th USENIX Security Symposium, USENIX Security 15, Washington, D.C., USA, August 12-14, 2015*, pages 223–238. USENIX Association, 2015.

66. Yoonseok Ko, Tamara Rezk, and Manuel Serrano. Securejs compiler: Portable memory isolation in javascript. In *Proceedings of the 36th Annual ACM Symposium on Applied Computing*, SAC '21, page 1265–1274, New York, NY, USA, 2021. Association for Computing Machinery.

67. Hugo Krawczyk, Kenneth G. Paterson, and Hoeteck Wee. On the security of the TLS protocol: A systematic analysis. In Ran Canetti and Juan A. Garay, editors, *Advances in Cryptology - CRYPTO 2013 - 33rd Annual Cryptology Conference, Santa Barbara, CA, USA, August 18-22, 2013. Proceedings, Part I*, volume 8042 of *Lecture Notes in Computer Science*, pages 429–448. Springer, 2013.

68. Sebastian Lekies, Ben Stock, and Martin Johns. 25 million flows later: large-scale detection of dom-based XSS. In Ahmad-Reza Sadeghi, Virgil D. Gligor, and Moti Yung, editors, *2013 ACM SIGSAC Conference on Computer and Communications Security, CCS'13, Berlin, Germany, November 4-8, 2013*, pages 1193–1204. ACM, 2013.

69. Xiaowei Li and Yuan Xue. BLOCK: a black-box approach for detection of state violation attacks towards web applications. In Robert H'obbes' Zakon, John P. McDermott, and Michael E. Locasto, editors, *Twenty-Seventh Annual Computer Security Applications Conference, ACSAC 2011, Orlando, FL, USA, 5-9 December 2011*, pages 247–256. ACM, 2011.

70. Benjamin Livshits and Stephen Chong. Towards fully automatic placement of security sanitizers and declassifiers. In Roberto Giacobazzi and Radhia Cousot, editors, *The 40th Annual ACM SIGPLAN-SIGACT Symposium on Principles of Programming Languages, POPL '13, Rome, Italy - January 23 - 25, 2013*, pages 385–398. ACM, 2013.

71. Tongbo Luo and Wenliang Du. Contego: Capability-based access control for web browsers. In *International Conference on Trust and Trustworthy Computing*, pages 231–238. Springer, 2011.

72. Moxie Marlinspike. New tricks for defeating SSL in practice. https://www.blackhat.com/presentations/bh-dc-09/Marlinspike/BlackHat-DC-09-Marlinspike-Defeating-SSL.pdf. Accessed: 22-06-2021.

73. Simon Meier, Benedikt Schmidt, Cas Cremers, and David A. Basin. The TAMARIN prover for the symbolic analysis of security protocols. In Natasha Sharygina and Helmut Veith, editors, *Computer Aided Verification - 25th International Conference, CAV 2013, Saint Petersburg, Russia, July 13-19, 2013. Proceedings*, volume 8044 of *Lecture Notes in Computer Science*, pages 696–701. Springer, 2013.

74. Leo A Meyerovich, Adrienne Porter Felt, and Mark S Miller. Object views: Fine-grained sharing in browsers. In *Proceedings of the 19th International Conference on World Wide Web*, pages 721–730, 2010.

75. Leo A. Meyerovich and Benjamin Livshits. Conscript: Specifying and enforcing fine-grained security policies for javascript in the browser. In *2010 IEEE Symposium on Security and Privacy*, pages 481–496, 2010.

76. Mark S. Miller, Mike Samuel, Ben Laurie, Ihab Awad, and Mike Stay. Caja - safe active content in sanitized javascript. http://math.ucr.edu/~mike/caja-spec-2008-06-06.pdf. Accessed: 25-10-2021.

77. Duncan Mitchell and Johannes Kinder. A formal model for checking cryptographic API usage in javascript. In Kazue Sako, Steve A. Schneider, and Peter Y. A. Ryan, editors, *Computer Security - ESORICS 2019 - 24th European Symposium on Research in Computer Security, Luxembourg, September 23-27, 2019, Proceedings, Part I*, volume 11735 of *Lecture Notes in Computer Science*, pages 341–360. Springer, 2019.

78. Mozilla. Javascript strict mode reference. https://developer.mozilla.org/en-US/docs/Web/JavaScript/Reference/Strict˙mode. Accessed: 25-10-2021.

79. Marius Musch, Marius Steffens, Sebastian Roth, Ben Stock, and Martin Johns. Scriptprotect: Mitigating unsafe third-party javascript practices. In *Proceedings of the 2019 ACM Asia Conference on Computer and Communications Security*, Asia CCS '19, page 391–402, New York, NY, USA, 2019. Association for Computing Machinery.

80. OWASP. Cross site scripting prevention cheat sheet. https://cheatsheetseries.owasp.org/cheatsheets/Cross˙Site˙Scripting˙Prevention˙Cheat˙Sheet.html. Accessed: 21-10-2021.

81. Lawrence C. Paulson. Inductive analysis of the internet protocol TLS. *ACM Trans. Inf. Syst. Secur.*, 2(3):332–351, 1999.

82. Phu H. Phung, David Sands, and Andrey Chudnov. Lightweight self-protecting javascript. ASIACCS '09, page 47–60, New York, NY, USA, 2009. Association for Computing Machinery.

83. Jonathan Protzenko, Benjamin Beurdouche, Denis Merigoux, and Karthikeyan Bhargavan. Formally verified cryptographic web applications in webassembly. In *2019 IEEE Symposium on Security and Privacy, SP 2019, San Francisco, CA, USA, May 19-23, 2019*, pages 1256–1274. IEEE, 2019.

84. Vineet Rajani, Abhishek Bichhawat, Deepak Garg, and Christian Hammer. Information flow control for event handling and the DOM in web browsers. In Cédric Fournet, Michael W. Hicks, and Luca Viganò, editors, *IEEE 28th Computer Security Foundations Symposium, CSF 2015, Verona, Italy, 13-17 July, 2015*, pages 366–379. IEEE Computer Society, 2015.

85. Charles Reis, John Dunagan, Helen J. Wang, Opher Dubrovsky, and Saher Esmeir. Browsershield: Vulnerability-driven filtering of dynamic html. *ACM Trans. Web*, 1(3):11–es, sep 2007.

86. Philippe De Ryck, Lieven Desmet, Wouter Joosen, and Frank Piessens. Automatic and precise client-side protection against CSRF attacks. In Vijay Atluri and Claudia Díaz, editors, *Computer Security - ESORICS 2011 - 16th European Symposium on Research in Computer Security, Leuven, Belgium, September 12-14, 2011. Proceedings*, volume 6879 of *Lecture Notes in Computer Science*, pages 100–116. Springer, 2011.

87. Andrei Sabelfeld and Andrew C. Myers. Language-based information-flow security. *IEEE J. Sel. Areas Commun.*, 21(1):5–19, 2003.

88. Mike Samuel, Prateek Saxena, and Dawn Song. Context-sensitive auto-sanitization in web templating languages using type qualifiers. In Yan Chen, George Danezis, and Vitaly Shmatikov, editors, *Proceedings of the 18th ACM Conference on Computer and Communications Security, CCS 2011, Chicago, Illinois, USA, October 17-21, 2011*, pages 587–600. ACM, 2011.

89. Prateek Saxena, David Molnar, and Benjamin Livshits. SCRIPTGARD: automatic context-sensitive sanitization for large-scale legacy web applications. In Yan Chen, George Danezis, and Vitaly Shmatikov, editors, *Proceedings of the 18th ACM Conference on Computer and Communications Security, CCS 2011, Chicago, Illinois, USA, October 17-21, 2011*, pages 601–614. ACM, 2011.

90. Sid Stamm, Brandon Sterne, and Gervase Markham. Reining in the web with content security policy. In Michael Rappa, Paul Jones, Juliana Freire, and Soumen Chakrabarti, editors, *Proceedings of the 19th International Conference on World Wide Web, WWW 2010, Raleigh, North Carolina, USA, April 26-30, 2010*, pages 921–930. ACM, 2010.

91. Emily Stark, Michael Hamburg, and Dan Boneh. Symmetric cryptography in javascript. In *Twenty-Fifth Annual Computer Security Applications Conference, ACSAC 2009, Honolulu, Hawaii, USA, 7-11 December 2009*, pages 373–381. IEEE Computer Society, 2009.

92. Ben Stock, Martin Johns, Marius Steffens, and Michael Backes. How the web tangled itself: Uncovering the history of client-side web (in)security. In Engin Kirda and Thomas Ristenpart, editors, *26th USENIX Security Symposium, USENIX Security 2017, Vancouver, BC, Canada, August 16-18, 2017*, pages 971–987. USENIX Association, 2017.

93. San-Tsai Sun and Konstantin Beznosov. The devil is in the (implementation) details: an empirical analysis of oauth SSO systems. In Ting Yu, George Danezis, and Virgil D. Gligor, editors, *the ACM Conference on Computer and Communications Security, CCS'12, Raleigh, NC, USA, October 16-18, 2012*, pages 378–390. ACM, 2012.

94. Ankur Taly, Úlfar Erlingsson, John C. Mitchell, Mark S. Miller, and Jasvir Nagra. Automated analysis of security-critical javascript apis. In *32nd IEEE Symposium on Security and Privacy, S&P 2011, 22-25 May 2011, Berkeley, California, USA*, pages 363–378. IEEE Computer Society, 2011.

95. Mike Ter Louw, Karthik Thotta Ganesh, and VN Venkatakrishnan. Adjail: Practical enforcement of confidentiality and integrity policies on web advertisements. In *USENIX Security Symposium*, pages 371–388, 2010.

96. Mike Ter Louw, Phu H Phung, Rohini Krishnamurti, and Venkat N Venkatakrishnan. Safescript: Javascript transformation for policy enforcement. In *Nordic Conference on Secure IT Systems*, pages 67–83. Springer, 2013.

97. Steven Van Acker, Philippe De Ryck, Lieven Desmet, Frank Piessens, and Wouter Joosen. Webjail: Least-privilege integration of third-party components in web mashups. In *Proceedings of the 27th Annual Computer Security Applications Conference*, ACSAC '11, page 307–316, New York, NY, USA, 2011. Association for Computing Machinery.

98. Steven Van Acker and Andrei Sabelfeld. Javascript sandboxing: Isolating and restricting client-side javascript. In Alessandro Aldini, Javier Lόpez, and Fabio Martinelli, editors, *Foundations of Security Analysis and Design VIII - FOSAD 2014/2015/2016 Tutorial Lectures*, volume 9808 of *Lecture Notes in Computer Science*, pages 32–86. Springer, 2016.

99. Lorenzo Veronese, Stefano Calzavara, and Luca Compagna. Bulwark: Holistic and verified security monitoring of web protocols. In Liqun Chen, Ninghui Li, Kaitai Liang, and Steve A. Schneider, editors, *Computer Security - ESORICS 2020 - 25th European Symposium on Research in Computer*

Security, ESORICS 2020, Guildford, UK, September 14-18, 2020, Proceedings, Part I, volume 12308 of *Lecture Notes in Computer Science*, pages 23–41. Springer, 2020.

100. Michael Walfish et al. Treehouse: Javascript sandboxes to help web developers help themselves. In *2012 {USENIX} Annual Technical Conference ({USENIX}{ATC} 12)*, pages 153–164, 2012.

101. Conrad Watt. Mechanising and verifying the webassembly specification. In June Andronick and Amy P. Felty, editors, *Proceedings of the 7th ACM SIGPLAN International Conference on Certified Programs and Proofs, CPP 2018, Los Angeles, CA, USA, January 8-9, 2018*, pages 53–65. ACM, 2018.

102. Conrad Watt, John Renner, Natalie Popescu, Sunjay Cauligi, and Deian Stefan. Ct-wasm: type-driven secure cryptography for the web ecosystem. *Proc. ACM Program. Lang.*, 3(POPL):77:1–77:29, 2019.

103. Lukas Weichselbaum, Michele Spagnuolo, Sebastian Lekies, and Artur Janc. CSP is dead, long live csp! on the insecurity of whitelists and the future of content security policy. In Edgar R. Weippl, Stefan Katzenbeisser, Christopher Kruegel, Andrew C. Myers, and Shai Halevi, editors, *Proceedings of the 2016 ACM SIGSAC Conference on Computer and Communications Security, Vienna, Austria, October 24-28, 2016*, pages 1376–1387. ACM, 2016.

104. Luyi Xing, Yangyi Chen, XiaoFeng Wang, and Shuo Chen. Integuard: Toward automatic protection of third-party web service integrations. In *20th Annual Network and Distributed System Security Symposium, NDSS 2013, San Diego, California, USA, February 24-27, 2013*. The Internet Society, 2013.

105. Mingxue Zhang and Wei Meng. Jsisolate: Lightweight in-browser javascript isolation. In *Proceedings of the 29th ACM Joint Meeting on European Software Engineering Conference and Symposium on the Foundations of Software Engineering*, ESEC/FSE 2021, page 193–204, New York, NY, USA, 2021.

106. Jean Karim Zinzindohoué, Karthikeyan Bhargavan, Jonathan Protzenko, and Benjamin Beurdouche. Hacl*: A verified modern cryptographic library. In Bhavani M. Thuraisingham, David Evans, Tal Malkin, and Dongyan Xu, editors, *Proceedings of the 2017 ACM SIGSAC Conference on Computer and Communications Security, CCS 2017, Dallas, TX, USA, October 30 - November 03, 2017*, pages 1789–1806. ACM, 2017.

7 Formal Methods for Quantum Algorithms

Christophe Chareton
Université Paris-Saclay, Palaiseau, France

Dongho Lee
Université Paris-Saclay and Inria, Palaiseau and Gif-sur-Yvette, France

Benoit Valiron
Université Paris-Saclay and Inria, Palaiseau and Gif-sur-Yvette, France

Renault Vilmart
Université Paris-Saclay and Inria, Palaiseau and Gif-sur-Yvette, France

Sébastien Bardin
Université Paris-Saclay Palaiseau, France

Zhaowei Xu
Université Paris-Saclay, Gif-sur-Yvette, France and University of Tartu, Estonia

CONTENTS

DOI: 10.1201/9781003090052-7

While the recent progress in quantum hardware opens the door for significant speedup in cryptography as well as additional key areas (biology, chemistry, optimization, machine learning, etc), quantum algorithms are still hard to implement right, and the validation of quantum programs is a challenge. Moreover, importing the testing and debugging practices at use in classical programming is extremely difficult in the quantum case, due to the destructive aspect of quantum measurement. As an alternative strategy, formal methods are prone to play a decisive role in the emerging field of quantum software. Recent works initiate solutions for problems occurring at every stage of the development process: high-level program design, implementation, compilation, etc. This chapter introduces both the requirements and challenges for an efficient use of formal methods in quantum computing and the current most promising research directions.

7.1 INTRODUCTION

7.1.1 CRYPTOGRAPHY AND QUANTUM INFORMATION

Quantum computing dates back to 1982, when Richard Feynman raised the idea [69] of simulating quantum mechanics phenomena by storing information in particles and controlling them according to the laws of quantum mechanics. In the brief history of quantum computing, the description in 1994 by Peter Shor of an algorithm [174], performing the decomposition of prime integers in polynomial time on the size of the input, plays a major role. Indeed, it was the first-ever described quantum algorithm with a practical utility–breaking the RSA public key cryptosystems in a tractable manner.

In an asymmetric cryptosystem such as RSA, information is encrypted via a key that is a solution for a given mathematical function–the decomposition of a given integer into prime factors for the case of RSA. The security of such a protocol is based on the fundamental assumption that no potential eavesdropper has the means to compute this solution efficiently. Shor's algorithm is based on (1) a reduction of the prime factor decomposition problem into the *order-finding* problem and (2) an adequate use of quantum parallelism to perform modular exponentiation of integers over many different inputs in a single row, enabling a polynomial resolution of the *order-finding*. Thus, the computation time for performing the prime decomposition is reduced from exponential to polynomial, and therefore breaks RSA's fundamental assumption. Shor's original article [174] also presents a variation of the order-finding resolution algorithm, solving the discrete logarithm problem with similar performances. Doing so, it brings a procedure for breaking elliptic curve cryptosystems.

Symmetric-key cryptosystems are also challenged by quantum computing [163]. As an example, Simon's quantum algorithm [176] brings an exponential speedup for computing the period of a function (given the promise that this period indeed exists). Several applications in public-key cryptosystems were described, providing exponential gain in, e.g., distinguishing a three-round Feistel construction [119, 163], key recovering in the Evan Mansour encryption scheme [120] and attacking the CBC-MAC message authentication scheme [163].

Finally, Grover search quantum algorithm [84] brings a quadratic speedup in the search for a distinguished element in unstructured databases. Hence, while in this case the complexity gain is less decisive than for the procedures introduced above, its potential for cryptography is significant as it weakens *any* symmetric-key encryption system.

Thus, quantum computing challenges current cryptographic uses and practices. Shor's algorithm opened a research program for cryptographic solutions resisting the power of quantum computation, called *post-quantum cryptography* [39].

Interestingly the induced challenge also received answers from quantum information theory itself. Indeed, one of the major distinctive features of quantum information is that it cannot be read without being affected. This entails that an eavesdropper trying to access a quantum information exchange cannot help betraying her attempt. Based on this feature, one can encode a cryptographic key in a quantum message and, in case of eavesdropping, detect it *a posteriori*, renounce this particular key and try another sending. The study of *Quantum key distribution protocols* is an active research area [164, 22, 129, 135].

7.1.2 QUANTUM COMPUTING AND QUANTUM SOFTWARE

These cryptographic aspects are one of many applications studied in the young research field of quantum computing. Others are, e.g., machine learning [26, 166, 134], optimization [64], solving linear systems [87], etc. In all these domains there are quantum algorithms beating the best known classical algorithms by quadratic or even exponential factors, complexity-wise.

These algorithms are based on laws and phenomena specific to quantum mechanics (such as quantum superposition, entanglement, unitary operations). Therefore, implementing them requires a framework consisting of both a *dedicated hardware* (quantum computers) and a *dedicated software* (quantum programming languages and compilation toolchains).

In the last 20 years, several such languages have been proposed, such as QISKIT [152], LIQ$Ui|\rangle$ [193], Q# [181], Quipper [83], PROJECTQ [180], *etc.* Still, the field is in its infancy, and many questions still need to be answered before we can reach the level of maturity observed for classic programming languages. Standing questions include, for example, introducing a foundational computing model and semantics for quantum programming languages, adequate programming abstractions and type systems, or the ability to interact with severely constrained hardware in an efficient way (optimizing compilers).

7.1.3 VERIFICATION OF QUANTUM PROGRAMS

While testing and debugging are the common verification practice in classical programming, they become extremely complicated in the quantum case. Indeed, debugging and assertion checking are *a priori* very complicated due to the destructive aspects of quantum measurement (see Section 7.2.3.1 below). Moreover, the probabilistic nature of quantum algorithms seriously impedes system-level quantum testing. Finally, classical emulation of quantum algorithms is (strongly believed to be) intractable.

On the other hand, nothing prevents *a priori* the *formal verification* [44] of quantum programs, i.e. *proving* by (more or less) automated reasoning methods that a given quantum program behaves as expected for any input, or at least that it is free from certain classes of bugs.

Interestingly, while formal methods were first developed for the classical case where they are still used with parsimony–mainly for safety-critical domains–as testing remains the main validation methods, their application to quantum computing could become more mainstream, due to the inherent difficulties of testing quantum programs.

7.1.4 GOAL OF THIS CHAPTER

This chapter introduces both the requirements and challenges for formal methods in quantum programs specification and verification, and the existing propositions to overcome these challenges.

The first sections give the general background. In Section 7.2 we introduce the main concepts at stake with quantum computing and quantum algorithms. We provide a state of the art introduction for formal methods, given in Section 7.3. The specific requirements for formal reasoning in the quantum case are then developed in Section 7.4. Then we come to concrete quantum programming and formal verification material. In Section 7.5 we introduce several existing solutions for the formal verification of quantum compilation and the equivalence of quantum program runs. Generating such runs requires specific programming languages. The formal interpretation of quantum languages is introduced in Section 7.6. Then in Section 7.7 we present the main existing solutions for formally verified quantum programming languages. In Section 7.8 we introduce references for further usage of formal methods linked with quantum information, and we conclude this chapter with a discussion in Section 7.9.

7.2 GENERAL BACKGROUND IN QUANTUM COMPUTING

By many aspects, quantum computing constitutes a new paradigm. Making great use of quantum superposition and quantum entanglement, it requires to define proper versions for such fundamental concepts as data structures at stake in computation, or the elementary logical operations at use. We introduce the well-known hybrid quantum computation model in Section 7.2.1.

Quantum computers are not intended to, and will not, replace classical ones. One should better see the opening of a new field, with possibilities to solve new problems. Section 7.2.2 presents these new problems and introduces quantum algorithms design.

As a new software technology, quantum computing comes with specific challenges and difficulties. These specificities are closely related to the particular needs for formal reasoning in quantum computing. They are introduced in Section 7.2.3.

7.2.1 HYBRID COMPUTATIONAL MODEL

Let us first introduce the main concepts at stake in quantum programming. They concern the architecture of quantum computers, the structure of quantum information and quantum programs, and their formal interpretation.

7.2.1.1 Hybrid Circuit Model

The vast majority of quantum algorithms are described within the context of the *quantum co-processor model* [114], i.e. an hybrid model where a *classical* computer controls a *quantum* co-processor holding a quantum memory, as shown in Figure 7.1. In particular, the classical computer performs control operations (**if ... else** statements, loops, etc). The co-processor can apply a fixed set of elementary operations (buffered as *quantum circuits*) to update and query (*measure*) the quantum memory. Importantly, while measurement allows retrieving classical (probabilistic) information from the quantum memory, it also modifies it (*destructive effect*).

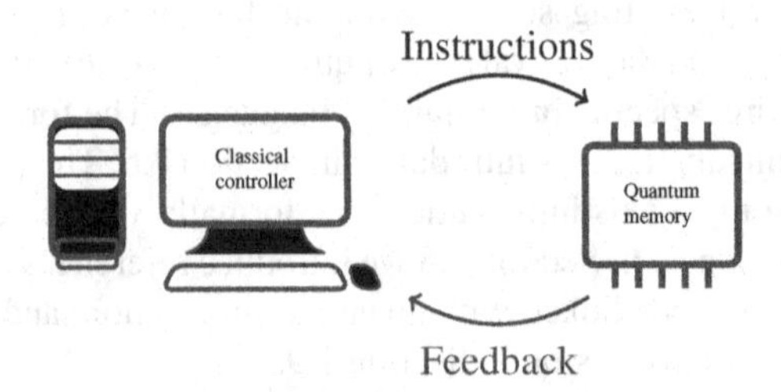

Figure 7.1 Scheme of the hybrid model.

Major *quantum programming languages* such as Quipper [83], LIQ$Ui|\rangle$[193], Q# [181], PROJECTQ [180], SILQ [27], and the rich ecosystem of existing quantum programming frameworks [153] follow this hybrid model.

7.2.1.2 Quantum Data Registers

The following paragraphs introduce several definitions and notations for quantum data registers. In particular, we follow the standard *Dirac notation.* For more details about this content, we refer the reader to the standard literature [140].

7.2.1.2.1 Kets and basis-kets.

While in classical computing the state of a bit is one between two possible states (0 or 1), in quantum computing the state of a *quantum bit* (or *qubit*) is described by *amplitudes* over the two elementary values 0 and 1 (denoted $|0\rangle$ and $|1\rangle$), i.e. linear combinations of vectors $\alpha_0|0\rangle + \alpha_1|1\rangle$ where α_0 and α_1 are any *complex values* satisfying $|\alpha_0|^2 + |\alpha_1|^2 = 1$. In a sense, amplitudes are generalization of probabilities.

More generally, quantum states are defined in complex finite-dimensional Hilbert spaces[1]: the state of a *qubit register* of n qubits (called a *ket* of length n–dimension 2^n) is a column vector with 2^n rows, formed as a *superposition* of the 2^n elementary basis vectors of length n (the "basis

[1]In the finite-dimensional case, Hilbert spaces are vector spaces equipped with an *inner* (scalar) product.

kets"), *i.e.* a ket is any linear combination of the form

$$|u\rangle_n = \sum_{k=0}^{2^n-1} \alpha_k |k\rangle_n \tag{7.1}$$

such that $\sum_{k=0}^{2^n-1} |\alpha_k|^2 = 1$.

7.2.1.2.2 Bit-vectors and basis kets.

Depending on the context, it may be more convenient to index the terms in the sum above with bit vectors instead of integers. We call *bit vector of length n* any sequence $x_0 x_1 \ldots x_{n-1}$ of elements in $\{0,1\}$. Along this chapter, we assume the implicit casting of these values to/from booleans (with the least significant bit on the right). For any positive n, we denote the set of bit vectors of size n by BV_n. We also surcharge notation $|j\rangle_n$ shown above with bit vector inputs. Formally, for any bitvector $\vec{x}$ of length n, $|\vec{x}\rangle_n = |\sum_{i=0}^{n-1} x_i * 2^{n-i-1}\rangle_n$. Hence, one can write state $|u\rangle_n$ from (7.1) as

$$|u\rangle_n = \sum_{\vec{x} \in \mathrm{BV}_n} \alpha_{\sum_{i=0}^{n-1} x_i \cdot 2^{n-i-1})} |\vec{x}\rangle_n$$

It may also be convenient to represent basis kets through their index's binary writing. For example, the two qubits kets basis is equivalently given as $\{|0\rangle_2, |1\rangle_2, |2\rangle_2, |3\rangle_2\}$ or as $\{|00\rangle, |01\rangle, |10\rangle, |11\rangle\}$.

We omit the length index n from notation $|u\rangle_n$ when it is either obvious from the context or irrelevant. We also adopt the implicit convention of writing basis kets with either integer indexes k, i, j or bit-vector $\vec{x}$ and general kets with indexes u, v, w. Hence, in the following $|u\rangle, |v\rangle, |i\rangle$ and $|\vec{x}\rangle$ all designate kets, the last two having the additional characteristics of being basis kets.

When considering two registers of respective size m and n, the state of the compound system lives in the *Kronecker product*[2] –or *tensor product*– of the original state spaces: a general state is then of the form

$$\sum_{\vec{x} \in \mathrm{BV}_m, \vec{y} \in \mathrm{BV}n} \alpha_{\vec{x},\vec{y}} |\vec{x}\rangle_m \otimes |\vec{y}\rangle_n.$$

In particular, the state of a qubit register of n qubits lives in the tensor product of n state-spaces of one single qubit.

[2]Given two matrices A (with r rows and c columns) and B, their Kronecker product is the matrix $A \otimes B = \begin{pmatrix} a_{11}B & \ldots & a_c B \\ \vdots & \ddots & \vdots \\ a_{r1}B & \ldots & a_{rc}B \end{pmatrix}$. This operation is central in quantum information representation. It enjoys a number of useful algebraic properties such as associativity, bilinearity or the equality $(A \otimes B) \cdot (C \otimes D) = (A \cdot C) \otimes (B \cdot D)$, where $\cdot$ denotes matrix multiplication.

7.2.1.2.3 *Adjointness.*

In the following we also use the adjoint transformation for matrices. The adjoint of matrix M with r rows and c columns is the matrix $M^\dagger$, with c rows and r columns and such that for any indexes $j,k \in [\![0,c[\![\times [\![0,r[\![$[3], cell $M^\dagger(j,k)$ holds the conjugate value $M(k,j)^*$ of $M(k,j)$ (for any complex number c, its conjugate c^* is the complex number with the same real part and the opposite imaginary part as c). The adjoint of a ket $|u\rangle_n$ is called a *bra*. It is a row vector with 2^n columns denoted $\langle u|_n$–or simply $\langle u|$–and with indices the conjugates of those of $|u\rangle_n$. This bra-ket notation is particularly convenient for representing operations over vectors. Given a ket $|u\rangle$ and a bra $\langle v|$, $|u\rangle\langle v|$ denotes their Kronecker product –or *outer product*. Furthermore, if $|u\rangle$ and $\langle v|$ have the same length, then $\langle v|u\rangle$ denotes their scalar product–also called *inner product*. In particular, in the case of basis states $|i\rangle$ and $\langle j|$, $\langle i|j\rangle = 1$ if $i = j$ and 0 otherwise and $|i\rangle\langle j|$ is the square matrix of width 2^n with null coefficient everywhere except for cell (i,j) with coefficient 1. If $i = j$, then $|i\rangle\langle j|$ operates as the projector upon $|i\rangle$.

7.2.1.2.4 *Quantum measurement and Born rule.*

The probabilistic law for measurement of kets is given by the so-called *Born rule*: for any $k \in [\![0,2^n[\![$, measuring state $|u\rangle_n$ from Equation 7.1 results in k with probability $|\alpha_k|^2$. The measurement is destructive: if the result were k, the state of the register is now $|k\rangle_n$ (with amplitude 1).

7.2.1.3 Separable and Entangled States

From Section 7.2.1.2, a quantum state vector of length n is a superposition of basis elements with coefficients whose squared moduli sum to one. Then, tensoring n quantum states $|u_j\rangle_1$ of length 1 results in a state $|u\rangle_n = \bigotimes_{j\in[\![0,n[\![} |u_j\rangle_1$ of length n. One can decompose back $|u\rangle_n$ into the family $\{|u_j\rangle\}_{j\in[\![0,n[\![}$: we say that $|u\rangle_n$ is a *separable* state. Note that the structure of quantum information introduced above contains states missing the property of being separable. As an example, the state

$$|\beta_{00}\rangle = \frac{1}{\sqrt{2}}(|00\rangle + |11\rangle)$$

cannot be written as a tensor product of two single-qubit states. This phenomenon is called *entanglement*, and $|\beta_{00}\rangle$ is an *entangled state*. It induces

[3] where, for any two integers i,j with $i < j$, $[\![i,j[\![$ denote the induced interval, that is the set of integers k such that $i \leq k < j$. Similarly, we use notation $[\cdot]$ for closed intervals of continuous values.

that one can store more quantum information in n qubits altogether than separately.

Example 7.2.1 (Bell states). *State* $|\beta_{00}\rangle$ *is a construction of particular interest in quantum mechanics and quantum computing. In their famous 1935 article [62], Einstein, Podolsky and Rosen argued for the incompleteness of quantum mechanics, based on considerations upon* $|\beta_{00}\rangle$. *In 1964 [21], J.S. Bell proposed an experiment to test the argument. It was based on statistics over experiments on the four following states:*

$$\begin{aligned} |\beta_{00}\rangle &= \tfrac{1}{\sqrt{2}}(|00\rangle_2 + |11\rangle_2) & \qquad |\beta_{01}\rangle &= \tfrac{1}{\sqrt{2}}(|01\rangle_2 + |10\rangle_2) \\ |\beta_{10}\rangle &= \tfrac{1}{\sqrt{2}}(|00\rangle_2 - |11\rangle_2) & \qquad |\beta_{11}\rangle &= \tfrac{1}{\sqrt{2}}(|01\rangle_2 - |10\rangle_2) \end{aligned}$$

These four states are now known as the Bell states *(notation* β *stands for the initial B) and are used in many quantum protocols, such as teleportation or superdense coding (see Section 7.5.1). We use them and their generation as a running example in the rest of this chapter.*

7.2.1.4 Quantum Circuits

Three kinds of operations may be applied to quantum memory, exemplified in Figure 7.2 with the circuit generating and measuring Bell states:

- the *initialization* phase allocates and initializes quantum registers (arrays of qubits) from classical data. In Figure 7.2 it is represented on the two first qubit wires as $\vdash$, indexed by value i_w. It *creates* a quantum register in one of the basis states (that is, in the case of a two-qubit register, in one of the four basis states $|00\rangle, |01\rangle, |10\rangle, |11\rangle$),
- the actual quantum computing part consists in transforming an initialized state. This is performed by applying a sequence of proper quantum operations, structured in a so-called *quantum circuit*. In Figure 7.2 this part is identified with a dashed box (itself sequenced with dotted boxes a. and b.),
- the extraction of useful information from a quantum computation is performed through the *measurement* operation, by which one probabilistically gets classical data from the quantum memory register. Measurement is represented, on each qubit it is applied to, as [meter symbol].
- In generalized circuits, not all wires in a register need to be initialized and measured. Hence in Figure 7.2 the transformations are performed over the two first wires of a wider register, and the additional wires are left untouched.

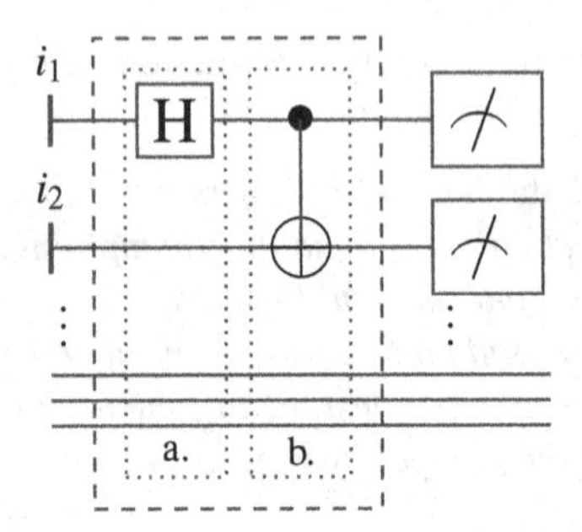

Figure 7.2 Generalized circuit to create and measure Bell states.

Note that we reserve the term circuit *for the pure quantum part (the dashed box in Figure 7.2). We call* generalized circuit *a process made of a circuit together with, possibly, initialization and measurements.*

Quantum circuits are built by *combining*, either in *sequence* or in *parallel*, a given set of elementary operations called *quantum gates*. In addition to sequence and parallelism, derived circuit combinators (controls, reversion, ancillas, etc) are often used in quantum circuit design (See Figure 7.9 in Section 7.6.1.1 for details). The circuit part of Figure 7.2 uses two different quantum gates, drawn in dotted boxes:

- the Hadamard gate H (a.), which induces a state superposition on a given qubit,
- the *control not* gate, often written *CNOT* (b.) and represented as $\oplus$. It is a binary gate, flipping the *target qubit* (in wire 2 in our case) when the *control qubit* (wire 1) has value 1.

7.2.1.5 Quantum Matrix Semantics and Density Operators

The transformation operated by a quantum circuit on a quantum register is commonly interpreted as a matrix. In this setting, the parallel combination of circuits is interpreted by the *Kronecker product* and the sequential combination by the *matrix multiplication*.

Quantum circuits.

Quantum circuits happen to operate as *unitary* operators (preserving the inner product between vectors). A set of elementary gates is (pseudo-) *universal* if, by a combination of parallel and sequential composition, one can synthesize (or approximate) all *unitary* operations. Examples for elementary gates are given in Table 7.1, with their matrix semantics interpretation. Apart from the already encountered gates H and $CNOT$, it features two additional families of gates, $Ph(\theta)$ and $R_Z(\theta)$, where θ is an angle inducing a so-called *phase factor* $e^{i\theta}$. $Ph(\theta)$ operates a simple scalar multiplication by a phase factor, while $R_Z(\theta)$ operates as a rotation. Table 7.1 is given with indexes ranging over any angle θ, making the set of gates universal. Usually, we restrict it to angles of measure $\frac{\pi}{2^n}$, with n ranging over integers. This restriction makes the resulting set of gates pseudo-universal.

Table 7.1
Elementary gates and their matrix semantics.

H	$CNOT$	$Ph(\theta)$	$R_Z(\theta)$
$\frac{1}{\sqrt{2}}\begin{pmatrix}1 & 1\\ 1 & -1\end{pmatrix}$	$\begin{pmatrix}1&0&0&0\\0&1&0&0\\0&0&0&1\\0&0&1&0\end{pmatrix}$	$\begin{pmatrix}e^{i\theta} & 0\\ 0 & e^{i\theta}\end{pmatrix}$	$\begin{pmatrix}e^{-i\theta} & 0\\ 0 & e^{i\theta}\end{pmatrix}$

Example 7.2.2 (Semantics for the Bell generating circuit). *Let us look at Figure 7.2 again. First, an Hadamard gate is applied to the first wire and nothing happens to the second wire (it stays untouched, which is represented by the identity matrix). The matrix for the first* column–*the dotted box indexed with a.–of Figure 7.2 is*

$$\frac{1}{\sqrt{2}}\begin{pmatrix}1 & 1\\ 1 & -1\end{pmatrix}\otimes\begin{pmatrix}1 & 0\\ 0 & 1\end{pmatrix}=\frac{1}{\sqrt{2}}\begin{pmatrix}1&0&1&0\\0&1&0&1\\1&0&-1&0\\0&1&0&-1\end{pmatrix}$$

Then gate CNOT is applied, with matrix $\begin{pmatrix}1&0&0&0\\0&1&0&0\\0&0&0&1\\0&0&1&0\end{pmatrix}$ *and the sequential combination of the two subcircuits translates, in the matrix semantics, as their usual product (mind the reverse ordering, wrt the figure):*

$$\boldsymbol{Mat}(\textit{Bell-circuit})=\begin{pmatrix}1&0&0&0\\0&1&0&0\\0&0&0&1\\0&0&1&0\end{pmatrix}\cdot\frac{1}{\sqrt{2}}\begin{pmatrix}1&0&1&0\\0&1&0&1\\1&0&-1&0\\0&1&0&-1\end{pmatrix}=\frac{1}{\sqrt{2}}\begin{pmatrix}1&0&1&0\\0&1&0&1\\0&1&0&-1\\1&0&-1&0\end{pmatrix}$$

Application to input initialized kets.

In the matrix formalism, we interpret $|0\rangle$ as the column vector $\binom{1}{0}$, $|1\rangle$ as $\binom{0}{1}$ and the concatenation $|ij\rangle$, where i and j both are sequences of 0 or 1, as the Kronecker product $|i\rangle \otimes |j\rangle$. For example, the two qubits basis kets $|00\rangle, |01\rangle, |10\rangle, |11\rangle$ are represented, respectively, as

$$\begin{pmatrix}1\\0\\0\\0\end{pmatrix}, \begin{pmatrix}0\\1\\0\\0\end{pmatrix}, \begin{pmatrix}0\\0\\1\\0\end{pmatrix} \text{ and } \begin{pmatrix}0\\0\\0\\1\end{pmatrix}.$$

The transformation performed by a quantum circuit C upon a quantum state $|\psi\rangle$ is interpreted as the matrix product $\mathbf{Mat}(C) \cdot |\psi\rangle$ of the matrix for this circuit by the column vector for this quantum state. By notation abuse, we also simply write it $C|\psi\rangle$.

Example 7.2.3. *One can now directly verify that the Bell generating circuit from Figure 7.2 generates the Bell states from Example 7.2.1: for any $a, b \in \{0,1\}$,*

$$\textit{Bell-circuit} \cdot |ab\rangle = |\beta_{ab}\rangle$$

Measurement.

Last, measurement is performed over an orthonormal basis of the Hilbert space. For sake of simplicity, we only consider the case of measurements in the computational basis. Hence, measuring a quantum register results in a basis state, with probabilities following the Born rule introduced in Section 7.2.1.2: measuring any state $|u\rangle_n$ results in basis state $|k\rangle_n$ with probability (written $\texttt{proba\_measure}(|u\rangle_n, |k\rangle_n)$) $|\alpha_k|^2$, where α_k is the amplitude of $|k\rangle_n$ in $|u\rangle_n$. Applying this rule to the Bell state, one easily state that for any $a, b, i, j \in \{0,1\}$,

$$\texttt{proba\_measure}(|\beta_{ab}\rangle, |ij\rangle) = \frac{1}{2}(\texttt{if } b=0 \texttt{ then } i \oplus j \texttt{ else } 1-(i \oplus j))$$

where $\oplus$ denotes addition modulo 2. Note, from Example 7.2.1, that index a in notation $|\beta_{ab}\rangle$ only accounts for a -1 factor in the second term of the state superposition. Hence, since measurement is ruled by the Born rule and since this rule ignores negation (see Section 7.2.1.2), then index a does not appear in the expression of $\texttt{proba\_measure}(|\beta_{ab}\rangle, |ij\rangle)$.

Discussion over the matrix semantics.

Matrix semantics is the usual standard formalism for quantum computing (see [140] for example). Still, the size of matrices grows exponentially with the width (number of qubits) of circuits, so it is often cumbersome when addressing circuits from non-trivial algorithm instances. Furthermore, algorithms usually manipulate parametrized families of circuits. The resulting parametrized families of matrices may not be conveniently writable.

Hence, a more compact interpretation for quantum circuits may be helpful. In particular, path-sum semantics [5, 4] directly interprets quantum circuits by the input/output function they induce over kets–corresponding, in matrix terms, for any circuit C of width n, to the function $|u\rangle_n \mapsto \mathbf{Mat}(C) \cdot |u\rangle_n$. To do so, it exhibits a generic form for quantum registers description, which is generated by a restricted number of parameters and composes nicely with sequence and parallel compositions. Path-sum semantics plays a growing role in formal specification and verification. It is introduced with further details in Section 7.5.2.

Density operators.

In the preceding paragraph, we described measurement as a non-deterministic operation over quantum states. Another strategy consists in dealing with a notion of states featuring this non-determinism. A *mixed state* (as opposed to a *pure state*) is a probability distribution over several states. Alternatively, it can be seen as an incomplete description of a state, featuring the incomplete knowledge one may have about it. Then, measurement can be characterized as a simple transition between mixed states.

In quantum processes, this view is formalized by density operators, that extends matrices formalism with the characterization of probabilistic states. For sake of brevity, in this paragraph we give only a short introduction to the density operator formalism. Our aim here is only to provide the required definitions and notations for this review. For further detail about density operators and for the related soundness proofs, we refer the interested reader either to [168] or [140] (Section 2.4).

Basically, the density operator for a pure state $|x\rangle$ is the reflexive outer product $|x\rangle\langle x|$. Given a set S of indices and a distribution of states $\{|x_k\rangle\}_{k\in S}$, each occurring with probability p_k, we represent the overall mixed state as the density operator

$$\rho := \sum_{k\in S} p_k |x_k\rangle\langle x_k|$$

By linearity, the result of applying a unitary U to $|x\rangle\langle x|$ is given by the product $U|x\rangle\langle x|U^\dagger$. A measurement of a quantum register q of size n may be described by the collection of possible projectors it realizes, that is the set $M = \{M_k := |k\rangle_n \langle k|_n\}_{k\in [\![0,2^n-1[\![}$. In the density operators formalism, the action of M over a state ρ may result in any state $M_k \rho M_k^\dagger$, with probability $tr(M_k M_k^\dagger \rho)$, where the trace $tr(M)$ of a square matrix M with n rows and columns is defined as the sum $\sum_{j\in [\![0,n[\![} M(j,j)$ of its diagonal cell values.

Then, the overall action of a measurement M over a density operator ρ results in $\rho' = \sum_{k\in[\![0,2^n-1[\![} M_k\rho M_k^\dagger$.

Now, measurement description generalizes to the case of *partial measurements*, where only a sub-register is measured. Let us consider the case of a quantum register q of size $n = n_1 + n_2$. We write q_1 and q_2 for the concatenated sub-registers and H, H_1, H_2 for the respectively induced Hilbert spaces. To simplify the notations we consider the case of measuring the last n_2 qubits. For any density operator ρ, if it is separable as $\rho = |x_1\rangle_{n_1} \langle y_1|_{n_1} \otimes |x_2\rangle_{n_2} \langle y_2|_{n_2}$, then the partial trace of ρ over H_2 is defined as $tr_2(\rho) = \langle y_2|x_2\rangle|x_1\rangle\langle y_1|$ and the definition generalizes by linearity to any density operator ρ. Then, $tr_2(\rho)$ equivalently represents the result of:

- (1) measuring register q_2 from the mixed state ρ and (2) forgetting the measured qubits while conserving memory of the unmeasured subregister q_1 state;
- or just forgetting about (*discarding*) register q_2 in the description of ρ. Then, $tr_2(\rho)$ is the description of the sub-system held by q_1. We call $tr_2(\rho)$ a *partial density operator* on H_1.

7.2.1.6 Other Models for Quantum Computations

Many alternatives are currently explored for physical implementations of quantum computing machines and worth mentioning. Some of them (such as Measurement-Based Quantum Computing [157, 32], topological quantum computations [76], linear optical networks [1], adiabatic quantum computing [65], *etc.*) differing on rather fundamental aspects (like, e.g., the elementary operations constituting computations). Nevertheless, currently, formal methods developments mainly address the standard circuit model introduced above.

The ZX-Calculus [47] also provides an alternative graphical formalism to reason about quantum processes. Basically, in this setting, quantum operations are represented by diagrams and their composition through sequence or parallelism corresponds to graphical compositions in the calculus. This language comes with a series of enabled transformations over graphs, preserving computational equivalence. ZX-Calculus is presented in Section 7.5.1.

7.2.2 ALGORITHMS

As previously introduced, quantum computers are meant to perform calculations that classical computers are *a priori* not able to perform in a *reasonable* time. We give the formal complexity theory characterization for this point in Section 7.2.2.1, then Section 7.2.2.2 discusses the usual conventions for quantum algorithms descriptions.

7.2.2.1 Quantum Algorithms and Complexity

It is commonly assumed that formal problems are *tractable* by a computer if there exists an algorithm to solve this problem in time (measured by the number of elementary operations it requires) that is bounded by a polynomial over the size of the input parameters. Formal problems satisfying this criterion for classical computers form a *complexity class* usually referred to as **P**. It is schematically represented in Figure 7.3.

As introduced in Section 7.2.1.2, extracting useful information from a quantum register requires a measurement, ruled by the Born law. Therefore, a quantum computation is an alternation of non-deterministic (measurement) and deterministic (circuit unitary application, classical post-treatment, etc) operations. Since such computations are probabilistic, the tractability criterion from above needs to be slightly adapted. Instead of considering problems for which a polynomial algorithm brings a solution with certainty, we consider those for which a polynomial algorithm brings a solution with an error probability of at most $\frac{1}{3}$. The corresponding class of problems for quantum computers is called *bounded error quantum polynomial time* (**BQP**).

In addition to **P** and **BQP**, Figure 7.3 represents the *non-deterministic polynomial time* class **NP**. It gathers formal problems $\mathscr{P}$ for which there is an algorithm that, given a candidate solution, checks whether this candidate is an actual solution for $\mathscr{P}$ in polynomial time. It is trivial that **P** is included in **NP** and it is also proved that $\mathbf{P} \subseteq \mathbf{BQP}$. There are good reasons to believe that these inclusions are strict. Nevertheless, strictness is not formally proved and there are a variety of problems that belong to **NP** without a known tractable resolution algorithm.

Hence, quantum algorithm performance is not to be evaluated against the best *possible* performance of any classical computation (which depends on whether $\mathbf{BQP} = \mathbf{P}$) but, more pragmatically, against the best classical *known* equivalent.

Then, quantum computing is relevant for problems that are polynomially solvable by a quantum computer (with a given probability of success) but *intractable* by a classical one. They appear in the gray zone in Figure 7.3. The question whether the dark gray part ($\mathbf{BQP} \setminus \mathbf{NP}$) is empty or not depends on whether $\mathbf{BQP} \subseteq \mathbf{NP}$, which is unknown, but several **BQP**-complete problems have been described through the literature [200]. These are neither easily computable nor verifiable with known classical means,

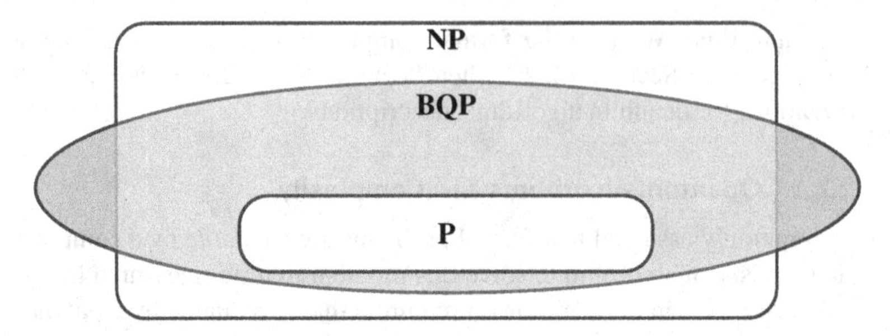

Figure 7.3 Complexity classes P, NP, BQP.

but may be computed with quantum means: since quantum computers output a right solution for them with probability $> \frac{2}{3}$, after several runs one can select the best represented output as the sought solution.

Let us stress that the correspondence between polynomial solvability and tractability is not strict. For a problem, not belonging to a polynomial class bounds the size of input parameters concretely tractable by a computer, but does not absolutely forbid computation for any instance of it. Hence, for some problems, the quantum advantage does not consist in providing a polynomial resolution, but in reducing computation time to extend the set of tractable inputs. A typical example is the Grover search algorithm [84], searching for a distinguished element in an unstructured data set, providing a quadratic acceleration against classical procedures.

7.2.2.2 Quantum Algorithm Design

Before introducing the challenges at stake with implementations of quantum algorithms and their formal solutions, we make a few observations on the usual format used to describe quantum algorithms, based on an example: Figure 7.4 reproduces the core quantum part for Shor's algorithm [140, p. 232] – certainly the most emblematic of all quantum algorithms.

The following observations generally hold for other quantum algorithms in the literature. We give them together with illustrations (within parenthesis) from the example of Figure 7.4.

1. The algorithm is structured in two main parts: a specification preamble and a **Procedure** description. The preamble indicates the minimal specification an implementation should satisfy. Note that it figures as an actual part of the algorithm itself. It contains three types of entries:
    - a description of the **Inputs**, giving a signature for the parameters (a black-box circuit U, integers x, N, L and two

Inputs: (1) A black-box $U_{x,N}$ which performs the transformation $|j\rangle|k\rangle \rightarrow |j\rangle|x^j k \bmod N\rangle$, for x co-prime to the L–bit number N,
(2) $t = 2L+1+\lceil\log(2+\frac{1}{2\varepsilon})\rceil$ qubits initialized to $|0\rangle$, and
(3) L qubits initialized to the state $|1\rangle$.

Outputs: The least integer $r > 0$ such that $x^r = 1 \pmod N$.

Runtime: $O(L^3)$ operations. Succeeds with probability $O(1)$.

Procedure:

1.	$|0\rangle|u\rangle$	initial state
2.	$\rightarrow \frac{1}{\sqrt{2^t}}\sum_{j=0}^{2^t-1}|j\rangle|1\rangle$	create superposition
3.	$\rightarrow \frac{1}{\sqrt{2^t}}\sum_{j=0}^{2^t-1}|j\rangle|x^j \bmod N\rangle$	apply $U_{x,N}$
	$\approx \frac{1}{\sqrt{r2^t}}\sum_{s=0}^{r-1}\sum_{j=0}^{2^t-1}e^{2\pi isj/r}|j\rangle|u_s\rangle$	
4.	$\rightarrow \frac{1}{\sqrt{r}}\sum_{s=0}^{r-1}\widetilde{|s/r\rangle}|u_s\rangle$	apply inverse Fourier transform to the first register
5.	$\rightarrow \widetilde{|s/r\rangle}$	measure first register
6.	$\rightarrow r$	apply continued fraction algorithm

Figure 7.4 Bird-eye view of the circuit for Shor's factoring algorithm [174] (as presented in [140, p. 232]).

quantum registers of sizes t and L) and some preconditions for these elements (e.g. x co-prime to L, L being N bits long, *etc*);

- a description of the **Outputs** of the algorithm. It contains, again, a signature (an integer) and a success condition for these **Outputs** (to be equal to the sought modular order);
- a **Runtime** specification, containing: (1) a probability of success for each run of the **Procedure** ($O(1)$), (2) resources specifications. In the example, the latter consists in bounding the number of required elementary operations. Further metrics are also often used (the maximal *width* of required quantum circuits–the number of qubits a circuits requires, the maximal depth of a circuit–the maximal number of operations performed on a given qubit, *etc*).

2. The **Procedure** itself consists of a sequence of declared operations, interspersed with formal descriptions of the state of the system along with the performance of these operations. (in Figure 7.4 these elements are given in parallel, declarations of operations constitute the right-hand side and intermediate formal assertions are on the left-hand side). These assertions serve as specifications for the declared operations. For instance, operation "create superposition" has precondition the formal expression of Line 1, left (framed in blue) and postcondition the one of Line 2, left (framed in red). They serve as arguments to convince the reader that the algorithm **Outputs** conditions are met at the end of the **Procedure** (notice that the ultimate such postcondition–the measured state being r– corresponds to the success condition for the overall algorithm). But we can also interpret them as *contracts* for the programmer, committing her to implement each function in any way provided that whenever its inputs satisfy the preconditions, then its outputs satisfy the postconditions.
3. The algorithm description is parametric, and so should be any program implementing it. Hence, the quantum programming paradigm is higher-order: a quantum program is a function from (classical data) input parameters to quantum circuits. Then, each instance of a quantum circuit behaves as a function from its (quantum data) inputs to its (quantum data) outputs.

Most of the current quantum programs [152, 83, 193, 181, 180, 27, 41] proceed as implementations for functions such as those declared in Figure 7.4 (right part of the **Procedure** part), providing no guarantee over the algorithm specifications, be it about their functional behavior or their resource specifications.

Based on the preceding comments, the purpose of formal verification can be summarized as completing algorithm implementations with their proved specifications. In other words:

Formal verification in quantum programming aims at providing solutions to furnish, in addition to quantum programs, evidence that these programs meet their specifications, in terms of both success probability and resource usage.

7.2.3 CHALLENGES FOR QUANTUM COMPUTATION

Let us now introduce some particularities of quantum programming with regards to classical computing. They raise design challenges that are particular to this programming paradigm.

7.2.3.1 Destructive Measurement and Non-Determinism

One of the main particularities of quantum programming is that the output produced by the quantum memory device follows the probabilistic Born rule (see Section 7.2.1.2). So, in the general case, the result of quantum computation is non-deterministic.

Furthermore, a computation in the model from Figure 7.1 contains both probabilistic quantum computations and classical control structures, performed by the classical controller. Hence, control itself may depend on the probabilistic data received from the quantum device and the execution flow itself is probabilistic.

7.2.3.2 Quantum Noise

Another particularity comes from the difficulty to maintain big quantum systems in a given state and to control the evolution of this state through time. Along with a quantum computation, uncontrolled modifications (bit or phase flip, amplitude damping, *etc.*) of the quantum state may occur.

To overcome this phenomenon, one solution consists in integrating error correction mechanisms into the compilation. Error correction design is an active research field [40, 130, 81, 75]. Many propositions have been developed. They mainly consist in designing redundant quantum circuits (one logical qubit is implemented by many different physical qubits). Then the main challenge is to design a solution for testing the reliability along with a computation without losing the state of the register due to destructive measurement.

Since error correction requires bigger quantum registers (due to redundancy), its possible implementation is conditioned by the design, elaboration and availability of large quantum processors.

An alternative strategy is to not correct quantum errors, but design computations to limit their effect. John Preskill introduced the notion of *Noisy-Intermediate Scale Quantum (NISQ)* technologies [151]. The formal analysis of error propagation requires the identification of possible errors together with rules specifying how their probability of occurrences propagates along with quantum circuits [97, 155].

7.2.3.3 Efficient Compilation on Constrained Hardware

Languages such as LIQ$Ui|\rangle$, Q#, Quipper, *etc.* enable the description and building of quantum circuits for so-called *logical qubits*. In practice, realizing a quantum circuit on an actual quantum machine (*physical qubits*) requires several compilation passes, in addition to the error correction mentioned in the preceding paragraph. Among others:

- the physical realization should respect the physical constraints of its target architecture, which concerns, e.g. connectivity of qubits or register size limits. Considering this point requires qubit reordering intermediary operations and an adequate mapping between theoretical and physical qubits;
- the set of possible quantum operations over physical qubits may not correspond to the set of elementary gates from the logical circuit description, which would require an adequate gate synthesis and circuit rewriting;
- last but not least, physical realization of an algorithm should be as resource frugal as possible, requiring the development of circuit optimization techniques.

Each of these steps consists of low-level operations over circuits. They must all preserve functional equivalence while reaching their proper purpose. For these low-level developing layers, one requires tools and languages formalizing functionally equivalent circuit transformation operations. ZX-Calculus [47] (Section 7.5.1) is particularly well-fitted for such design, other propositions include the proof of path-sum semantics equivalence [5, 4] (Section 7.5.2) and formally verified circuit optimization [91] (Section 7.5.4).

7.3 GENERAL BACKGROUND ON FORMAL METHODS

We now present a brief overview of formal methods. While the domain is old and has led to rich literature, we try to highlight the underlying main principles and to quickly describe the most popular classes of techniques so far.

7.3.1 INTRODUCTION

Formal methods and formal verification [44] denote a wide range of techniques aiming at *proving* the correctness of a system with a *mathematical guarantee*—reasoning over *all* possible inputs and paths of the system, with methods drawn from logic, automated reasoning and program analysis. The last two decades have seen an extraordinary blooming of the field, with significant case studies ranging from pure mathematics [80] to complete software architectures [113, 125] and industrial systems [20, 108]. In addition to offering an alternative to testing, formal verification has in principle the decisive additional advantages to both enable parametric proof certificates and offer once-for-all absolute guarantees for the correctness of programs.

7.3.2 PRINCIPLES

Formal methods' main principles were laid mostly in the 1970s. Pioneers include Floyd [74], Hoare [94], Dijkstra [59], Cousot [49] and Clark [42]. While there is a wide diversity of approaches in the field, any formal method builds upon the following three key ingredients:

- a formal semantics M representing the possible behaviors of a system or program–M is typically equipped with an operational semantics and behaviors are often represented as a set of traces $L(M)$;
- a formal specification φ of the acceptable or correct behaviors–φ is typically a logical formula or an automaton representing a set of traces L_φ;
- a (semi-)decision procedure verifying that possible behaviors are indeed correct, denoted $M \models \varphi$–typically a semi-algorithm to check whether $L(M) \subseteq L_\varphi$ holds or not.

Regarding the complexity of realistic systems and programs, the verification problem is usually undecidable, hence the impossibility to have a fully automated and perfectly precise decision procedure for it. The different formal method communities bring different responses to get around this fundamental limitation, yielding different trade-offs in the design space, favoring either restriction of the classes of systems under analysis, restrictions of the classes of properties, human guidance or one-sided answer (over-approximations or under-approximations).

Overall, after two decades of maturation, formal methods have made enough progress to be successfully applied to (mostly safety-critical) software [89, 191, 12, 31, 50, 117, 108].

7.3.3 THE FORMAL METHOD ZOO

We present now in more detail the main classes of formal methods. While recent techniques tend to blur the lines and combine aspects from several main approaches, this classification is still useful to understand the tradeoff at stake in the field.

- **Type checking and unification:** at the crossroad of programming language design and formal methods, type systems [149] allow forbidding by design certain classes of errors or bad code patterns (such as trying to add together a number and a boolean in Java or, in a quantum setting, trying to apply a unitary operation over a classical data register). Traditionally, type systems focus on simple "well-formedness" properties (good typing), but they scale very well (modular reasoning) and require only a little manual annotation effort (type inference). While first type systems were based on basic unification [93, 137], advanced type systems with dependent types or flow-sensitivity come closer and closer to full-fledged verification techniques;
- Model checking and its many variants: while initially focused on finite-state systems [42] (typically, idealized protocols or hardware models) and complex temporal properties–with essentially graph-based and automata-based decision procedures, model checking [43] has notably evolved along the year to cope with infinite-state systems, either through specific decidable classes (e.g., Petri Nets or Timed Automata) or through abstractions. The current approaches to software model checking include notably symbolic bounded verification [117, 35] for bug finding and counter-example guided abstraction refinement [89] for bug finding and proof of invariants (but it may loop forever). Usually, model checking relies on specifications expressed in a variant of modal logic such as temporal logic [42, 150], dynamic logic [86, 123, 72] or mu-calculus [115, 167];
- **Abstract Interpretation:** Generally speaking, Abstract Interpretation [49] is a general theory of abstraction for fixpoint computations. Abstract Interpretation-based static analysis [161] builds over Abstract Interpretation to effectively compute sound (i.e. overapproximated) abstractions of all reachable states of a program. Hence, these techniques are well suited for proving invariants. More precisely, Abstract Interpretation provides a systematic recipe to design sound abstract computation over sets of program states, by connecting the concrete domain (e.g., sets of states) to a given abstract domain (e.g., interval constraints) through a Galois Connection between an abstraction and a concretization functions. In practice, Abstract Interpretation comes down to computing the fixpoint over the abstract domain, ensuring termination but losing precision. Different abstract domains provide different trade-offs between cost and precision. Historically

speaking, the approach targets full automation and implicit properties (e.g., runtime error);

- **Deductive verification and first-order reasoning:** Deductive program verification [17, 71, 94, 182] is probably the oldest formal method technique, dating back to 1969 [94] and the development of Hoare logic. In this approach, programs are annotated with logical assertions, such as pre- and postconditions for operations or loop invariants, then so-called proof obligations are automatically generated (e.g., by the weakest precondition algorithm) in such a way that proving (a.k.a. discharging) them ensures that the logical assertions hold along any execution of the program. These proof obligations are commonly expressed in first-order or separation logic [158] and proven by the help of proof assistants [145, 141] or automatic solvers lying on Satisfiability Modulo Theory [18] or Automated Theorem Proving [73];
- **Interactive proof and second-order reasoning:** some techniques completely drop the hope for automation in favor of expressivity, relying on 2nd order or even higher-order specification and proofs languages–typically in Coq [145] or ISABELLE/HOL [141]. Once programmed and proved in the language, a certified functional program can then often be extracted. This family of approaches is very versatile and almost any problem or specification can be encoded, yet it requires lots of manual effort, both for the specification and the proofs– higher-order proofs can be automatically checked but not found. Still, the technique has been for example used for certified compilers or operating systems [125, 113].

7.4 OVERVIEW OF FORMAL METHODS IN QUANTUM COMPUTING

As presented in Section 7.3, formal methods for proving properties of classical algorithms, programs and systems are well-developed and versatile. In this section, we present the needs for formal methods in the realm of quantum computation, while the later sections are devoted to answering them.

7.4.1 THE NEED FOR FORMAL METHODS IN QUANTUM COMPUTING

As introduced in Section 7.2, the data structures at stake in quantum computing make the computations hard to represent for developers.

Intermediary formal languages are of great help for understanding what quantum programs do and describing their functional behavior.

Furthermore, as introduced in Section 7.1, directly importing the testing and debugging practices at use in classical programming is extremely difficult in the quantum case[4], due to the destructive aspect of quantum measurement. Moreover, the probabilistic nature of quantum algorithms seriously impedes system-level quantum testing. As a consequence, test-based programming strategies do not seem adequate in the quantum case and quantum computing needs alternative debugging strategies and methodologies.

So far, existing quantum processors were small enough that their behavior could be entirely simulated on a classical device. Hence, a short-term solution for overcoming the debugging challenge relied on classical simulations of quantum programs. Since quantum computer prototypes are now reaching the size limit over which this simulation will not be possible anymore (among others, [9] is often referred to as the milestone for this context change, referred to as *quantum supremacy*), more robust solutions must be developed.

On the other hand, nothing prevents *a priori* the formal verification of quantum programs. In addition to constituting alternative debugging strategies, formal methods have several additional decisive advantages. In particular:

- They enable parametrized reasoning and certification in the higher-order quantum programming context introduced in Section 7.2.2.2: formal certification of a parametrized program holds for any circuit generated by this program, whatever the value of its parameters. In contrast, testing certification holds for the particular values of these parameters that are used in a test. Formal certification is not limited by the size of the parameters;
- It provides once for all an absolute, mathematically proven, certification of a program's specifications, whereas testing furnishes at best only statistical arguments based on bounded-size input samples.

7.4.2 TYPOLOGY OF PROPERTIES TO VERIFY

In this section we detail the different properties one has to mind for developing correct quantum programs. The goal of formal certification is to provide solutions for their verification.

[4] It requires major adaptations and redefinitions, see Section 7.8.4 for details.

7.4.2.1 Functional Specifications

A major challenge is to give assurance on the input/output relationship computed by a given program, that is verifying whether a given program implements an intended function f. Functional specifications are two-layered:

High-level specifications.

We give a circuit an overall specification independent from the concrete implementation. It is made of a success condition and a minimum probability, for any run of this circuit, to result in an output satisfying the success condition.

In the hybrid model, circuits are run on a quantum co-processor but controlled by a classical computer, performing control operations (such as **if** and **while** instructions, simple sequence, *etc*). As a simple example, an algorithm such as the one from Figure 7.4 outputs a success with probability p. It can be included in a control structure including k iterations of it. This higher-level procedure has probability $[1-(1-p)^k]$ to output a success at least once, which can be made arbitrarily close to 1.

More generally, a high-level quantum verification framework [132, 194] considers an algorithm as a controlled sequence of quantum functions. There, one considers quantum operations as primitives and composes them together via controlled sequence operations. These operations are interpreted as functions in the semantical formalism. For example, [168, 132, 194] formalize quantum programs in the *density operators formalism.* This view is introduced in detail in Section 7.7.1 together with the Quantum Hoare Logic (QHL) [132, 194].

Intermediate specifications.

In the **Procedure** section from Figure 7.4, each mid-level step of the algorithm is given a formal specification, that is a description of the state of the system. These intermediate specifications are deterministic and concern quantum data.

In a lower-level verification approach view, instead of inputting quantum operations as primitive functions, one builds quantum circuit implementations of these operations, by adequately combining quantum gates. Such a framework [91, 156, 38] relies on a circuit description language such as Quipper or QWIRE. Then, an adequate semantics characterization for the built circuits enables to reason about the quantum data received as inputs and delivered as outputs. A certification solution for quantum circuits enables us to reason compositionally about their semantics. This programming view is explored in Sections 7.7.2 and 7.7.3.

7.4.2.2 Complexity Specifications

The major reason for developing quantum computers and quantum algorithms is to lower computing complexity specifications, *w.r.t.* classical computing solutions (see Section 7.2.2.1 for precisions). Therefore, the relevance of a quantum implementation relies on the fact it satisfies low complexity specifications. As introduced in Section 7.2.2.2, they may be formulated through different metrics, such as the width and/or depth of quantum circuits, their number of elementary gates or more complex metrics such as *quantum volume* [126].

The complexity specification is also crucial for another reason: remind from Section 7.2.3.2 that quantum computation is subject to noise: the bigger a quantum circuit is, the most prone to error it is. Functional specifications introduced so far reason about the theoretical output of quantum computations, in the absence of errors. The risk of error in a circuit is closely related to the structural characteristics of this circuit, among which are the different measures of complexity. Therefore, the information provided by these measures is also crucial to appreciate the functional trustfulness of an implementation.

7.4.2.3 Structural Constraints

Quantum circuit design must also consider various *structural constraints* that discriminate through several criteria:

1. They can be either relative to a target architecture or absolute (induced by quantum physics laws). The first category comprises, for example, the number of available qubits in a processor, the connectivity between physical qubits, the set of available elementary gates, *etc.* The second category mainly deals with aspects induced by quantum Calculus unitarity (no cloning theorem, ancilla management, quantum control, *etc.*);
2. Now, depending on the programming language at stake, absolute structural constraints may either be taken into charge by the language design or left to the user's responsibility. For example, the *no-cloning rule* is derived from the unitarity of quantum processes. It forbids using the same quantum data register twice:
   - in languages where quantum data registers are full right objects (eg: Quipper, QWIRE, *etc*), caring for the respect of no-cloning is left to the user. In this case, formal verification may help her to do so. Solutions like ProtoQuipper [162] or QWIRE [147, 156] tackle this problem through linear type systems (see Section 7.6.3);
   - another possibility is to reduce the expressivity of the language (eg: QFC/QPL [168], SQIR [91], QBRICKS [38]), to

prevent any possible violation of no-cloning. In SQIR or QBRICKS, quantum data registers are addressed via integer indexes, but the quantum data they hold are not directly accessible from the programming language itself. These data concern the semantics of the language and they are formalized only in the specification language. Hence, the respecting conditions for the no-cloning theorem are reduced to simple indexing rules for quantum circuits.

3. Last, structural program constraints can be either syntactic or semantic. The first category contains, for example, all constraints that are linked with qubit identification (eg. : *do not control an operation by the value of a qubit it is acting on*). The most representative example for semantic constraints concerns a particular aspect of quantum computing that we do not detail in this chapter: the management of ancilla qubits. Ancilla qubits provide additional memory for some sections of quantum circuits, the content of which is then discharged at some stage of the computation. Discharging a part of a register is possible (without affecting the rest of the memory) only if there is no interaction between the memory to discharge and the rest of the memory (See [140] for further details). Hence, ancilla management is possible modulo some non-entanglement specifications, regarding the semantics of quantum circuits.

7.4.2.4 Circuit Equivalence

Compilation of quantum programs contains many circuit rewriting operations (see Section 7.2.3.3). They concern the implementation of logical qubits in a physical framework and require certification for functional behavior preservation. Concretely, given a logical circuit C, compiling C traces as a chain of circuits, starting from C and each obtained from the precedent by applying a circuit rewriting operation. Each such rewriting must preserve the input/output relation, to ensure that, provided C fits its functional specifications, then so does the final physical qubits circuit. In Section 7.5 we present two tools enabling the verification of circuit equivalence: the ZX-Calculus (Section 7.5.1) and the path-sum equivalence verification (Section 7.5.2).

Further formal comparisons between quantum processes.

Different notions of equivalence between quantum processes are also at stake with further uses of quantum information, such as communication protocols. Recent developments [185, 19] generalize the equivalence specification to further comparison predicates between quantum processes. Since they are not designed for the formalization of algorithms, which is the scope of the present chapter, we do not detail these propositions in the present chapter.

7.5 LOW-LEVEL VERIFICATION: COMPILATION AND EQUIVALENCE

Realizing logical circuits into physical devices (*circuit compilation*) requires to deal with severe constraints: the number of available qubits, their connectivity, the set of elementary operations, the instability of quantum information–requiring the insertion of error correction mechanisms, *etc.* As mentioned in Section 7.4.2.4, the underlying circuit transformations must preserve functional equivalence with the initial circuit representation, all along the compilation process. In the present section we introduce formal tools for checking such equivalences and certifying compilation correctness.

7.5.1 ZX-CALCULUS AND QUANTOMATIC/PYZX

ZX-Calculus [46] is a powerful graphical language for representing and manipulating quantum information. This language historically stems from category theory applied to quantum mechanics, through the program Categorical Quantum Mechanics initiated by Samson Abramsky and Bob Coecke [2].

For our purposes, it is interesting to see ZX-diagrams as a lax version of quantum circuits. This laxness on the one hand implies that not all ZX-diagrams are implementable with physical qubits, but on the other hand, it allows formalism to get powerful results on the underlying equational theory (rewriting rules, pseudo-normal forms).

The level of abstraction provided by the language allows the user to reason about quantum programs or protocols while significantly alleviating the "bureaucracy checks" typically coming with circuit-level reasoning, in particular, checking sub-circuit equivalence in the presence of ancillas. It also allows unifying different models of quantum computation (circuits, measurement-based quantum computing, lattice surgery, *etc.*), as well as to provide optimization strategies for these models. Last but not least, it can be used to formally (yet, graphically) verify properties on protocols or programs – all that based on simple graph-based manipulations.

7.5.1.1 Semantical Model

The ZX-diagrams are generated from a set of primitives:

$$\left\{ \ldots, \; \overset{n}{\underset{m}{\ldots}} \alpha, \; \overset{n}{\underset{m}{\ldots}} \alpha, \; \ldots \right\}_{\substack{n,m\in\mathbb{N}\\ \alpha\in\mathbb{R}}}$$

which can be composed either:

- sequentially: D_1 over D_2
- or in parallel: D_1 D_2

where D_1 and D_2 are both ZX-diagrams (themselves composed of the above primitives). We denote by **ZX** the set of ZX-diagrams. In these, information flows from top to bottom, which is in contrast with quantum circuits where it flows from left to right. This is only a matter of convention, as string diagrams, on which the ZX-Calculus formalism relies upon, are oriented vertically.

These diagrams are used to represent linear maps, thanks to the so-called *standard interpretation* of ZX-diagrams as complex number matrices $[\![.]\!] : \mathbf{ZX} \to \mathcal{M}(\mathbb{C})$[5]. It is inductively defined as:

$$[\![D_1 \text{ over } D_2]\!] = [\![D_2]\!] \circ [\![D_1]\!]$$

$$[\![D_1 \; D_2]\!] = [\![D_1]\!] \otimes [\![D_2]\!]$$

$$[\![\,|\,]\!] = id_{\mathbb{C}^2} = |0\rangle\langle 0| + |1\rangle\langle 1| \qquad [\![\times]\!] = \sum_{i,j\in\{0,1\}} |ji\rangle\langle ij|$$

$$[\![\frown]\!] = [\![\smile]\!]^\dagger = |00\rangle + |11\rangle \qquad [\![H]\!] = |+\rangle\langle 0| + |-\rangle\langle 1|$$

$$[\![\text{green node } \alpha, n \text{ inputs}, m \text{ outputs}]\!] = |0^m\rangle\langle 0^n| + e^{i\alpha}|1^m\rangle\langle 1^n|$$

$$[\![\text{red node } \alpha, n \text{ inputs}, m \text{ outputs}]\!] = |+^m\rangle\langle +^n| + e^{i\alpha}|-^m\rangle\langle -^n|$$

where $|+\rangle := \frac{|0\rangle+|1\rangle}{\sqrt{2}}$ and $|-\rangle := \frac{|0\rangle-|1\rangle}{\sqrt{2}}$ and $|u\rangle\langle v|$ is the ket bra outer product from Section 7.2.1.2. For example, $id_{\mathbb{C}^2} = |0\rangle\langle 0| + |1\rangle\langle 1| = \begin{pmatrix}1 & 0\end{pmatrix} \otimes \begin{pmatrix}1\\0\end{pmatrix} + \begin{pmatrix}0 & 1\end{pmatrix} \otimes \begin{pmatrix}0\\1\end{pmatrix} = \begin{pmatrix}1&0\\0&0\end{pmatrix} + \begin{pmatrix}0&0\\0&1\end{pmatrix} = \begin{pmatrix}1&0\\0&1\end{pmatrix}$.

Notice that the green (light) and red (dark) nodes only differ from the basis in which they are defined (as $(|+\rangle, |-\rangle)$ defines an orthonormal basis

[5] To be more precise, the standard interpretation associates to any ZX-diagram in $\mathbf{ZX}[n,m]$ (i.e. with n inputs and m outputs) a complex matrix of dimension $2^m \times 2^n$ i.e. in $\mathcal{M}_{2^m\times 2^n}(\mathbb{C})$.

of $\mathbb{C}^2$) and that they can have an arbitrary number of inputs and outputs. It often happens that a green or red node has a parameter of value 0. In this case, by convention, this angle 0 is omitted. Finally, notice that represents exactly the Hadamard gate of quantum circuits. This is not a coincidence, as ZX-diagrams can be seen as a generalization of quantum circuits. In particular, we can map any quantum circuit to a ZX-diagram that represents exactly the same quantum operator:

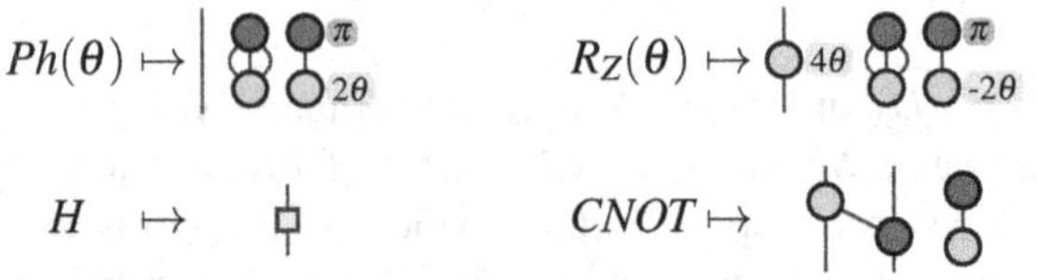

$Ph(\theta) \mapsto$ π 2θ $\qquad$ $R_Z(\theta) \mapsto$ 4θ π -2θ

$H \mapsto$ $\qquad$ $CNOT \mapsto$

and that preserves sequential and parallel compositions. The elementary gates given above are the ones detailed in Table 7.1.

We can actually map any *generalized* quantum circuit (i.e. circuit including measure) into a ZX-diagram. Indeed, initializations of qubits are easy to represent: $|0\rangle \mapsto$ 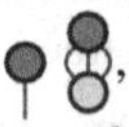 , and there exists an extension of the ZX-Calculus [48, 36] that allows the language to represent measurements. In this extension, we represent the environment as $\top$, which becomes an additional generator of the diagrams (we denote by $\mathbf{ZX}^{\top}$ this updated set of ZX-diagrams). This generator can also be understood as discarding a qubit. However, contrary to classical data, this action affects the rest of the system. Introducing $\top$ forces us to change the codomain of the standard interpretation, but we will not give the details here. Simply keep in mind that the measurement in the computational basis $(|0\rangle, |1\rangle)$ is represented by 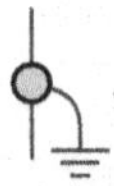 .

In this way, we can (fairly) easily represent any generalized quantum circuit as a ZX-diagram. But we can actually represent more, and this is an active field of research to try and characterize diagrams that can be put in circuit form (we talk about "extracting a circuit"). First was introduced the notion of causal flow [52] which was then extended to that of "gflow" (for generalized flow) [33]. Some other variations exist [11].

Quantum circuits, however, are not the only computational model one might want ZX-diagrams to compile to. Indeed, it so happens that the primitives of the ZX-Calculus quite naturally match those of *lattice surgery* [55], a scheme for error correction [75, 95]. In particular, ZX-diagrams implementing a (physical) lattice surgery procedure features a special notion of flow, the *PF flow* (for Pauli fusion flow) [54].

7.5.1.2 Verified Properties

The strength of ZX-Calculus comes from its powerful equational theory. This equational theory allows to define equivalence classes of ZX-diagrams and to conveniently decide whether two different diagrams represent the same quantum operator.

This question can be asked for quantum circuits as well, as two different circuits may represent the same operator (e.g. $H^2 = Id$). Some such equational theories exist for quantum circuits [169, 6], but none, as of today, for a universal fragment of quantum mechanics (notice that in [179] a completeness theorem is given for the interaction between measurements and pure parts of the circuit, but crucially one for the pure part itself is not provided but only assumed).

The main difference between the two formalisms is that the equational theory of the ZX-Calculus allows for a powerful result in this language, aggregated under the paradigm *"only connectivity matters"*. This result states that we can treat any ZX-diagram as an undirected open graph, where the Hadamard box and the green and red nodes are considered as vertices. In particular, any (open) graph isomorphism is an allowed transformation.

Example 7.5.1. 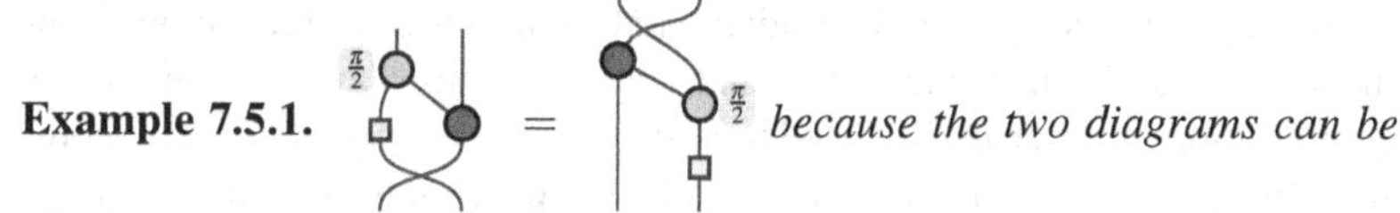 *because the two diagrams can be obtained from one another by simply "moving their nodes around" (while keeping inputs and outputs fixed).*

This result also allows us to unambiguously represent a horizontal wire. For instance:

This "meta"-rule, that all isomorphisms of open graphs are allowed, constitutes the backbone of the ZX-Calculus. In what follows, different sets of axioms, that satisfy different needs, will be presented, but this meta-rule will always be there implicitly.

When two diagrams D_1 and D_2 are proven to be equal using the equational theory $\mathscr{T}$, we write $\mathscr{T} \vdash D_1 = D_2$. The axiomatization **zx** for the ZX-Calculus can be found in Figure 7.5, and it was recently proven to be complete for the standard interpretation $[\![.]\!]$ [189]:

Theorem 1. **ZX**/*zx is complete with respect to* $[\![.]\!]$*:*

$$\forall D_1, D_2 \in \mathbf{ZX},\ [\![D_1]\!] = [\![D_2]\!] \iff zx \vdash D_1 = D_2$$

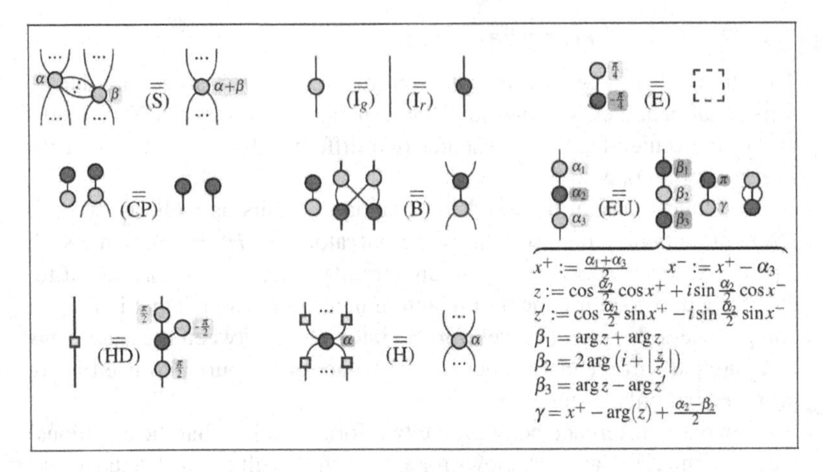

Figure 7.5 The equational theory zx[6]. All rules – provably – hold in their upside-down and color-swapped (between green and red) versions.

Here **ZX**/zx represents the quotient of **ZX** by the equational theory zx. The completeness property is fundamental. It allows us to reason on quantum processes through diagrammatic transformations rather than by matrix computations. In particular, it tells us that whenever two diagrams represent the same quantum operator, they can be turned into one another using only the rules of zx.

It is customary in quantum computing to work with particular (restricted) sets of gates. For a lot of such restrictions, there exist complete axiomatizations [85, 102, 103, 101, 104, 105]. The ZX-Calculus with measurements has a similar completeness result for $\mathbf{ZX}^{\div}/\mathrm{zx}^{\div}$ [36] with an updated set of rules $\mathrm{zx}^{\div}$ which we will not give here for conciseness purposes.

Some properties of quantum protocols or algorithms can then be verified by diagram transformations. To give the reader the flavor of such verifications, we detail the example of superdense coding.

Example 7.5.2 (Superdense Coding)**.** *The idea of the superdense coding protocol is to transmit two classical bits using a single qubit. This is not possible in general, but it is when the two parties initially share an*

[6]The way we denote the equations relates to their names in broader literature if such exists, and more informal names if not. (S) is the *spider* rule, (I_g) and (I_r) are the green and red *identity* rules, (E) is the *empty diagram* rule, (CP) the *copy* rule, (B) the *bialgebra* rule, (EU) the *Euler angles* rule, (HD) the *Hadamard decomposition* and (H) the *Hadamard color change* rule.

entangled pair of qubits. The protocol goes as follows:

- *Alice and Bob initially share the (previously defined) EPR pair* $\beta_{00} = \frac{|00\rangle + |11\rangle}{\sqrt{2}}$*, and Alice moreover has two bits she wants to send to Bob*
- *Alice applies* $\sigma_X = \begin{pmatrix} 0 & 1 \\ 1 & 0 \end{pmatrix}$ *to her qubit if her first bit is* 1*, then* $\sigma_Z = \begin{pmatrix} 1 & 0 \\ 0 & -1 \end{pmatrix}$ *if her second bit is* 1
- *Alice sends her qubit to Bob*
- *Bob applies a CNOT between his qubit and the one he received from Alice, then a H gate on his qubit, and finally measures his two qubits*

This protocol can be represented with a ZX-diagram as follows.

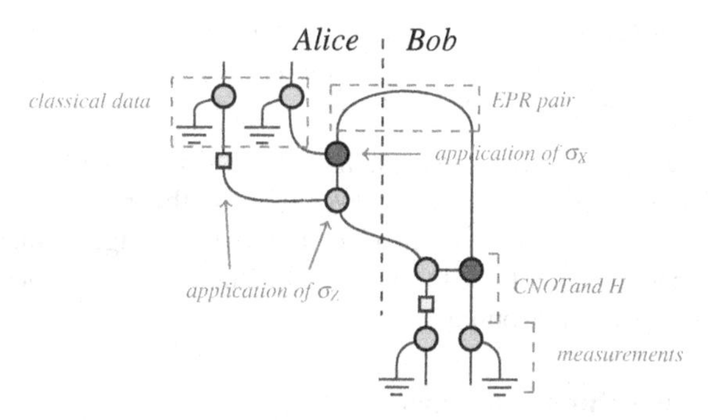

It is then possible to verify that Bob eventually does get (copies of) Alice's bits, using the equational theory (although the whole derivation is not given here):

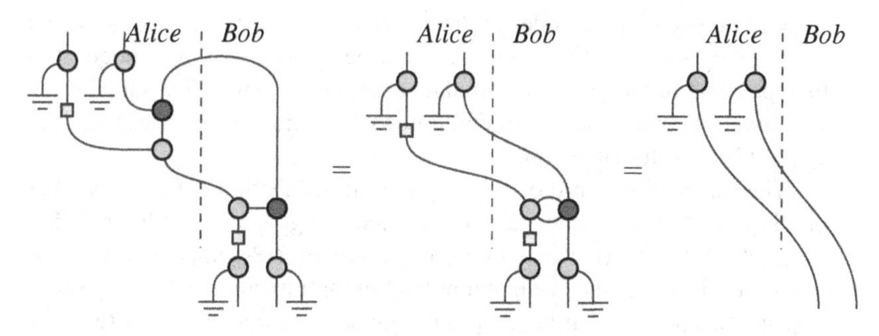

We can hence see that data is transmitted from Alice to Bob, without any loss. Interestingly, this protocol can be extended for secure communication between the two parties [192]. The larger protocol uses instances of the

smaller one to also check whether an eavesdropper has tried intercepting or copying data.

If Bob is aware that the data he received was compromised, he can abort everything by simply discarding his qubits, so that the eavesdropper (Eve) gets absolutely no information:

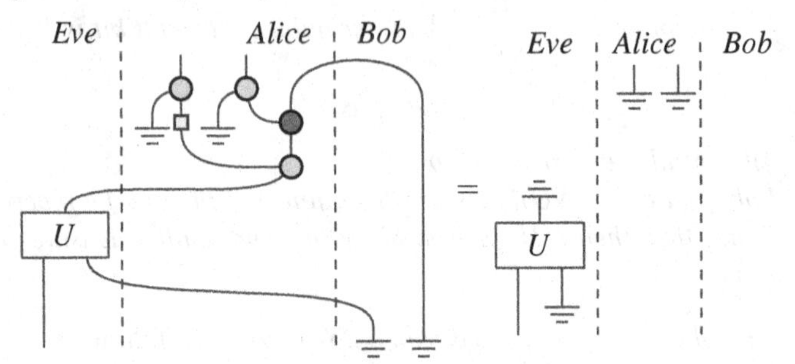

where U denotes an unknown operator applied by Eve to the qubit she intercepted. Notice here how no information can pass from Alice to Eve. No information is retrieved by the latter.

A plethora of quantum protocols have been verified with ZX-Calculus in a similar manner [92]. Note however that the theory is ever evolving, and in particular, ⏚ was not introduced in the language at that time, so the author had to use a trick to make up for the absence of measurement (namely case-based reasoning).

7.5.1.3 Algorithms and Tools

It is possible to manipulate ZX-diagrams in a computer-verified way. For instance, Quantomatic [111] allows the users to define at the same time diagrams in graphical form and equational theories. It is also possible to work with user-defined nodes in the diagram so that even though ⏚ is not part of the "vanilla" ZX-Calculus, it can be defined as a new node. It is then possible in the tool to manipulate diagrams in a way that satisfies the equational theory, and even to define rewriting strategies that can be then applied in an automated way.

The verification of protocols and programs using the ZX-Calculus relies on diagrammatic equivalence. This problem, in general, is at least QMA-complete, [29, 98] (the quantum counterpart of NP-completeness). This problem is linked to the one of simplification/optimization, which asks how a quantum operator can be simplified, given a particular metric (e.g. the number of non-Clifford gates). Indeed, for instance if D_1 and D_2 are two diagrams representing the same *unitary* (i.e. $[\![D_1]\!] = [\![D_2]\!]$), then simplifying $D_2^\dagger \circ D_1$ should ideally get us to the identity.

In the case of the Clifford fragment (obtained when the angles in 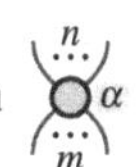

and 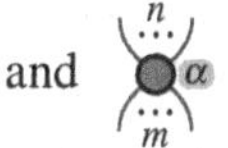 are restricted to multiples of $\frac{\pi}{2}$), there exists a strategy that reduces the diagram in (pseudo-)normal form [10]. When this algorithm terminates, the resulting diagram is of size $O(n^2)$ where n is the number of inputs and outputs in the diagram. The algorithm is polynomial in the overall size of the diagram it is applied on.

Turning an arbitrary diagram into a normal form can be done in principle [104]; however, the complexity of this algorithm is EXPSPACE for universal fragments. So this approach is obviously not preferred in general. However, one can use the ideas of the algorithm for the Clifford fragment as a starting point to get a rewriting strategy for the general case. Applications to quantum circuits and improvements on this strategy can be found in the literature [110, 53, 60, 11], and implementations in the PyZX tool [109]. This tool can in particular be used to tackle circuit equivalence verification, using a different but related approach to that of Section 7.5.2 below. The formalism used later is that of *path-sums*, where morphisms were showed in [124, 190] to be essentially equivalent to ZX-diagrams, allowing us to apply strategies for path-sums to the ZX-Calculus and vice-versa. Next Section is devoted to we introducing the path-sum formalism and its use for circuits verification.

7.5.2 PATH-SUM CIRCUIT EQUIVALENCE VERIFICATION

Path-sums are a recent alternative direction for verifying the equivalence between quantum circuits [4, 5]. In this section we briefly present it, together with the main verification related achievements.

Note that a generalization of path-sum semantics is introduced in Section 7.7.2, for parametrized families of circuits. For sake of readability, conciseness and coherence with this further content, in the coming paragraphs we slightly simplify path-sums related notations. We refer the reader to the original definitions [4, 5] for the full formalism and underlying mathematical structures.

7.5.2.1 Semantical Model

The standard semantics for quantum circuits is the matrix formalism, introduced in Section 7.2.1.5. It associates to each quantum circuit C a matrix $\mathbf{Mat}(C)$ and it interprets the behavior of this circuit as a function $|x\rangle \to \mathbf{Mat}(C) \cdot |x\rangle$ from kets to kets, where $\cdot$ stands for the usual matrix product.

Notice that this standard semantics builds on an intermediary object–the matrix–to derive and interprets the functional behavior of circuits. And it does so by use of a higher-order function–the matrix product. Contrarily, path-sums are a straight construction of the *input/output* function performed by circuits, enabling compositional reasoning.

Concretely, a path-sum $PS(x)$ is a quantum register state (a ket), parametrized by an input basis ket $|x\rangle$ and defined as the sum of kets

$$PS(x) ::= \frac{1}{\sqrt{2}^n} \sum_{k=0}^{2^n-1} e^{\frac{2\cdot\pi\cdot i\cdot P_k(x)}{2^m}} |\phi_k(x)\rangle \tag{7.2}$$

where the $P_k(x)$ are called *phase polynomials* while the $|\phi_k(x)\rangle$ are *basis-kets*. This representation is *closed* under functional composition and Kronecker product. For instance, if

$$C : |x\rangle \mapsto PS(x) = \frac{1}{\sqrt{2}^n} \sum_{k=0}^{2^n-1} e^{\frac{2\cdot\pi\cdot i\cdot P_k(x)}{2^m}} |\phi_k(x)\rangle,$$
$$C' : |y\rangle \mapsto PS'(y) = \frac{1}{\sqrt{2}^{n'}} \sum_{k=0}^{2^{n'}-1} e^{\frac{2\cdot\pi\cdot i\cdot P'_k(y)}{2^{m'}}} |\phi'_k(y)\rangle,$$

then their parallel combination `parallel`(C, C') sends $|x\rangle \otimes |y\rangle$ to:

$$\frac{1}{\sqrt{2}^{n+n'}} \sum_{j=0}^{2^{n+n'}-1} e^{\frac{2\cdot\pi\cdot i\left(2^{m'}\cdot P_{j/2^n}(x)+2^m\cdot P'_{j\%2^n}(y)\right)}{2^{m+m'}}} |\phi_{j/2^n}(x)\rangle \otimes |\phi'_{j\%2^n}(y)\rangle$$

The sequential combination of quantum circuits C and C' receives a similar compositional definition, parametrized by path-sums components for circuits C and C'.

7.5.2.2 Path-Sums Reduction

While path-sums compose nicely, a given linear map (eg. the input/output function for a quantum circuit) does not have a unique representative path-sum. Hence, given two different path-sum, how to decide whether they both encode the behavior of a given circuit? To tackle this problem, an equivalence relation is defined with a few, simple rules that can be oriented. As an example, the *HH* rule enables to simplify a path-sum expression over the reduction from a sequence of two consecutive Hadamard gates to the identity–see Figure 7.6. All these rules transform a path-sum into an equivalent one, with a lower number of *path variables* (parameter n in the notation of Equation 7.2).

$$\frac{PS(x) = \frac{1}{\sqrt{2}^{n+1}} \sum_{j=0}^{2} \sum_{k=0}^{2^n} e^{\frac{1}{2} \cdot \pi \cdot i \left(\frac{1}{2} j (k_i + Q_k(x)) + P_k(x) \right)} |\phi_k(x)\rangle}{PS(x) = \frac{1}{\sqrt{2}^{n+1}} \sum_{k=0}^{2^n} e^{\frac{1}{2} \cdot \pi \cdot i (P[i := Q_k(x)]_k(x))} |\phi[i := Q_k(x)]_k(x)\rangle} \; HH$$

Figure 7.6 The *HH* path-sum transformation rule.

We refer the desirous reader to [5, 4] for an exhaustive exposition of the corresponding proof system. It was proved strongly normalizing, meaning that every sequence of reduction rules application terminates with an irreducible path-sum. Furthermore, finding and applying such a normalizing sequence is feasible in time polynomial in the width of the circuit at stake, which makes the overall reduction procedure tractable.

7.5.2.3 Verified Properties

Hence, path-sums provide a human-readable formalism for the interpretation of quantum circuits as ket data functions. Furthermore, it is given a polynomial normalization procedure, based on a restricted set of rewriting rules. The method was probed against both circuit equivalence and functional specifications verification. More precisely:

Translation validation consists, for a given quantum algorithm, in (1) computing the path-sums for both a non-optimized and an optimized circuit realization and (2) using the normalization procedure for the automatic checking of their equivalence. It was performed on various quantum routine instances (Grover, modular adder, Galois field multiplication, *etc*) of various size (up to several dozens of qubits). Interestingly, the methods proved as efficient for identifying non-equivalence (over erroneous instances) as for checking equivalence.

Quantum algorithms verification consists in verifying whether a given circuit instance respects its functional description. It was performed for instances of similar case studies as for the translation validation (QFT, Hidden Shift [187]), with up to a hundred qubits.

As conclusion, the path-sum formalism provides a fully automatized procedure for verifying the equivalence between two circuits. Hence, given two quantum circuits, the latter being a supposed optimized version of the former, path-sums treatment enables us to verify that they implement the same quantum function. Since path-sums perform internal complexity reduction, an open direction is a search for efficient heuristics extracting an optimized quantum circuit from reduced path-sums. As mentioned in Section 7.5.1, this problem is closely linked to the reduction of ZX-Calculus diagrams. So, in its general form, it faces the same complexity limitations. The search for efficient reduction procedure applying to identified useful fragments could then both benefit from and feed advances in the ZX setting.

In its present state of development, using path-sums for quantum algorithm verification is restricted to compilation time, when program parameters are instantiated. Furthermore, a new run of the path-sum reduction is required at each new call of a given quantum function. In Section 7.7.2 we introduce the QBRICKS language, whose semantics is based on a parametrized extension of path-sums. It enables the verification, once for all, of parametrized programs, holding for any possible future parameter instances–yet at the price of full automation as the manipulation of parametrized path-sums requires first-order logic reasoning.

7.5.3 QUANTUM ABSTRACT INTERPRETATION

The techniques presented so far target an exhaustive functional description of a quantum circuit. Because of the intrinsic complexity of quantum computing, their use is limited, in particular by the size of circuits. As illustrated in the benchmarks summed up in Section 7.5.2.3, path-sums reduction enabled, for instance, the verification of quantum circuits up to a hundred of qubits.

To push the boundary, a possible strategy comes from Abstract Interpretation (see Section 7.3.3). There, one does not target an exhaustive description of the functional behavior of a circuit, but an over-approximation of it. Such a framework relies on:

- the identification of a conveniently structured set of properties of interest (the *abstract domain*);
- sound abstraction and concretization to/from this abstract domain;
- reasoning tools for the abstract domain.

In [148], the author introduces such an abstract interpretation of quantum states based on their entanglement structure. The technique enables to identify mutually separable sub-registers. It is useful, for example, for uncomputation, qubit discarding or identifying a convenient decomposition for further analysis of a state.

A more recent development on abstract interpretation applied to quantum process is [199], in which the authors define abstract domains made of tuples of partial projections over quantum sub-registers. Intuitively, the idea is to overcome the exponential complexity of quantum states by decomposing them into sub-spaces. Interestingly, the method was implemented and evaluation results are provided. In particular, this abstraction enabled to characterize the invariant in the main loop structuring the Grover's search algorithm and to prove it for instances of width up to 300 qubits.

7.5.4 TOWARD INTEGRATED VERIFIED OPTIMIZATION: VOQC

A noticeable effort for an integrated verified quantum optimization was recently led through the development of a *Verified Optimizer for Quantum Circuits* (VOQC–pronounced "vox") [91]. As main aspects, in comparison to ZX- and path-sum calculus, VOQC:

- is integrated into a core programming environment and applies on circuits issued from parametrized programs;
- not only validates the equivalence between an input quantum circuit and a candidate optimized version of it, but also provides the formally verified optimization procedure, directly generating this optimized version.

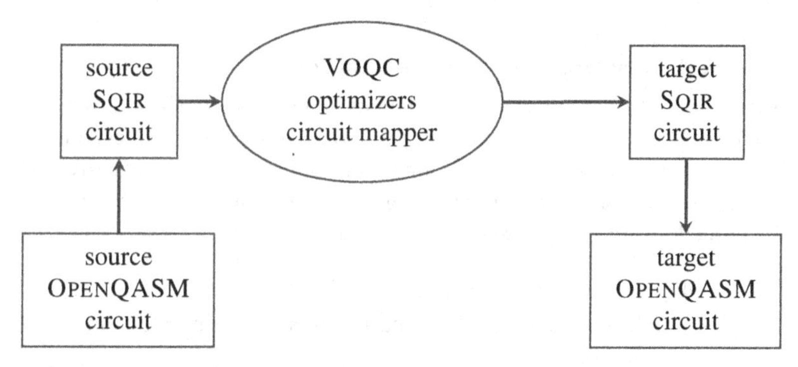

Figure 7.7 A simplified view of VOQC architecture.

7.5.4.1 Architecture

In Figure 7.7 we give a simplified view of VOQC architecture: it relies on an Intermediate Representation Language named SQIR. Since it may also be used as a verified programming environment, SQIR is introduced and developed *per se* in a dedicated section below (Section 7.7.3). In the present context, we can think of it as a core language generating parametrized quantum circuits. Then, for any instance of the parameters, VOQC extracts the corresponding sequence of operations, applies an optimization procedure upon this sequence of operations and extracts back an optimized proved equivalent SQIR quantum circuit. Furthermore,

- in addition to optimization, VOQC also contains some circuit mapping functionalities. They perform further circuit transformation so that the output circuits fit to specific quantum architecture qubit connectivity constraints. This functionality is a preliminary address to the problem raised as first bullet in Section 7.2.3.3,
- VOQC environment also provides both ways compilations between SQIR and the standard assembly language OPENQASM [51]. Hence, it opens the way for a modular easy integration in any standard programming environment, in particular QISKIT [152], which uses OPENQASM as assembly language.

7.5.4.2 Optimization Procedure

VOQC optimization process provides two functionalities, one is deterministic (optimization by propagation and cancellation) and the other one requires a replacing circuit input.

Optimization by propagation and cancellation

is based on local circuit rewriting schemes and self composition properties of elementary gates, borrowed from [139] (eg., sequences of an even number of either *H*, *CNOT* or X gate annihilate as the identity, successive occurrences of R_z gates melt by summing their angle parameters, *etc*). Hence, the procedure consists of two successive steps:

- propagate: for any elementary gate, it
  - considers several identified patterns enabling gate commutations,
  - finds all occurrences of these patterns,
  - and performs the related commutation, pushing any occurrence of this elementary gate to the end of the computation,
- cancellation then consists in deleting the resulting repetitive occurrences of the elementary gate at stake.

Optimization by circuit replacement

consists in substituting a part of a quantum circuit (a subcircuit) by another one that is proven to be functionally equivalent. In this case, the equivalence proof is led by help of the path-sums semantics (see Section 7.5.2). Hence, in its most general form, the process requires an external equivalence proof oracle. Nevertheless, VOQC provides some instances of such proved equivalence patterns (eg, rotation merging), whose application is automatic.

Performance and Achievements.

VOQC performance has been evaluated against several standard quantum computation routines, and compared with several existing optimizers [7, 152, 139, 178, 109]. Note that in VOQC, since optimization is performed as a succession of rewriting operations, the formal verification consists in assessing the equivalence between the input and the output of the optimization, it does not address the optimization performance. Still, on reported experiments, VOQC performance competes with other existing non-verified optimizers, both regarding computation time and circuit complexity reduction–precise performance comparison tables appear in [91]. Thus, in the current state of the art, the benefits of formally verified circuit optimization comes for free.

7.5.5 FORMALLY VERIFIED QUANTUM COMPILATION IN AN IMPERATIVE SETTING: GIALLAR

Giallar [183] (precedently know as CertiQ [173]) is another noticeable effort for verified compilation. Interestingly, it applies to QISKIT language, which is certainly *the most widely used quantum compiler*. Giallar proceeds similarly as VOQC, by applying successive optimization *passes*, each consisting in applying local circuit equivalence transformations over quantum circuits. Giallar has been evaluated against the very compilation environment provided by IBM QISKIT framework. And verified optimization went through 44 out of the 56 compilation passes identified from different versions of the framework. Hence, the current limitations of formally verified compilation are mainly inherited from the state of the art in (non-verified) quantum compilation.

The application to QISKIT also inevitably brings an additional drawback: while Giallar can bring assurance that compilation respects the functional equivalence between an input and an output quantum circuit, the initial circuit building brick still lacks formal verification. Indeed, the environment is not provided formal means to assess that the input circuit meets a given functional specification. An open work direction then is to develop formal circuit building verification methods similar to SQIR or QBRICKS but applying to (mostly imperative) widely used development environments such as QISKIT, so as to complete the verified development chain.

7.6 FORMAL QUANTUM PROGRAMMING LANGUAGES

For reasoning on concrete *runs* of quantum algorithms solving specific problem *instances*, the notion of quantum circuit is natural. Yet, a quantum circuit is only a by-product of a specific run of a quantum algorithm, holding only for a specific instantiation of the algorithm parameters.

Hence, a quantum algorithm is not reducible to a quantum circuit. To run quantum algorithms and reason about them, one needs *quantum programming languages* (QPL): this is the topic of the current section.

7.6.1 QUANTUM PROGRAMMING LANGUAGES DESIGN

In Section 7.5, circuits were merely seen as sequences of elementary gates. However, in most quantum algorithms circuits follow a more complex structure: they are built compositionally from smaller sub-circuits and circuits combinators. Circuits are usually *static* objects, buffered until completion before being flushed to the quantum co-processor. Still, in some algorithms, they are *dynamically* generated: the tail of the circuit depends on the result of former measurements.

In this section, we discuss the high-level structure of quantum algorithms, the requirements for a quantum programming language, and review some of the existing proposals.

7.6.1.1 Structure of Quantum Algorithms

The usual model for quantum computation was depicted in Figure 7.1: a classical computer controls a quantum co-processor, whose role is to hold a quantum memory. A programmatic interface for interacting with the co-processor is provided to the programmer sitting in front of the classical computer. The interface gives methods to send instructions to the quantum memory to allocate and initialize new quantum registers, apply unitary gates on qubits, and eventually perform measurements. Even though the set of instructions is commonly represented as a circuit, it is merely the result of a *trace of classical execution* of a classical program on the classical computer.

Figure 7.8 presents two standard workflows with a quantum co-processor. In Figure 7.8a, the classical execution inputs some (classical) parameters, performs some pre-processing, generates a circuit, sends the

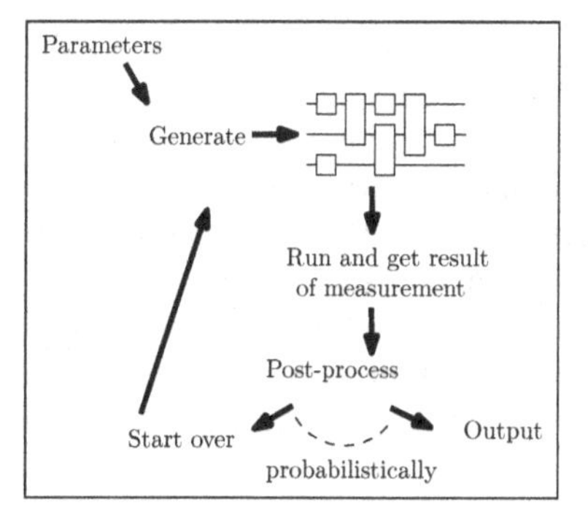

(a) Static scheme

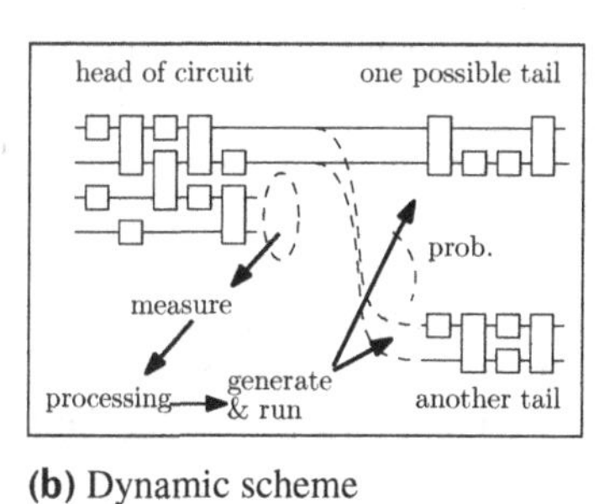

(b) Dynamic scheme

Figure 7.8 Workflows for quantum algorithm.

circuit to the coprocessor, collects the result of the measurement, and finally performs some post-processing to decide whether an output can be produced or if one needs to start over. Shor's factoring algorithm [175] or Grover's algorithm [84] fall into this scheme: the circuit is used as a fancy probabilistic oracle. Most of the recent variational algorithms [136] also fall into this scheme, with the subtlety that the circuit might be updated at each step. The other, less standard workflow is presented in Figure 7.8b. In this scheme, the circuit is built "on the fly", and measurements might be performed on a sub-part of the memory along the course of execution of the circuit. The latter part of the circuit might then depends on the result of classical processing in the middle of the computation.

Understanding a quantum circuit as a by-product of the execution of a classical program shines a fresh view on quantum algorithms: it cannot be identified with a quantum circuit. Instead, in general, at the very least a quantum algorithm describes a *family* of quantum circuits. Indeed, consider the setting of Figure 7.8a. The algorithm is fed with some parameters and then builds a circuit: the circuit will depend on the shape of the parameters. If for instance, we were using Shor's factoring algorithm, we would not build the same circuit for factoring 15 or 114,908,028,227. The bottom line is that a quantum programming language should be able to describe *parametrized* families of circuits.

The circuits described by quantum algorithms are potentially very large–a concrete instance of the HHL algorithm [87] for solving linear systems of equations has been shown [165] to count as much as $\sim 10^{40}$ elementary gates, if not optimized. Unlike the circuit-construction schemes hinted at in Section 7.5, this circuit is not uniquely given as a list of elementary gates: it is built from sub-circuits –possibly described as a list of elementary gates but not only–and from high-level circuit combinators. These combinators build a circuit by (classically) processing a possibly large sub-circuit. Some standard such combinators are shown in Figure 7.9 (where we represent inverse with reflected letters). There is a distinction to be made between the combinator, applied on a sub-circuit, and its semantics, which is an action on each elementary gate. Combinators are abstractions that can be composed to build larger combinators, such as the one presented in Figure 7.10, built from inversion, controlling and sequential composition.

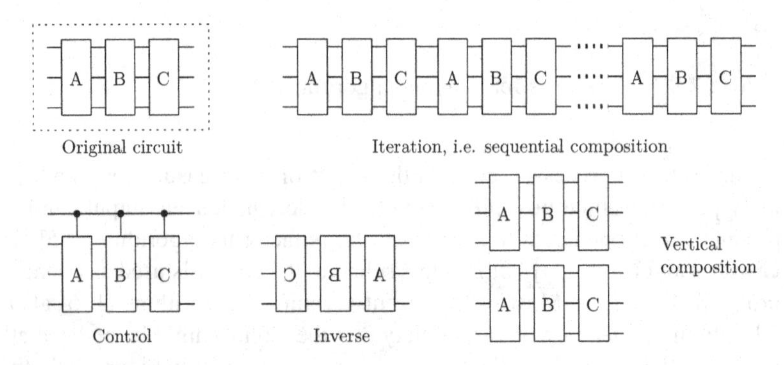

Figure 7.9 Standard circuit combinators.

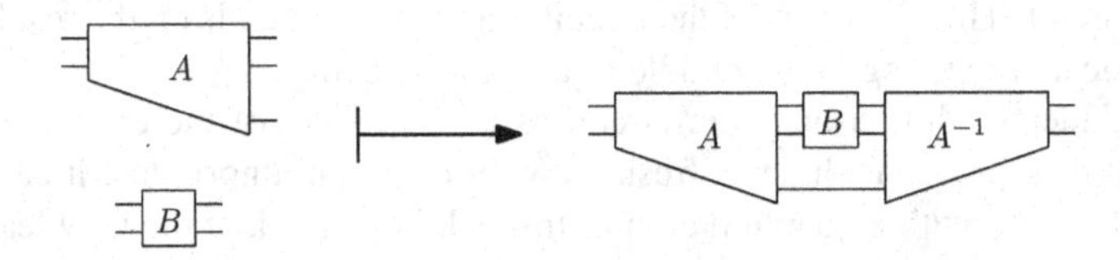

Figure 7.10 Example of a derived circuit combinator.

7.6.1.2 Requirements for Quantum Programming Languages

Any scalable quantum programming language should therefore allow the following operations within a common framework:

- Manipulation of quantum registers and quantum circuits as first-class objects. The programmer should both be able to refer to "wires" in a natural manner and handle circuits as independent objects;

- Description of *parametric families* of quantum circuits, both in a procedural manner as a sequence of operations–gates or subcircuits–and in an applicative manner, using circuit combinators;
- Classical processing. In our experience [83], quantum algorithms mostly consists of classical processing–processing the parameters, building the circuits, processing the result of the measurement.

This broad description of course calls for refinements. For instance some of the classical processing might be performed on the quantum co-processor–typically the simple classical controls involved in quantum error correction. The level of classical processing performed on the classical computer and performed on the quantum co-processor is dependent on the physical implementation. Even though some recent proposals such as Quingo [154] discuss the design of quantum programming languages aware of the two levels of classical processing–in and out of the co-processor, this is still a work in progress.

7.6.1.3 Review of the Existing Approaches

Most of the current existing quantum programming languages follow the requirements discussed in Section 7.6.1.2. In this section, we review some typical approaches followed both in academic and industrial settings. This review is by no means meant to be exhaustive: its only purpose is to discuss the possible strategies for the design of QPLs.

When designing a realistic programming language from scratch, the main problem is access to existing libraries and tools. In the context of quantum computation, one would need for instance to access the file system, make use of specific libraries such as Lapack or BLAS, *etc*. One can also rely on the well-maintained and optimized compiler or interpreter of the host language. To quickly come up with a scalable language, the easiest strategy consists in *embedding* the target language in a host language. Indeed, a quantum programming language can be seen as a domain-specific language (DSL), and it can be built over a regular language.

Even though the advantages of working inside a host language are clear, there are two main drawbacks, The first one is the potential rigidity of the host language: there might be constructs natural to the DSL that are hardly realizable inside the host language. The second drawback has to do with the compilation toolchain: the shallow embedding of the DSL makes it impossible to access its abstract syntax tree, rendering specific manipulation thereof impossible.

Embedded QPLs.

The first scalable embedded proposal is Quipper [83, 82]. Embedded in Haskell, it capitalizes on *monads* to model the interaction with the quantum co-processor. Quipper's monadic semantics is meant to be easily abstracted and reasoned over: it is the subject of Section 7.6.2.2. Since Quipper, there has been a steady stream of embedded quantum programming languages, often dedicated to a specific quantum co-processor or attached to a specific vendor, and mostly in Python: QISKIT [152] and PROJECTQ [180] for IBMQ, CirQ [41] for Google, Strawberry Fields [107] for Xanadu, AQASM for Atos, *etc.* From a language-design point of view, most of these approaches heavily rely on Python objects to represent circuits and operations: the focus is on usability and versatility rather than safety and well-foundedness.

Standalone QPLs.

On the other side of the spectrum, some quantum programming languages have been designed as standalone languages, with their own parser and abstract syntax tree. Maybe the first proposed scalable language was Ömer's QCL [142]. Ömer experimented several features such as circuit-as-function, automatic inversion and oracle generation. However, due to its non-modular approach, the language did not have successors.

LIQ$Ui|\rangle$ [193] and its sequel Q# [181] developed by Microsoft are good examples of an attempt at building a standalone language while keeping a tight link with an existing programming environment, as LIQ$Ui|\rangle$and Q# are tightly linked with the F# framework–which can be used as a host language for F# and is itself embedded in the whole .NET framework), making it possible to easily "reuse" library functions from within a Q# piece of code. On the other hand, Q# has its own syntax and type system, to capture run-time errors specific to quantum computation.

Scaffold [99] is another example of a standalone QPL. Even though the language is rather low-level its compiler–ScaffCC [100]–has been heavily optimized and experimented over, and it serves as a support for a long stream of research on quantum compiler optimizations.

The last noteworthy language to cite in the series is SILQ [27], as it serves as a good interface with the next paragraph: aimed at capturing most of the best practices in terms of soundness and safety, it is nonetheless targeted toward usability.

Formal QPLs.

The last line of works on QPLs we would like to mention here is formal languages aimed at exploring and understanding the design principles and the semantics of quantum algorithms. We shall be brief as this is the topic of the remainder of this chapter. The initial line of work was initiated by Selinger [168] with the study of a small flow-chart language with primitive constructs to interact with the quantum co-processor: qubit initialization, elementary gate application and measurement. This language was later extended to a simple Lambda-Calculus with similar primitive quantum features [170, 172]. Even though the language is not aimed at full-scale quantum algorithms, it is nonetheless enough to serve as a testbed for experimenting type systems and many operational [170, 57, 121] and denotational [143, 171, 58, 88] semantics.

The study of formal QPLs took a turn toward circuit-description languages *à la* Quipper with the development of scalable quantum languages. One of the first proposals of formalization is QWIRE [147], embedded in the proof assistant Coq. QWIRE uses Coq expressive system to encode the sophisticated typing rules of QWIRE. In a sense, Coq type system is expressive enough to use Coq as a host language *and* still be able to manipulate the abstract syntax tree of a program. The main design choice for QWIRE is to separate pure quantum computation with its constraints such as no-cloning, from classical computation.

Albeit disconnected from Coq, the formalization of Quipper has followed a similar root. This development is based on the formal language ProtoQuipper [162], which extracts the critical features out of Quipper: the creation and manipulation of circuits using a minimal Lambda-Calculus. The language is equipped with a linear type system and a simple operational semantics based on circuit construction. The simple core proposed by ProtoQuipper has stirred a line of research on the topic, including the formalization of inductive datatypes, recursion and dependent types in this context [160, 159, 131, 77].

The last class of formal programming language we want to mention focuses on the specification and verification of high-level properties of programs and are solely based on circuit manipulation: unlike QWIRE or Quipper, qubits are not first-class objects and circuits are simple "bricks" to be horizontally or vertically stacked. In this class of languages, one can mention qPCF [144], mainly a theoretical exploration of dependent type systems in this setting, and QBRICKS, presented in Section 7.7.2.

7.6.2 FORMALIZING THE OPERATIONAL SEMANTICS

In order to reason on quantum programming languages, one needs to have a formal understanding of their operational semantics.

7.6.2.1 Quantum Lambda-Calculi

The Lambda-Calculus [16] is a formal language encapsulating the main property of higher-order functional languages: functions are first-class citizens that can be passed as arguments to other functions. Lambda-Calculus features many extensions to model and reason about side-effects such as probabilistic or non-deterministic behaviors, shared memory, read/write, *etc.*

One of the first formal proposals of a quantum, functional language has precisely been a *quantum* extension of Lambda-Calculus [170]. On top of the regular Lambda-Calculus constructs, the *quantum Lambda-Calculus* features constants to name the operations of qubit initialization, unitary maps and measurements. A minimal system consists of the following terms:

$$\begin{array}{rcl} M,N & ::= & x \mid \lambda x.M \mid MN \mid \\ & & \texttt{tt} \mid \texttt{ff} \mid \texttt{if}\, M\, \texttt{then}\, N_1\, \texttt{else}\, N_2 \mid \\ & & \texttt{qinit} \mid \texttt{U} \mid \texttt{meas}. \end{array}$$

Terms are represented with M and N, while variables x range over an enumerable set of identifiers. The term $\lambda x.M$ is an *abstraction*: it stands for a function of argument x and of body M. The application of a function M to an argument N is represented with MN. To this core Lambda-Calculus, we can add constructs to deal with booleans: tt and ff are the boolean constant values, while $\texttt{if}\, M\, \texttt{then}\, N_1\, \texttt{else}\, N_2$ is the usual test. Finally, qinit stands for qubit initialization, meas for measurement and U ranges over a set of unitary operations. These three constants are functions: for instance, qinit tt corresponds to $|1\rangle$ and qinit ff to $|0\rangle$, while meas applied to a qubit stands for the measurement of this qubit. A fair coin can then be represented by the term

$$\texttt{meas}\,(\texttt{H}\,(\texttt{qinit}\,\texttt{tt})), \tag{7.3}$$

where H stands for the Hadamard gate.

The question is now: how do we formalize the evaluation of a piece of code? In the regular Lambda-Calculus, evaluation is performed with *substitution* as follows:

$$(\lambda x.M)N \to M[x := N]$$

where $M[x := N]$ stands for M where all free occurrences of x –i.e., those corresponding to the argument of the function–have been replaced by N. Even though we can still require such a rule in the context of the quantum Lambda-Calculus, this does not say how to deal with the term qinit tt.

In order to give an operational semantics to the Lambda-Calculus, a naive idea could be to add yet another construction: a set of constants $c_{|\phi\rangle}$, once for every possible qubit state $|\phi\rangle$. If–as shown by van Tonder [188]–this can somehow be made to work, a more natural presentation consists in mimicking the behavior of a quantum co-processor, in the style of Knill's QRAM model [114]: we define an *abstract machine* $(|\phi\rangle_n, L, M)$ consisting of a finite memory state $|\phi\rangle_n$ of n qubits, a term M with n free variables $x_1,\ldots,x_n$, and a *linking function* L, bijection between $\{x_1,\ldots,x_n\}$ and the qubit indices $\{1,\ldots,n\}$. Variables of M captured by L are essentially pointers to qubits standing in the quantum memory. The fair coin of Eq. (7.3) then evaluates as follows.

$$
\begin{aligned}
&(|\rangle_0, \{\}, \texttt{meas}\,(\texttt{H}\,(\texttt{qinit tt})))\\
&\quad \to (|1\rangle_1, \{x \mapsto 1\}, \texttt{meas}\,(\texttt{H}x))\\
&\quad \to (\tfrac{1}{\sqrt{2}}(|0\rangle_1 - |1\rangle_1), \{x \mapsto 1\}, \texttt{meas}\,x)\\
&\quad \to \begin{cases} (|0\rangle_1, \{\}, \texttt{ff}) & \text{with prob. } 0.5\\ (|1\rangle_1, \{\}, \texttt{tt}) & \text{with prob. } 0.5. \end{cases}
\end{aligned}
$$

In this evaluation, most of quantum computation has been exemplified: initialization of qubits, unitary operations and measurements. Handling the latter in particular requires a *probabilistic evaluation*, and this requires some care–we invite the interested reader to consult for example Selinger & Valiron [170] for details.

7.6.2.2 Monadic Semantics

The operational semantics of the quantum Lambda-Calculus is very limited. Indeed, as discussed in Section 7.6.1, quantum algorithms do not in general send operations one by one to the quantum co-processor: instead, a quantum program must build circuits (or pieces thereof) before sending them to the co-processor as batch jobs. The quantum Lambda-Calculus does not allow to build circuits: operations can only be sent one at a time. In particular, there is no possibility to create, manipulate and process circuits: circuit generation in the quantum Lambda-Calculus is a *side-effect* that is external to the language. One cannot interfere with it, and embedding the quantum Lambda-Calculus as it stands inside a host language, as suggested in Section 7.6.1.3, would not help.

The solution devised by Quipper consists in relying on a special language feature from Haskell called *monad*. A monad is a type operator encapsulating a side effect. Consider for instance a probabilistic side effect. There are therefore two classes of terms: terms without side-effect, with types e.g. `Bool`, or `Int`, and terms with side-effect, with types e.g. `P(Bool)` or `P(Int)` standing for "term evaluating to a boolean/integer, possibly with a probabilistic effect". The operator `P(-)` captures the probabilistic side effect.

A monad comes with two standard maps. In the case of `P` we would have:

```
return :: A -> P(A)
bind :: P(A) -> (A -> P(B)) -> P(B)
```

The `return` operation says that an effect-free term can be considered as having an effect–in the case of the probabilistic effect, it just means "with probability 1". The `bind` operation[7] says how to compose effectful operations: given a function inputting `A` and returning an object of type `B` with a probabilistic side-effect, how to apply this function to a term of type `A` also having a probabilistic effect? We surely get something of type `P(B)`, but the way to construct it is described by `bind`. A few equations have to be satisfied by `return` and `bind` for them to describe a monad. For instance, `bind return` is the identity on `P(A)`. There can of course be more operations: for instance, we can add to the signature of `P` an operator `coin` of type `() -> P(Bool)`[8].

A nice property of monads is that effectful operations can be written with syntactic sugar in an imperative style:

```
do
  x <- coin ()
  if x then return 0 else return 1
```

is a term of type `P(Int)` equals to

```
bind (coin ()) (λx.if x then return 0 else return 1)
```

once the syntactic sugar has been removed.

Following this approach, quantum computation can be understood as side-effect: it combines both (1) Read/Write effect, since gates are sent to the coprocessor, and results of measurements are received; (2) Probabilistic effects, since measurement is a probabilistic operation. The first attempt at formalizing this monad is Green's quantum IO monad [3]: it has then been further developed in Quipper [83].

Internally, circuits in Quipper are represented using a simple inductive datatype akin to a list of gates[9]. The interaction with the quantum-coprocessor is modeled using a specific I/O monad `Circ`. This monad encapsulates the construction of circuits featuring wires holding qubits but also wires holding bits —results of measurements. A bit is however

[7] In Haskell, this map is denoted with >>=. For the sake of legibility, here we denote it with `bind`.

[8] In Haskell, the unit type is denoted with `()`.

[9] Technically a tree structure, as measurements entails branching.

uniquely useable "inside" the monad: to use it in Haskell —in an if-then-else construct for instance— we need to "lift" it into a regular Boolean. The signature of the monad in particular includes

```
qinit :: Bool -> Circ(Qubit)
measure :: Qubit -> Circ(Bit)
hadamard :: Qubit -> Circ(Qubit)
```

A fair coin can be implemented as a circuit returning a bit, of type `() -> Circ Bit`.

```
bitcoin () = do
  q <- qinit True
  q' <- hadamard q
  r <- measure q'
  return r
```

The function `bitcoin` will merely generate a computation —producing a circuit— waiting to be executed. To implement the toss-coin of Section 7.6.2.2, we then need to "run" the circuit for lifting the bit into a Boolean. Provided that we have a function

```
run :: Circ Bit -> P Bool
```

one can then implement `coin ()` as `run (bitcoin ())`.

Thanks to the monadic encapsulation, circuits can be manipulated within Haskell. For instance, inversion and control can be coded in Haskell as circuit combinators with the following types.

```
inverse :: (a -> Circ b) -> (b -> Circ a)
control :: (a -> Circ b) -> ((a,Qubit) -> Circ (b,Qubit))
```

The compositionality of the monadic semantics also makes it possible to automatically construct oracles out of classical specification [83, 186].

Compared with the quantum λ-Calculus discussed in Section 7.6.2.1, where the program can only send gates one by one to the co-processor, Quipper gives to the programmer the ability to manipulate circuits. Although both the quantum λ-Calculus and Quipper represent quantum computations mathematically, Quipper provides a richer model, better-suited for program specification and verification than the plain λ-Calculus.

7.6.3 TYPE SYSTEMS

In Section 7.6.2, we discussed how to model the operational behavior of a quantum program. We have however not mentioned yet the run-time errors inherent to quantum computation. In the classical world, type systems are a standard strategy to catch run-time errors at compile-time. Several run-time errors specific to quantum computation can also be caught with a type system, with a few specificities that we discuss in this section.

7.6.3.1 Quantum Data and Type Linearity

The main problem with quantum information is that it is *non-duplicable* (a.k.a. *non clonable*, see Section 7.4.2.3). In terms of quantum programming language, this means that a program cannot duplicate a quantum bit: if U is a unitary map acting on two qubits, the function $\lambda x.U(x \otimes x)$ trying to feed U with two copies of its argument makes no sense. Similarly, it is not possible to control a gate acting on a qubit with the same qubit. The Quipper code of Figure 7.11 is, therefore, buggy.

```
exp :: Circ Qubit
exp = do
  q1 <- qinit True
  q2 <- qinit True
  r <- qnot q1 `controlled` q1
  return r
```

Figure 7.11 An example of a buggy Quipper program.

Type systems, in a broad sense, provide a predicate that says that a well-typed program does not have a certain class of bugs: in the case of quantum programming languages, a large class of bugs comes from duplicating non-duplicable objects. This calls for a *linear* type system, enforcing at least the non-duplication of qubits. This has been taken into account in recent scalable implementations such as Silq [27].

On the theoretical side, type systems for quantum Lambda-Calculi and ProtoQuipper [162]–the formalization of Quipper–are typically based on linear logic. Originally designed by Girard [79] (as a continuation of [122]), linear logic assumes that formulas are linear–i.e. non-duplicable and non-erasable–by default, and the logic comes equipped with a logic constructor "!" to annotate duplicable and erasable formulas. Linear logic also proposes a special pairing constructor $\otimes$ replacing the usual product and compatible with both the linearity constructs and the (linear) implication $\multimap$.

A core type system for a quantum Lambda-Calculus with pairing therefore consists of the following grammar:

$$A,B \quad ::= \quad \texttt{qubit} \mid \texttt{bool} \mid A \multimap B \mid A \otimes B \mid !A.$$

Type $A \multimap B$ represents the type of (linear) functions, using their argument only once. Type $A \otimes B$ represents the pair of a term of type A and a term of type B. Type $!A$ stands for a duplicable term of type A. We give a few examples as follows.

- The identity function $\lambda x.x$ is of type $A \multimap A$, but also of type $!(A \multimap A)$ as it is duplicable (since it does not contain any non-duplicable object);
- If the pairing construct is represented with $\langle -, - \rangle$, the function $\lambda x.\langle x, x \rangle$ is of type $!A \multimap (!A \otimes !A)$: it asks for a duplicable argument;
- The operator `qinit` is of type $!(\texttt{bool} \multimap \texttt{qubit})$: it is duplicable but it does *not* generate a duplicable qubit;
- The operator `meas` can however be typed with $!(\texttt{qubit} \multimap !\texttt{bool})$ as a boolean should be duplicable;
- Provided that U is a unitary acting on two qubits, one can type it in a functional manner with $\texttt{qubit} \otimes \texttt{qubit} \multimap \texttt{qubit} \otimes \texttt{qubit}$: it inputs two (non-duplicable) qubits and outputs the (still non-duplicable) modified qubits;
- In particular, provided that we assume implicit dereliction, casting duplicable elements of type $!A$ to A, the term $\lambda x.U\langle x, x \rangle$ can only be typed with $!\texttt{qubit} \multimap \texttt{qubit} \otimes \texttt{qubit}$: its argument has to be duplicable. The fact that this program can never be actually used on a concrete qubit is a property of the type system (intuitively, `qinit` only generates non-duplicable qubits).

7.6.3.2 Example: Quantum Teleportation

Mixing quantum computation and higher-order objects can yield non-trivial objects. The scheme of teleportation is given in Figure 7.12. It consists of three steps: (A) creation of a Bell state, (B) measure in the Bell basis to retrieve two booleans and (C) application of a gate $U_{b_1 b_2}$ dependent on the result of the measure. The state of the top wire is then "teleported" to the bottom wire. It is possible to understand the pieces of the quantum teleportation protocol as three (duplicable) functions:

$$\begin{array}{ll} (A) & !(() \multimap \texttt{qubit} \otimes \texttt{qubit}) \\ (B) & !(\texttt{qubit} \multimap (\texttt{qubit} \multimap \texttt{bool} \otimes \texttt{bool})) \\ (C) & !(\texttt{qubit} \multimap (\texttt{bool} \otimes \texttt{bool} \multimap \texttt{qubit})) \end{array}$$

The parts (B) and (C) are duplicable functions producing two non-duplicable functions of type $\texttt{qubit} \multimap \texttt{bool} \otimes \texttt{bool}$ and $\texttt{bool} \otimes \texttt{bool} \multimap$

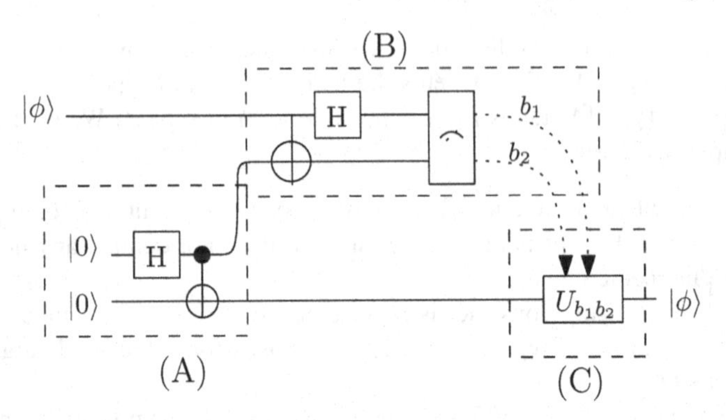

Figure 7.12 Teleportation algorithm.

qubit. The teleportation algorithm then feeds the two qubits emitted by (A) to (B) and (C); this gives a general type

$$!(() \multimap (\texttt{qubit} \multimap \texttt{bool} \otimes \texttt{bool}) \otimes (\texttt{bool} \otimes \texttt{bool} \multimap \texttt{qubit}))$$

for the protocol. It can be used several times (as it is duplicable). Each time it is run, it generates a pair of *non-duplicable* functions $\langle f, g \rangle$, and the specification of the protocol states that these two functions are inverse one of the other.

These two functions f and g of the pair are *non-duplicable*. Indeed, each of them holds a (non-duplicable) qubit coming from (A). Moreover, in a sense, these functions are entangled, since the Bell state from (A) is entangled.

7.6.3.3 Extending the Type System to Support Circuits

Let us assume that the type system of the quantum Lambda-Calculus is extended with lists: $[A]$ stands for lists of elements of type A (see e.g. [170] to see how to do this). A function $[\texttt{qubit}] \multimap [\texttt{qubit}]$ inputs a list of qubits. It can apply unitary gates to these qubit arguments: It can in fact describe different circuits, depending on the size of the list. Such a function , therefore, describes a *family* of circuits. The quantum Lambda-Calculus is not expressive enough to extract one circuit out of this family of circuits and operate on it (e.g. by inversing or controlling it).

The ProtoQuipper language [162] and its successors [160, 159, 77, 131] are formalized fragments of the programming language Quipper. They enforce structural properties of quantum programs using the linear type system of the quantum Lambda-Calculus, yet extending it to support circuit manipulation. ProtoQuipper comes with a new type construct $Circ(A,B)$: the type of circuits from A to B, and two functions:

- box sends $A \multimap B$ to $Circ(A,B)$. It takes a function $A \multimap B$, partially evaluates it, and stores the emitted circuit in an object of type $Circ(A,B)$.
- unbox sends $Circ(A,B)$ to $A \multimap B$. It takes a circuit from A to B and reads it as a function of input A and output B.

The behavior of box and unbox is specified by the operational semantic of the language [162].

One of the subtleties is the fact that box turns a function – possibly representing a family of many circuits – to *one* circuit. In the case of a function of type $[\mathtt{qubit}] \multimap [\mathtt{qubit}]$, this corresponds to choosing one size of list and building the circuit for this input size. Whenever the type system supports inductive types such as lists, the operator box then also takes a *shape* as a second argument, for deciding on the shape of the circuit to build. In recent works [77], ProtoQuipper's type system has been extended to very expressive *dependent types*, in order to characterize with a very fine-grain the shape structure of a family of circuits.

For instance, suppose that the program P sends $[\mathtt{qubit}]$ to $[\mathtt{qubit}]$. It corresponds to a family of circuits, but if we pick a choice of input size n, the type gives no information on the output shape of the circuit – that is, the number of output wires. Maybe P duplicates each input wire? With dependent types, we can for instance index list-types with size and type P with

$$\forall n.[\mathtt{qubit}]_n \multimap [\mathtt{qubit}]_{2n}.$$

This type tells us that P corresponds to a family of circuits of even output wires. This makes it possible to catch errors when using circuit combinators: for instance, the inverse operator can be typed with

$$\forall n\ m.Circ([\mathtt{qubit}]_m, [\mathtt{qubit}]_n) \multimap Circ([\mathtt{qubit}]_n, [\mathtt{qubit}]_m)$$

The inverse of P then becomes a function of type

$$\forall n.[\mathtt{qubit}]_{2n} \multimap [\mathtt{qubit}]_n.$$

In particular, this function can only be applied to lists of even sizes. This run-time error cannot be checked without shape information.

Although such a type system becomes very expressive, in general, it fails to feature a type inference algorithm, as this would require to be able to solve arbitrary arithmetic equations.

7.6.3.4 Dependent Types and Proofs of Programs

To build a dependently typed language, an alternative approach is to embed it inside an existing host language with this feature: QWIRE [147, 146] follows this route and relies on the language and proof-assistant Coq [25]. While Quipper uses Haskell's monads to encode circuits, QWIRE capitalizes on Coq's inductive types within the formalism of the Calculus of Inductive Constructions (CiC) [145], the logical framework of Coq.

In Haskell, inductive types are limited in expressivity: without dependent types, it is not possible to impose constraints on the content of a datastructure. As circuits in Quipper are internally made of elements of an inductive type, it makes it impossible to forbid ill-defined constructions such as re-using an erased wire, or using a wire twice on a controlled-not, as shown in Fig. 7.11.

QWIRE can instead rely on dependent, inductive types to enforce such constraints on circuit constructions: instead of simply considering circuits as lists of gates, a circuit in QWIRE consists of a list of gates together with proofs that the gates are added in a sensible way–in other words, the constructors of the inductive type of circuits in QWIRE corresponds to a set of *typing rules* for writing valid circuits.

Interestingly enough, QWIRE does not have to rely on "!" type constructor to distinguish between duplicable and non-duplicable data. The idea is that instead of working in a linear-logic-based type-system, QWIRE considers an equivalent linear/nonlinear model [23]. In this paradigm, there are two intertwined languages:

- A linear language aimed at qubit manipulation and gate application: a program in this language is a circuit. The linearity of the type system enforces the necessary constraints so that e.g. Fig. 7.11 is indeed invalid;
- A high-level, regular language–typically a lambda-calculus, with a regular type system. This language represents the "usual" programming paradigm where classical, conventional computation happens;
- Then there are two operations, akin to `box` and `unbox`, to move from the linear language to the classical language.

In the context of QWIRE, the classical, regular language is Coq and the linear language is encoded using the inductive type of circuits. As Coq features dependent types, QWIRE can then be regarded as a dependently-typed quantum programming language. However, due to the sophisticated type system, QWIRE does not feature a type inference algorithm.

7.6.3.5 Discussion

Type systems for quantum programming languages provide very efficient ways to encode and–whenever featuring a type inference algorithm–*automatically* verify some important properties of programs, and in particular to rule out at compile-time large classes of run-time errors specific to quantum computation. In particular, type systems have been used to characterize and enforce

- structure of parametric families of circuits;
- linearity of non-duplicable elements;
- control and inversion of only purely quantum circuits.

However, to be able to go further and characterize functional correctness with respect to specification, or validate the number of gates of a circuit, or catch subtle bugs involving concatenation of inverted circuits, one needs to move away from the simple linear type systems of quantum lambda-calculi and shift toward sophisticated dependent types, such as the extension ProtoQuipper-D [77] of ProtoQuipper, or the approach of QWIRE. The gain in expressiveness is then at the expense of automation [177, 147].

The quest for a finer trade-off, permitting automation while capturing some of what is currently only available with dependent type system is an active research area in the community.

7.7 HIGH- AND MID-LEVEL VERIFICATION: ALGORITHMS AND PROGRAMS

Most *quantum programming languages* (QISKIT [152] Quipper [83], LIQ$Ui|\rangle$ [193], Q# [181], PROJECTQ [180], SILQ [27], *etc*) embed features for quantum circuit manipulations within a standard classical programming language. Such *circuit-building quantum languages* is the current consensus for high-level executable quantum programming languages. A current major challenge is to link this language design paradigm with formally verified programming. In the present section, we introduce the main existing propositions in that direction.

7.7.1 QUANTUM HOARE LOGIC

Quantum Hoare logic (QHL) [37, 66, 106, 194, 185, 184, 19] is a general framework for reasoning about the classical control instructions over unitary operations in quantum algorithms. Referring again to Figure 7.1, the focus is on the interaction between the classical computer and the quantum co-processor, instead of the circuit analysis as in Section 7.5, or the gate to gate circuit building functions as in Sections 7.7.2 and 7.7.3. Therefore, we consider it as a high-level description of quantum algorithms.

It is based on the assertion method of Floyd and Hoare [74, 94]–attach each program point with an assertion and whenever the data flow reaches a program point the attached assertion should be satisfied–which was originated with Alan Turing [8]. Hoare's approach enables (interactive) theorem proving for high-level algorithmic description verification that proceeds at the same abstraction level as the language itself. This makes verification more human-friendly than lower-level (machine-friendly) verification.

7.7.1.1 Quantum Programming Language: Quantum WHILE-Programs

The guideline for the hybrid model introduced in Section 7.2.1.1 is summed up by the slogan "*quantum data and classical control*" [168]: quantum data can be superposed and entangled, they are manipulated by basic quantum operations–unitary evolution and measurement, but the high-level control is still classical (e.g. branch, loops, *etc*).

In light of this slogan, QHL introduces a minimal programming language for describing quantum algorithms [195]. We follow [194] for the introduction of QHL technical environment. Let q (resp. $\bar{q}$) be a quantum variable (resp. a list of quantum variables); let U be a unitary operator acting on the qubits $\bar{q}$ and let $M \triangleq \{M_m\}_m$ with $\sum_m M_m^\dagger M_m = I$ be a measurement on the qubits $\bar{q}$, each M_m corresponding to a measurement result m (see Section 7.2.1.5). As a special case, let $M' \triangleq \{M_0, M_1\}$ with $M_0^\dagger M_0 + M_1^\dagger M_1 = I$. Then quantum WHILE-programs are generated by the following syntax.

S	$\triangleq$	**skip**	No operation
	$\mid$	$q := \lvert 0\rangle$	Initialization
	$\mid$	$\bar{q} \mathrel{*}= U$	Unitary operation
	$\mid$	$S_1; S_2$	Sequential composition
	$\mid$	**if** $\square m \cdot M[\bar{q}] = m \rightarrow S_m$ **fi**	Probabilistic branching
	$\mid$	**while** $M'[\bar{q}] = 1$ **do** S_0 **od**	Probabilistic while loop

The intended semantics of language constructs above is similar to that of their classical counterparts. To illustrate the quantum features contained in these constructs, we make the following comments:

(i) in the initialization, the choice of a fixed state $|0\rangle$ is without loss of generality, since any known quantum state can be prepared by applying a unitary operator to $|0\rangle$;

(ii) according to Born rule, measurement results follow a probabilistic law. Since they lie on measurement result observations, branching (resp. while loops) is therefore probabilistic. It creates different branches $\{S_m\}_m$ (resp. $\{\textbf{skip}, S_0\}$), chosen according to outcomes of the measurement M (resp. M') on the qubits $\bar{q}$.

We refer the reader to [194] for a detailed exposition of the syntax above.

Example 7.7.1 (Preparation of the Bell state $|\beta_{00}\rangle$). *Let p and q be quantum variables, each denoting one qubit. Then the following program initiates them to* $|0\rangle$ *and implements the circuit from Figure 7.2, preparing state* $|\beta_{00}\rangle$*from Example 7.2.1.:*

$$\beta_{00} \quad \triangleq \quad p := |0\rangle;\ q := |0\rangle;\ p *= H;\ (p,q) *= CNOT$$

Note that the quantum programming language defined above is in the spirit of the hybrid circuit model presented in Section 7.2. Indeed, the basic sequence of quantum operations (initialization, unitary operation, and measurement) are meant to be interpreted as a generalized quantum circuit to be executed on a quantum co-processor; and post-measurement branchings (in, e.g., probabilistic branching and while loop) are meant to be controlled by a classical computer.

7.7.1.2 Quantum States, Operations and Predicates

Measuring a quantum state transforms it, following the Born rule (see Section 7.2.1.2). The resulting probability distribution over quantum states is formalized as a *mixed state* (as opposed to *pure states*, see Section 7.2.1.5). For example, a measurement on any pure quantum state $|+\rangle \triangleq \frac{1}{\sqrt{2}}(|0\rangle + |1\rangle)$ or $|-\rangle \triangleq \frac{1}{\sqrt{2}}(|0\rangle - |1\rangle)$ will result in the mixed state $\mathbb{E} = \{(\frac{1}{2}, |0\rangle), (\frac{1}{2}, |1\rangle)\}$, with states $|0\rangle$ and $|1\rangle$ occurring with an equal probability of $\frac{1}{2}$ (notice that this observation makes both states $|+\rangle$ $|-\rangle$ impossible to distinguish by simple measurement).

In this way, the representation of the final state of applying a series of measurements to a quantum state could expand exponentially. To address this issue, a square-matrix representation of quantum states, i.e. partial density operator, is adopted instead. For example, pure quantum state $|+\rangle$ is represented as $|+\rangle\langle+|$, and mixed quantum state $\mathbb{E}$ as $\frac{1}{2}|0\rangle\langle 0| + \frac{1}{2}|1\rangle\langle 1|$. See Section 7.2.1.5 for a brief introduction to partial density operators, and [168, 195, 140] for further details.

If we see the matrix representation of a quantum state (partial density operator) as a linear operator, then a quantum operation– initialization, unitary evolution and measurement–can be thought of as a super operator, *i.e.* a function from linear operators to linear operators. What's interesting is that every quantum WHILE-program defined above can be interpreted as a super operator, and partial density operators are closed under super operators. This justifies the success of representing quantum states as partial density operators and defining the denotational semantics of quantum programs as super operators [168, 195].

Following [56], a quantum predicate on vector space H is defined as a Hermitian operator M between the zero operator 0_H (representing the contradiction) and the identity operator I_H (representing the tautology). Instead of the usual binary satisfaction judgment, QHL evaluates the satisfaction of a predicate by a state as a real value between 0 (false) and 1 (true). It is defined as the trace $tr(M\rho)$ of the product $M\rho$. Intuitively, it represents the expectation for the truth value of M in the mixed state ρ (which is, again, a probability distribution over pure states).

Then, the intuition of implication between predicates is also probabilistic. It is filled by the *Löwner order* $M \sqsubseteq N$, relating operators M and N, if and only if, for any state ρ, the expectation truth value of N in ρ is more or equal to that of M in ρ. This condition is formalized as $tr(M\rho) \leq tr(N\rho)$ for all states ρ (See, e.g., [194, Lemma 2.1]).

Adopting such quantum predicates as assertions, among many others (e.g., interpreted as physical observables), provides simple expression means for many properties of quantum effects. For example, quantum predicate $|+\rangle\langle+|$ expresses that a state ρ is in the equal superposition $|+\rangle$ with probability $tr(|+\rangle\langle+|\rho)$; quantum predicate $|\beta_{00}\rangle\langle\beta_{00}|$ expresses that a state ρ is in the maximal entanglement $|\beta_{00}\rangle$ with probability $tr(|\beta_{00}\rangle\langle\beta_{00}|\rho)$, *etc.*

7.7.1.3 Quantum Program Verification

For now, a quantum (partial) correctness formula can be the Hoare's triple $\{P\}\ S\ \{Q\}$, where S is a quantum WHILE-program, and P, Q are quantum predicates. To define the partial-correctness semantics of quantum Hoare's triples, in the sequel, let $[\![S]\!]$ denote the semantic function of S (Note that $[\![S]\!]$ is a quantum operation defined by induction on S, cf. [195]), and $[\![S]\!](\rho)$ the output of S on the input ρ.

Definition 7.7.1 (Semantics of partial correctness, cf. [194]). *Let P, Q be quantum predicates and S a quantum* WHILE*-program. We say that S is (partially) correct w.r.t. precondition P and postcondition Q, written* $\models \{P\}\ S\ \{Q\}$, *if*

$$\forall \rho, \quad tr(P\rho) \;\leq\; tr(Q[\![S]\!](\rho)) + \left[tr(\rho) - tr([\![S]\!](\rho))\right]. \tag{7.4}$$

Note that Inequality (7.4) can be seen as a probabilistic version of the following statement: if state ρ satisfies predicate P, then, executing program S on input ρ, either S fails to terminate or the resulting state $[\![S]\!](\rho)$ satisfies predicate Q.

Table 7.2
Proof system for partial correctness.

(Skip Axiom)	$\{P\}$ **skip** $\{P\}$
(Init Axiom)	$\{\sum_i \vert i\rangle_q \langle 0\vert P\vert 0\rangle_q \langle i\vert\}\ q := \vert 0\rangle\ \{P\}$
(Unit Axiom)	$\{U^\dagger PU\}\ \bar{q} \mathrel{*}= U\ \{P\}$
(Comp Rule)	$\dfrac{\{P\}\ S_1\ \{Q\} \quad \{Q\}\ S_2\ \{R\}}{\{P\}\ S_1;S_2\ \{R\}}$
(If Rule)	$\dfrac{\{P_m\}\ S_m\ \{Q\} \text{ for all } m}{\{\sum_m M_m^\dagger P_m M_m\}\ \textbf{if}\ \Box m{\cdot}M[\bar{q}]{=}m{\rightarrow}S_m\ \textbf{fi}\ \{Q\}}$
(Par Loop Rule)	$\dfrac{\{P\}\ S_0\ \{M_0^\dagger QM_0{+}M_1^\dagger PM_1\}}{\{M_0^\dagger QM_0{+}M_1^\dagger PM_1\}\ \textbf{while}\ M'[\bar{q}]{=}1\ \textbf{do}\ S_0\ \textbf{od}\ \{Q\}}$
(Order Rule)	$\dfrac{P{\sqsubseteq}P' \quad \{P'\}\ S\ \{Q'\} \quad Q'{\sqsubseteq}Q}{\{P\}\ S\ \{Q\}}$

The axiom system for proving partial correctness of quantum WHILE-programs is composed of axioms and inference rules manipulating quantum Hoare's triples [194]. It is shown in Table 7.2 $\big($where $\{|i\rangle\}_q$ is the computational basis for quantum variable $q\big)$. Remark that each of these rules and axioms follows the assertion method. Here we only show how to derive the most complex rule (Par Loop Rule). The derivation of other proof rules can be done similarly.

Intuition of (Par Loop Rule).

To derive (Par Loop Rule), by assertion method, we attach each program point, say l_1, l_2, l_3, of a WHILE-statement with an assertion, say R, P, Q, respectively:

$$\{l_1 : R\}\ \textbf{while}\ M'[\bar{q}] = 1\ \textbf{do}\ \{l_2 : P\}\ S\ \textbf{od}\ \{l_3 : Q\}$$

Fix the input ρ at the program point l_1 satisfying the assertion R. By semantics of a WHILE loop, after the measurement M', one part $M_1 \rho M_1^\dagger$

of the input will go to the loop body through the program point l_2 where the assertion P will be satisfied; the other part $M_0 \rho M_0^\dagger$ will leave the while loop through the program point l_3 in which the assertion Q will be satisfied. Hence:

$$tr(R\rho) \quad \leq \quad tr\big(Q(M_0 \rho M_0^\dagger)\big) + tr\big(P(M_1 \rho M_1^\dagger)\big)$$

Due to the arbitrariness of ρ, by properties of the trace function and Löwner order, we have that $R \sqsubseteq M_0^\dagger Q M_0 + M_1^\dagger P M_1$. Then, by weakening R to $M_0^\dagger Q M_0 + M_1^\dagger P M_1$ and lifting the above reasoning process into an inference rule, (Par Loop Rule) follows.

The following example illustrates how to derive a partially correct quantum Hoare triple using the axioms and inference rules presented above.

Example 7.7.2 (Specification and correctness proof for the $|\beta_{00}\rangle$ state construction program)**.** *Recall from Example 7.7.1 the definition of quantum program* β_{00}*:*

$$\beta_{00} \quad \triangleq \quad p := |0\rangle;\ q := |0\rangle;\ p *= H;\ (p,q) *= CNOT$$

To show (partial) correctness of this program, it suffices to prove

$$\{I_p \otimes I_q\}\ \beta_{00}\ \{|\beta_{00}\rangle_{p,q}\ \langle\beta_{00}|\} \tag{7.5}$$

This can be done as follows. For the sake of space we need to decompose the derivation tree. We first derive specifications for the initialization instructions:

$$\dfrac{\dfrac{}{\{I_p \otimes I_q\}\ p := |0\rangle\ \{|0\rangle_p\ \langle 0| \otimes I_q\}}\text{Init Axiom} \qquad \dfrac{}{\{|0\rangle_p\ \langle 0| \otimes I_q\}\ q := |0\rangle\ \{|0\rangle_p\ \langle 0| \otimes |0\rangle_q\ \langle 0|\}}\text{Init Axiom}}{\{I_p \otimes I_q\}\ p := |0\rangle;\ q := |0\rangle\ \{|0\rangle_p\ \langle 0| \otimes |0\rangle_q\ \langle 0|\} \quad (i)}\text{Comp Rule}$$

Then come the unitary application instructions. We set the following abbreviations for Hoare triples:

$(ii) \ := \ \{|0\rangle_p\ \langle 0| \otimes |0\rangle_q\ \langle 0|\}\ p *= H\ \{|+\rangle_p\ \langle +| \otimes |0\rangle_q\ \langle 0|\}$

$(iii) \ := \ \{|+\rangle_p\ \langle +| \otimes |0\rangle_q\ \langle 0|\}\ (p,q) *= CNOT\ \{|\beta_{00}\rangle_{p,q}\ \langle\beta_{00}|\}$

They are instances of axioms and we can combine them:

$$\dfrac{\dfrac{}{(ii)}\text{Init Axiom} \qquad \dfrac{}{(iii)}\text{Init Axiom}}{\{|0\rangle_p\ \langle 0| \otimes |0\rangle_q\ \langle 0|\}\ p *= H;\ (p,q) *= CNOT\ \{|\beta_{00}\rangle_{p,q}\ \langle\beta_{00}|\} \quad (iv)}\text{Comp Rule}$$

Finally, the two preceding proof tree branch together via the sequential composition rule, which achieves the derivation:

$$\dfrac{(i) \qquad (iv)}{\{I_p \otimes I_q\}\ \beta_{00}\ \{|\beta_{00}\rangle_{p,q}\ \langle\beta_{00}|\}}\text{Comp Rule}$$

7.7.1.4 Implementations and Extensions

Several works have taken advantage of extended Quantum Hoare Logic, e.g. algorithmic analysis of termination problem [128] or characterization and generation of loop invariants $\big($i.e. $M_0^\dagger Q M_0 + M_1^\dagger P M_1$ in (Par Loop Rule)$\big)$ [198].

The practical illustration of QHL can be found in Liu et al. paper [133], containing an implementation in ISABELLE/HOL together with a formalization of Grover [84] and Quantum Phase Estimation (QPE [112]) algorithms. Nevertheless in these examples, the central verification part is assumed through Python libraries uses.

More recent work [132] includes full proof for a parametrized version of Grover's search algorithm. It constitutes an illustration of QHL use on a non-trivial example.

7.7.1.5 Other Quantum Hoare Logics

In addition to the framework introduced above, applying Hoare Logic methods to quantum field brought several additional developments, focusing on different aspects of quantum computations. We introduce a few of these QHL-related framework in the following paragraphs.

7.7.1.5.1 Quantum Hoare Logic with ghost variables

One of the principal shortcuts of [194] comes from the limitations of the specification language. In [184], the author in particular targets the possibility to characterize probabilistic distributions of values. His proposition introduces *ghost* variables in the specification language: a *ghost* variable does not occur in the program but only in its specifications. In [184], a ghost variable is interpreted under an implicit existential quantification. Ghost variables enable, in particular, to explicitly refer to discarded, measured or overwritten qubits. In addition to probability distribution definitions, it brings several features to the expressive power of the specifications language, such as separability–unentanglement–of variables or the fact that a given variable holds a classical data value.

7.7.1.5.2 Quantum Hoare Logic with classical variables

With similar concern, [67] extends QHL specifications with classical variables. In the specification and verification of algorithms of practical use, holding classical information is indeed crucial. It enables, for example, describing and specifying an algorithm holding classical parameters or a hybrid program intertwining classical and quantum instructions (see Section 7.2.1.1).

Hence, the semantics relies on so-called *cq-states* (classical/quantum states) made of both:

- a classical variable assignment for the interpretation of classical variables;
- a density operator interpreting quantum variables.

This extension of the semantics induces an extension of the proof system, which is proved sound and complete–with respect to partial correctness. Interestingly, the paper presents detailed specified case studies, including Grover's and Shor's algorithms.

7.7.1.5.3 Robustness analysis

is another important line of work, initiated in [97] as a continuation of QHL. The different solutions presented so far rely on the implicit assumption that quantum gates are applied deterministically as indicated by their matrix semantics. Still, as stated in Section 7.2.3.2, this noise-free modeling is not perfectly accurate in the NISQ era.

A more realistic description of the behavior of a gate would consider several different possible behaviors, weighed with their respective probabilities of occurrence: the *intended* one, expected by the semantics, and one or several additional erroneous behaviors[10].

Interestingly, in this setting, the application of a quantum gate is formally represented as a probability distribution over unitaries. Hence, a non-deterministic gate application naturally formalizes as super operators acting over density operators.

Then, a metrics is defined for measuring the difference between the behavior of a quantum system under a given error scheme (for each gate application, the mention of a possible erroneous behavior together with its probability of occurrence) and its intended behavior. It is called the *trace distance* and serves as an evaluation for the robustness of the implementation.

Among other case studies, the method is illustrated by evaluating a minimal quantum error correcting scheme.

7.7.1.5.4 Quantum separation logic

[201] aims at simplifying quantum programs specifications. The leading observation is that, while quantum programs often manipulate big

[10]Strictly speaking, the language in 7.2.3.2 is limited to considering up to one possible error per gate application. But this is without lost of generality since more sophisticated scenarios can be encoded through, eg., the replacement of gate application intervals by potentially erroneous identity transformations.

quantum circuits, with matrix semantics growing exponentially over their width, many quantum algorithms proceed via sequences of local manipulations over quantum sub-registers. The authors exhibit the examples of Quantum Machine Learning [26] and Variational Quantum Algorithms (VQA [136]), which are among the most promising classes of algorithms in the NISQ era (see Section 7.2.3.2).

To efficiently reason about state evolution in such implementations, a Quantum Separation Logic is proposed, together with a dedicated proof system. This logic allows the expression of local manipulation on separated quantum registers while maintaining the state of the rest of the register. The approach is probed with case studies from both quantum programming analysis (VQA) and communication protocols security checks (one-time pad and secret sharing).

7.7.1.5.5 Quantum relational Hoare logic

[185, 19] allows to reason about how the outputs of two quantum programs relate to each other given a relation between their inputs, which can be used to analyze security of post-quantum cryptography and quantum protocols.

7.7.1.5.6 Quantum Hoare type theory

[177] is inspired by classical Hoare type theory and extends the Quantum IO Monad [3] by indexing it with pre- and postconditions that serve as program specifications, which has the potential to be a unified system for programming, specifying, and reasoning about quantum programs.

7.7.1.5.7 Quantum dynamic logic.

We end up this section by mentioning a line of work that is not a strict extension or application of Hoare Logic [94], but shares similar concerns and related solutions. Just as QHL and its extensions, dynamic logic formalizes the evolution of a state along the execution of a process acting over it.

Dynamic logic inherits from the modal logic apparatus where, in addition to propositional logic connectives, modalities $\Box$ and $\Diamond$ are introduced. Intuitively, given a formula $\varphi, \Box\varphi$ means that φ is *necessarily* true and $\Diamond\varphi$ means that φ is *possibly* true. The standard semantics, based on Kripke models [116], helps catching this intuition. A Kripke model $\mathscr{K}$ is made of a set of states S holding, each, a valuation $P : S \to \{\textbf{true}, \textbf{false}\}$ for a set of propositional variables P. And the modality is interpreted through an accessibility relation $R \subseteq S \times S$ For example, given a propositional variable p, $\Box p$ (resp. $\Diamond p$) is true in state s, written $\mathscr{K}, s \models \Box p$, iff p (resp. $\mathscr{K}, s \models \Diamond p$) is true in every (resp. at least one) state s' such that $(s, s') \in R$.

In dynamic logic, several modalities coexist, formalizing a set of *actions*. Hence, given an action a, formula $[a]\varphi$ is true iff φ is always true after performing action a. Actions can combine together, for example $[a;a']\varphi$ means that φ is always true after the successive performance of actions a and a'.

In their *Logic of Quantum Programs* (LQP [14, 15]), Baltag and Smets designed a quantum version of dynamic logic. Here, states correspond to one-dimensional subspaces of a Hilbert space for a quantum register. Actions are of two kinds:

- a *test* encodes a measurement through the set of projection transitions corresponding to the different possible outcomes (for example, given formulas φ and ψ, $\varphi?\psi$ is satisfied in the state of evaluation if and only if any state satisfying φ after measurement–any successful test for φ–also satisfies ψ),
- name *action* is reserved for the second kind. An *action* deterministically encodes a unitary state transformation.

In [14], the framework is illustrated by a correction proof for the teleportation and quantum secret sharing protocols. Baltag and Smets link their work to similar quantum logics as in [34].

Note that to treat measurement, LQP formalizes information about the set of states that are possibly reachable, but the language does not hold any notion of probability. This shortcut is overcome in [13], where the authors introduce a probabilistic test modality characterizing, for any formula φ and rational $x \in [0,1[$, that a test for φ has probability at least x to succeed from the evaluation state. In [13], the authors also provide a decidability proof for their setting and illustrate its expressive power by formally specifying Grover's search algorithm and a leader election protocol. Further work in the community has used this probabilistic logic setting for the verification of the BB84 Quantum Key Distribution algorithm [24].

7.7.2 Qbricks

QBRICKS [38] is a recently proposed circuit description language together with a deductive verification framework. It enables automated proof support for program specifications, reducing the required human effort for the development of verified programs.

QBRICKS object language (QBRICKS-DSL) consists in a minimal functional language with features for the design of circuit families. Similarly to the formal contract style of algorithm descriptions (see Section 7.2.2.2), QBRICKS functions are written with explicit pre-and post-conditions, specifying their complexity and the parametrized input-output quantum data registers function they implement. These specifications are written in a dedicated formal language, called QBRICKS-SPEC.

To support proofs, QBRICKS is given a Hoare style derivation rules system including rules for each parametrized circuit constructor. These rules are enriched with equational theories enabling, in particular, reasoning about measurement and probabilities.

QBRICKS is a domain-specific language, embedded in the Why3 [70] deductive verification framework : programs are written in ML language and annotated with specifications in QBRICKS-DSL (pre- and postconditions, loop invariants, calls for lemmas, *etc*). Compiling a QBRICKS program interprets these specifications as proof obligations. Then, a dedicated interface enables to directly access these proof obligations and either send them to a set of automatic SMT-solvers (CVC4, Alt-Ergo, Z3, *etc.*), or enter some interactive proof transformation commands (additional calls for lemmas or hypotheses, term substitutions, *etc.*) or even to proof assistants (Coq, ISABELLE/HOL).

7.7.2.1 Writing Quantum Circuits Functions in Qbricks: Qbricks-DSL

QBRICKS-DSL makes use of a regular inductive datatype for circuits, where the data constructors are elementary gates, sequential and parallel composition, and ancilla creation. In particular, unlike in e.g. Quipper or QWIRE, a quantum circuit in QBRICKS is not a function acting on qubits: it is a simple, static object. Nonetheless, for the sake of implementing quantum circuits from the literature, this does not restrict expressiveness as they are usually precisely represented as sequences of blocks.

The core of QBRICKS-DSL is presented in Figure 7.13. It is a small first-order functional, call-by-value language. To the elementary gates presented in Section 7.2.1.5, QBRICKS adds the qubit swapping gate `SWAP` and the identity `ID`. The constructors for high-level circuit operations are sequential composition `SEQ`, parallel composition `PAR` and ancilla creation/termination `ANC`.

$$
\begin{array}{rrcl}
\text{Expression} & e & ::= & x \mid c \mid f(e_1,\dots,e_n) \mid \texttt{let}\,\langle x_1,\dots,x_n\rangle = e\,\texttt{in}\,e' \mid \\
 & & & \texttt{if}\,e_1\,\texttt{then}\,e_2\,\texttt{else}\,e_3 \mid \texttt{iter}\,f\,e_1\,e_2 \\
\text{Data Constructor} & c & ::= & \underline{n} \mid \texttt{tt} \mid \texttt{ff} \mid \langle e_1,\dots,e_n\rangle \mid \texttt{CNOT} \mid \texttt{SWAP} \mid \texttt{ID} \mid \texttt{H} \mid \texttt{Ph}(e) \mid \\
 & & & \texttt{R}_z(e) \mid \texttt{ANC}(e) \mid \texttt{SEQ}(e_1,e_2) \mid \texttt{PAR}(e_1,e_2) \\
\text{Function} & f & ::= & f_{\text{d}} \mid f_{\text{c}} \\
\text{Declaration} & d & ::= & \texttt{let}\,f_{\text{d}}(x_1,\dots,x_n) = e
\end{array}
$$

Figure 7.13 The syntax for QBRICKS-DSL.

The term constructs are limited to function calls, `let`-style composition, test with the ternary construct `if-then-else` and simple iteration: `iter` $f\ n\ a$ stands for $f(f(\cdots f(a)\cdots))$, a succession of n calls to f.

Even though the language does not feature measurement, it is nonetheless possible to *reason* on probabilistic outputs of circuits, if we were to measure its output. This is expressed in a regular theory of real and complex numbers in the specification language (see Section 7.7.2.4 below for details).

7.7.2.2 Parametrized Path-Sums

To interpret circuit description functions, QBRICKS uses parametrized path-sums (pps), that is an extension of path-sums [5] (Section 7.5.2).

In QBRICKS setting, a path-sum P is an object from an opaque type with four parameters, presented in Table 7.3 with their types and identifier shortcuts: two integer constants `pps_width`(P) (the width of the target circuit) and `pps_range`(P) (the *range*, meaning that the output sum of kets has a term for each bit vector $\vec{y}$ of length `pps_range`(P)–written $\vec{y} \in \mathrm{BV}_{\texttt{p\_r}(P)}$) and two functions `pps_angle`$(P)$ and `pps_ket`(P) : for any input bit vector $\vec{x}$ of length `pps_width`(p) (standing for a basis ket input to the target circuit) and for any index bit vector $\vec{y}$ of length `pps_range`(p), functions `pps_angle`(P) and `pps_ket`(P) respectively define a real scalar and a bit vector of length `pps_width`(p) (standing for a basis ket output to the target circuit)[11].

Table 7.3

Pps **accessors and types.**

Identifier	Type	Id-abbreviation
`pps_width`	`int`	`p_w`
`pps_range`	`int`	`p_r`
`pps_angle`	`bit_vector` → `bit_vector` → `real`	`p_a`
`pps_ket`	`bit_vector` → `bit_vector` → `bit_vector`	`p_k`

Then for any bit vector $\vec{x}$ of size `p_w`(P), the expression

$$Ps(h, |\vec{x}\rangle) = \frac{1}{\sqrt{2^{\texttt{p\_r}(P)}}} \sum_{\vec{y} \in \mathrm{BV}_{\texttt{p\_r}(P)}} e^{2\cdot\pi i.\texttt{p\_a}(P)(\vec{x},\vec{y})} |\texttt{p\_k}(P)(\vec{x},\vec{y})\rangle_{\texttt{p\_w}(P)} \quad (7.6)$$

[11] In the rest of this section, type `bit_vector` corresponds to bit vectors $\vec{x}$. They are encoded in an abstract type that provides a positive integer `length`$(\vec{x})$ (the length of the vector) and a value function `get_bv`$(\vec{x})$: `int` → `int`. For any integer i, we commonly abbreviate `get_bv`$(\vec{x})i$ as $\vec{x}_i$; if $i \in [\![0, \texttt{length}(\vec{x})[\![$ (that is, if i is actually in the range of the bit vector) then it is such that $0 \le \vec{x}_i < 2$.

combines these different element together to define a linear application for quantum state vectors. Function Ps is extended by linearity to any ket $|u\rangle$ of length $\mathtt{p\_w}(p)$[12].

A path-sum P is said to *correctly interpret a given circuit* C (written $(C \triangleright P)$) if and only if C has width $n = \mathtt{pps\_width}(P)$ and for any bit vector $\vec{x}$ of length n, $\mathbf{Mat}(C) \cdot |\vec{x}\rangle = Ps(P, |\vec{x}\rangle)$. The relation $(^{\cdot} \triangleright {}^{\cdot})$ enjoys nice composition properties along QBRICKS-DSL circuit constructors (see [38]).

QBRICKS-SPEC generalizes path-sums by introducing Parametrized path-sums (`pps`). A `pps` is a function that inputs a set of parameters and outputs a path-sum. Then, it can be seen as a family of path-sums (one for each possible value of its parameters) describing the effects of the different members in a family of quantum circuits. Hence, it is well-fitted for the specification of parametrized algorithms such as Shor order finding (`Shor-OF` see Figure 7.4).

The main strength of `pps` semantics, with regards to formal verification, is that each of the path-sum parametrized accessors (see Table 7.3) combines compositionally along with circuit constructors.

Hence, it enables reasoning about parametrized quantum circuits and their semantics without manipulating sum terms or other higher-order objects. Thanks to this tool, the automatic generation of proof obligations for QBRICKS specifications results in only first-order formulas, enabling a high level of automation when sent to SMT-solvers.

7.7.2.3 From Quantum Circuits to Path-Sums

The specification language for QBRICKS is a first-order predicate language, equipped with various equational theories. For any quantum circuit family C, QBRICKS-DSL enables to identify a `pps` $\mathtt{circ\_to\_pps}(C)$. Each of its parametrized accessors is defined inductively upon the structure of C and it is proved that $(C \triangleright \mathtt{circ\_to\_pps}(C))$, for any instance of circuit. These accessors are listed and given abbreviations in Table 7.4.

7.7.2.4 Probabilistic Reasoning

QBRICKS-DSL does not contain any constructor for the measurement of quantum registers. Nevertheless, QBRICKS-SPEC provides reasoning tools about it. In particular, function

$$\mathtt{proba\_measure} : \mathtt{circ} \times \mathtt{ket} \times \mathtt{int} \to \mathtt{real}$$

[12] That is, to any linear combination of basis kets $|\vec{x}\rangle$ of length $\mathtt{p\_w}(p)$.

Table 7.4
Function `circ_to_pps` and accessors.

Accessor	Abbreviation
pps_width(circ_to_pps(C))	C_width(C)
pps_range(circ_to_pps(C))	C_range(C)
pps_angle(circ_to_pps(C))	C_angle($C,-,-$)
pps_ket(circ_to_pps(C))	C_ket($C,-,-$)

inputs a circuit C, a quantum data register $|v\rangle_n$ and an index $j \in [\![0, 2^n[\![$. It outputs the probability, for one measuring the quantum register resulting from applying circuit C to $|v\rangle$, to get the bit vector representing integer j as a result. Function `proba_measure` is defined, for any input ket $|u\rangle$ of length pps_width(C), by application of the Born rule (see Section 7.2.1.2), as

$$\texttt{proba\_measure}(C, |i\rangle_n, j) = |(Ps(\texttt{circ\_to\_pps}, |i\rangle_n)(j)|^2.$$

QBRICKS-SPEC also provides similar reasoning ghost functions for discussing the effect of partial measurement over quantum sub-registers.

7.7.2.5 Verified Properties

The QBRICKS framework aims at providing tools for writing and verifying the standard format of quantum algorithm specifications as they appear in algorithms (see, e.g., Figure 7.4): to perform a given computation task with a given amount of resources. Hence, as illustrated above, QBRICKS-SPEC is designed for the formalization of both:

- parametrized input/output relations for families of circuits. For a family of circuits, they typically consist in characterizing their parametrized output ket vector. Thanks to function `proba_measure`, QBRICKS enables to identify the probability to get a given result after measurement and derivated probability reasoning (such as bounding the parametrized probability of success of a computation, *wrt* a pre-defined success condition). Proof support for these specifications is processed through the pps formalism,
- complexity specifications: QBRICKS enables to specify the parametrized width, number of required ancilla qubits and number of elementary gates of a circuit family.

Example 7.7.3 (Pps specifications for the Bell generating circuit). *For example, a specification for the Bell generating circuit can be written using accessors of* `pps circ_to_pps` *(Bell-circuit), as below*

$$\Gamma, \overrightarrow{x}, \overrightarrow{y} : \mathtt{bit\_vector}, j : \mathtt{int} \vdash$$
$$\{ \ \mathtt{bv\_length}(\vec{x}) = 2 \ \wedge \ \mathtt{bv\_length}(\vec{y}) = 1 \ \wedge \ j \in [\![0,2[\![\ \}$$
$$\textit{Bell-circuit}$$
$$\left\{ \begin{array}{l} \mathtt{C\_width}(\mathtt{result}) = 2 \ \wedge \ \mathtt{C\_ket}(\mathtt{result}, \overrightarrow{x}, \overrightarrow{y}) = \overrightarrow{x(0) \cdot (1 - x(1))} \ \wedge \\ \mathtt{C\_range}(\mathtt{result}) = 1 \ \wedge \ \mathtt{C\_angle}(\mathtt{result}\overrightarrow{x}, \overrightarrow{y}) = x(0) * y(0) \ \wedge \\ \mathtt{width}(\mathtt{result}) = 2 \ \wedge \ \mathtt{size}(\mathtt{result}) = 2 \ \wedge \ \mathtt{ancillas}(\mathtt{result}) = 0 \end{array} \right\}$$

This Hoare style notation, $\{Pre\}p\{Post\}$ states that whenever *Pre* is satisfied, then running p ensures that *Post* is satisfied. Formula *Post* uses `result` as a variable standing for program p. The specification uses free–*ghost*–bit vectors variables $\vec{x}$ and $\vec{y}$ and integer variable j. It requires $\vec{x}$ and $\vec{y}$ to have respective lengths 2 and 1 and j to be in $[\![0,2[\![$. Given these preconditions, the specifications ensures that the angle `C_angle` outputs $\overrightarrow{x}_0 * \overrightarrow{y}_0$ for inputs $\vec{x}$ and $\vec{y}$ and the ket function `C_ket` outputs $\overrightarrow{x}_0 \cdot (1 - \overrightarrow{x}_1)$ for inputs $\vec{x}, \vec{y}$ and j. Then, one easily verifies that, applying equation 7.6 on each bit-vector $\overrightarrow{a \cdot b}$ of length 2 results in the corresponding output $|\beta_{ab}\rangle$, formally:

$$Ps(\mathtt{circ\_to\_pps}(\textit{Bell-circuit}), |ab\rangle) = \mathbf{Mat}(\textit{Bell-circuit}) \cdot |ab\rangle)$$

So the postcondition, by use of `C_width`, `C_range`, `C_angle` and `C_ket`, enables a complete characterization of the input/output function performed by the Bell circuit. In addition, the specification brings some complexity related postcondition: the circuit has width 2 and size 2 (*i.e.*, length of the required quantum register and number of performed elementary operations), and does not use any additional ancilla qubit.

7.7.2.6 Deduction and Proof Support

Proof support in QBRICKS strongly relies on the compositional structure of quantum circuits, enabling compositional reasoning on both pps and complexity features. As an example, in Figure 7.14 we give some of the rules used for the characterization of circuits size. Gates `ID` and `SWAP` are considered as free, so they count for null. The other elementary gates have size 1, both sequence and parallel compositions sum the size of their components and ancilla creation/termination does not affect circuit size.

Similar deduction rules are defined for circuit width, number of ancilla qubits and pps accessors. We do not introduce them here for the sake of concision but we refer the desirous reader to [38].

$$\frac{C \in \{\texttt{ID},\texttt{SWAP}\}}{\texttt{size}(C) = 0}\ (\texttt{Id-SWAP-size}) \qquad \frac{C \in \{\texttt{H},\texttt{Ph}(n),\texttt{R}_z(n)\}}{\texttt{size}(C) = 1}\ (\texttt{H-Ph-R}_z\texttt{-size})$$

$$\frac{\Gamma \vdash \texttt{size}(C_1) = n_1 \qquad \Gamma \vdash \texttt{size}(C_2) = n_2 \qquad \Gamma \vdash \texttt{width}(C_1) = \texttt{width}(C_2)}{\Gamma \vdash \texttt{size}(\texttt{SEQ}(C_1,C_2)) = n_1 + n_2}\ (\texttt{seq-size})$$

$$\frac{\Gamma \vdash \texttt{size}(C_1) = n_1 \qquad \Gamma \vdash \texttt{size}(C_2) = n_2}{\Gamma \vdash \texttt{size}(\texttt{PAR}(C_1,C_2)) = n_1 + n_2}\ (\texttt{par-size})$$

$$\frac{\Gamma \vdash \texttt{size}(C) = n}{\Gamma \vdash \texttt{size}(\texttt{ANC}(C)) = n}\ (\texttt{anc-size})$$

Figure 7.14 Deduction rules for QBRICKS: size (number of gates).

7.7.2.7 Implementation and Case Studies

QBRICKS enabled to implement, specify and formalize parametrized versions for Quantum Fourier Transform, Grover search algorithm, QPE, Shor order finding, *etc*. The main specificity of these QBRICKS implementations is to hold a high level of proof support automation. This is largely due to the reduction of proof obligations to first-order logic predicates by use of pps characterizations.

7.7.3 Sqir

The SQIR language [91, 90] is the representation language used by the VOQC optimizer (see Section 7.5.4). On its own, it also constitutes a solution for formally proved correct quantum programs. It is developed concurrently with QBRICKS and holds similar concerns: basically, to reduce the expressivity of programming languages such as Quipper and QWIRE so as to (1) still enable the whole implementation of emblematic algorithms (2) enable formal proof of specifications.

The development of SQIR followed that of QWIRE (see Section 7.6.1.3) when the authors observed the difficulty to hold formal verification, mainly linked to the management of memory wires. Hence SQIR and QWIRE have overlapping author and developer ship, they are both deeply embedded in the Coq proof assistant and they share the same mathematical libraries for, e.g., matrices and complex numbers. Schematically, compared to QWIRE, SQIR has a reduced expressivity (disabling, eg., the identification of qubit held values), making tractable the formal verification of functional program properties.

7.7.3.1 Programming Language

Just as QBRICKS, the programming part of SQIR is reduced to the minimum enabling implementation of main quantum programming features. It is two-layered: the *unitary part* corresponds to the design of unitary circuits and a generalized circuit layer adds branching measurement and ket initialization.

In SQIR, quantum circuits have type $\mathbb{N} \to$ `Set`, with a positive integer parameter corresponding to their width. Quantum memory wires are identified by integer indexes, bounded by a global register size parameter `d`. Then, the type for unitary operators in SQIR (`ucom`) features the application of a circuit to a quantum register with `d` wires. It is defined inductively as follows.

```
Inductive ucom (U: ℕ → Set) (d: ℕ) : Set :=
| useq : ucom U d → ucom U d → ucom U d
| uapp1 : U 1 → ℕ → ucom U d
| uapp2 : U 2 → ℕ → ℕ → ucom U d
| uapp3 : U 3 → ℕ → ℕ → ℕ → ucom U d
```

This definition holds two kinds of operations:

- the sequential composition, `useq`, which inputs two circuits and outputs their sequential composition,
- the application of an elementary gate to (a) given wire(s), depending on the width of this gate. There are three different versions of this operations, for gates of width 1,2 or 3, named respectively `uapp1`, `uapp2` and `uapp3`. As an example, `uapp1` inputs a gate `U` of width 1 and a parameter `i`. It outputs the result of applying `U` on wire `i` in a register of size `d`[13].

Then, SQIR provides a generalized circuit building layer, enabling the sequential composition of unitary commands, their initialization and branching measurement. Again, a circuit is an object `com` of type $\mathbb{N} \to$ `Set` applied on a register of specified size `d`. It is defined, inductively, as follows.

```
Inductive com (U: ℕ → Set) (d: ℕ) : Set :=
| uc : ucom U d → com U d → ucom U d
| skip : com U d
| meas : ℕ → com U d → com U d → com U d
| seq : com U d → com U d → com U d
```

[13]Note that, for this construction to make sense, the parameter `i` should not be greater than `d`. This condition is encoded by the semantics of SQIR.

A generalized circuit is built as either the lifting `uc` of a circuit into a generalized circuit, the empty `skip` operation, the branching measurement `meas` (which inputs a wire identifier for a qubit to measure and two generalized circuits to execute, depending on the measurement result) and the sequence `seq` of two different generalized circuits.

7.7.3.2 Matrix Semantics and Specifications

The semantics for SQIR programs is based on the standard matrix apparatus. For the unitary fragment, it uses the matrix semantics presented in Section 7.2.1.5. It is extended for generalized circuits by density operators semantics, similarly as in Section 7.7.1.

In its present state of development, SQIR enables to specify functional properties, describing the input-output relation, similarly to the one for QBRICKS introduced in Section 7.7.2.5

7.7.3.3 Implementation and Case Studies

SQIR is implemented as an embedded DSL into the Coq proof assistant. It was illustrated with specified and proved parametrized implementations of Simon's algorithm, QPE and Grover's Search Algorithm.

7.7.3.4 Comparison between Qbricks and Sqir

QBRICKS and SQIR are being developed concurrently, with very similar objectives. In particular they both trade-off between offering user-friendly programming features and reducing the language expressivity to the minimal, to enable functional formal verification. The solutions they provide share many common points. We discuss their main design differences.

- SQIR elementary operations consist in applying quantum gates on given wires of a quantum register, whereas QBRICKS proceeds by assembling quantum gates together into a quantum circuit, just as bricks of a wall. Both views are inter-simulable: QBRICKS provides a macro `place` with integer parameters specifying the wire identifier a given sub-circuit should be applied to and the size of the overall circuit (corresponding to the size of the available quantum register). This macro is built by the parallel combination of its sub-circuit arguments with the appropriate number of occurrences of `ID` gate. It is of similar use as SQIR function `uapp`. On the other hand, QBRICKS gates assemblage is trivially simulable through SQIR `uapp`;
- On the other hand, SQIR provides a generalized circuit building layer, including measurement and classical control. Nevertheless,

this upper layer is formalized through density operators, which are cumbersome objects for formal reasoning. So far, this part of the language only received illustrations with toy examples, such as superdense coding or quantum teleportation. In more involved implementations (such as QPE or Grover), SQIR authors followed a specification and proof strategy similar to that of QBRICKS: by reasoning on the quantum data outputs of circuits, specifying over the probability distribution of result if a measurement were performed. Hence, designing a generalized circuit building language semantics probed against actual implementations of real usage algorithms is still an open challenge;

- As introduced in Section 7.4.2.2, complexity properties of circuits constitute a fundamental aspect of quantum certification, decisive with regards to both the physical reliability of a computation and the quantum advantage it may provide. In the present state of development, SQIR does not offer a solution for this type of specifications. Still, this could be merely implemented in SQIR as additional functionality.

7.7.4 CONCLUSION ABOUT FORMAL VERIFICATION OF QUANTUM PROGRAMS

In Table 7.5 we sum up the main concrete case study realizations of formally verified quantum algorithms: instances of Grover algorithm from QBRICKS, SQIR and QHL, Deutsch-Jozsa and QPE instances from QBRICKS and SQIR, and `Shor-OF` implementation from QBRICKS. For each of these implementations, we give the length of the code (column LoC) and a measure of the human proof effort required for the specification proofs. It was obtained by adding the length of the program specifications (Spec stands for the number of lines of specifications and intermediary lemmas) and the number of proof commands that were required to prove these specifications (column Cmd).

To the best of our knowledge Table 7.5 is comprehensive regarding parametrized formally proved quantum algorithm [14]. Let us stress out how

[14] An additional formalization of Deutsch Jozsa algorithm is presented by Bordg et al. [30]. We do not include it in Table 7.5 since it is not generated by a programming language but directly led as an algebraic proof. The total length of the proof is over 1 700 lines. Additionally, the online SQIR repository contains a Shor algorithm folder, but to the best of our knowledge, it was never explicitly presented nor described. It seems, by the way, hardly comparable with the QBRICKS implementation since it focuses on different aspects by bringing additional classical post-processing functions but lacking the oracle implementation.

Table 7.5
Compared implementations of formally verified quantum algorithms.

	QBRICKS[38]		SQIR [90, 91]		QHL[132]	
	LoC	Spec+Cmd	LoC	Spec+Cmd	LoC	Spec+Cmd
DJ	11	85	10	261		
Grover	39	279	15	926	90	2975
QPE	23	246	40	812		
Shor-OF	132	1212				

young the field is (in complement to the reduced number of concrete realizations, note that none of them is dated earlier than 2019). Nevertheless, it has already brought promising results.

One of the main challenges for formal verification is to reduce the human proof effort that is required for the certification of programs. As Table 7.5 shows, comparing this effort to the length of effective programs, QBRICKS offers a quite stable ratio $\simeq 10$. In a quite regular way, SQIR adds a $\simeq 3.5$ factor to this ratio and QHL, for the case of Grover algorithm, requires $\simeq 10$ times more human effort [15]

7.8 DISCUSSION AND BIBLIOGRAPHICAL NOTES

We end up this chapter by providing some additional references for usage of formal methods in quantum information and quantum computing.

7.8.1 DEDUCTIVE VERIFICATION

Deductive verification appears to be the most promising direction for the development of formal methods in quantum computing. In particular, it is particularly adequate for the formal verification of functional specifications, which is crucial for quantum programming and prone to play, there, a role similar to that of testing and debugging in classical computing (see Section 7.4.1). It is worth noting that, currently, all existing formally verified quantum algorithms descriptions [132] or implementations [38, 91] are based either on deductive verification or interactive proofs.

[15] Note that the QHL implementation of Grover algorithm concerns a restricted case, with regards to the two others figuring in Table 7.5. Furthermore, it does not contain the gate-to-gate circuit building but uses large circuit portions as primitives instead. Therefore, factor $\simeq 10$ is actually an underestimation.

7.8.2 MODEL CHECKING

Attempts for functional verification of quantum algorithm with model checking techniques were also led before these developments [78, 197, 196]. They enabled to verify toy examples of quantum processes in a completely automated way. Nevertheless, this direction is limited by its high scale sensitivity, which is specifically problematic for quantum programs, since they are designed to tackle large problems instances.

7.8.3 TYPE CHECKING

Apart from functional specification, specialized type systems for quantum programming languages also facilitate programming and debugging. In Section 7.4.1 we introduced the verification of structural constraints and the non-duplicability of quantum information. Type checking may also have further use in quantum computing.

Recently, the SILQ language was proposed. It is based on a linear type system which, upon other features, enables to verify, for any quantum circuit, whether it can be *uncomputed*, a computation feature required at many stages of quantum implementations. Based on this type system, SILQ enables to automatically generate the uncomputation steps of circuits. This partial automation of the development lowers the expertise requirements for developers and the length of programs with regards to languages such as Q# or Quipper.

7.8.4 RUNTIME ASSERTION CHECKING

First, recall from Section 7.1 that proving quantum programs is mainly meant to replace the standard classical debugging method of testing and assertion checking. Apart from the development of formal proofs as an alternative, efforts are led to adapt this classical strategy to the quantum case. There, we still decorate programs with formal specifications (called *assertions* in this context) describing the evolution of the system state through an execution. But instead of mathematical proofs, these assertions are probed by statistical testing over program fragments. The challenges faced by such methods are mainly twofold:

Destructive measurement : memory reading destructs the superposition of a quantum state, therefore one cannot continue the execution after checking. Hence, assertion checking can be applied only to fragments of an execution.

Non-determinism : what we aim to check is a superposition of states, which induces a probability distribution of outcomes when a measurement is performed. Then, checking an assertion requires a number of testing runs large enough to build up a representative statistical distribution.

To overcome these difficulties, a first strategy is to reduce the specifications so as to express only properties that may be handled by assertion checking. Huang and Martonosi propose a "runtime-monitoring like" verification method for quantum circuits [96]. The annotation language is reduced so as to specify, for a quantum register, to be either in a classical state, in a superposition or entangled, without any concern about further description of the state.

More recently, Li et al. [127] developed an assertion-checking-based method for the verification of fine quantum registers states properties, including functional descriptions of circuit behaviors. This method is based on:

- an assertion language, based on QHL projections (see Section 7.7.1), enabling functional specifications about computations at stake;
- the use and formalization of *gentle measurements*: quantum registers are not measured in the usual computation basis, but in a basis containing an output that is very close to the expected state;
- an appropriate formalization of the notion of distance between quantum states, enabling verification in terms of confidence interval between the current state of the system and its expected value.

Hence, gentle measurement enables to test executions over full functional state specifications. In addition to bringing expressivity to the specifications, it lowers the undesired effects of destructive measurements: since a gentle measurement operator contains an eigenstate that is an approximation of the expected quantum state, in most cases the measurement effect on the system state can be considered negligible and the execution can be pursued. Furthermore, the test result probability distribution is centered on a specific value. Therefore, a much-reduced set of runs bring valuable statistical conclusions.

However, verification following this strategy only holds for a particular instance of a circuit, instead of a family of quantum circuits with unassigned parameters as in propositions from Section 7.7. Furthermore, in the general case, gentle measurements are implemented by applying (1) a unitary U uncomputing the system state into a ket of the computational basis (2) a measurement in the computational basis (3) unitary $U^\dagger$, to recover the initial state. Then, assertion checking inputs unitaries U and $U^\dagger$ that are themselves prone to error. More precisely, to test against the exact expected value of the state, operator $U^\dagger$ should be equivalent to the computation under test. Practically, gentle measurements approximate the state under test. This enables a simplification of the measurement operator, at the cost of the robustness of the procedure.

7.8.5 VERIFICATION OF QUANTUM COMMUNICATION PROTOCOLS

Another challenge for formal methods is the verification of quantum information processes concerns quantum protocols.

Quantum key distribution protocols [164, 22, 129, 135] enable secured information exchanges between two parties. These protocols exploit the fact that, due to the destructive measurement, physics laws prevent almost any [16] possibility for a potential eavesdropping in the exchange of quantum information. In particular, several formally verified implementations of the BB84 quantum key distribution protocol [22] have been proposed in the literature, based either on process calculus [138, 118], formalization in Coq [28] or model checking [63, 68].

Bordg et al. [30] propose a formalization of quantum information in the ISABELLE/HOL proof assistant. They illustrate their methods through the cases of quantum teleportation and the quantum prisoner dilemma. This work also contains a formally verified implementation of the Deutsch-Jozsa algorithm in ISABELLE/HOL.

In a more fundamental prospect, Echenim et al. [61] provide an ISABELLE/HOL proof for the CHSH inequalities [45]. These are probability distributions about crossed measurement results of quantum observables. They provide a proof for the Bell theorem [21], inducing that no classical theory could account for the entanglement phenomenon (hence that quantum physics cannot be reduced to local classical theories).

7.9 CONCLUSION

7.9.1 SUMMARY

Throughout this chapter, we introduced the context, the main challenges and the most promising results in formally verified quantum programmatic.

[16]These protocols are based on the fact that, in the general case, a potential eavesdropper destructs a quantum message he attempts to intercept, so that the parties can detect the attempt. Nevertheless, it is based on probabilities and there is always a chance for an eavesdropper to perform only conservative measurements. The corresponding probability is bounded by r^n, with $r \in]0,1[$ and n the length of the sent message.

The current state of affairs in this emerging domain can be summed up as follows:

- Quantum computing is an emerging domain, with huge potential application fields and promises. Progresses in the development of concrete machines are reaching the practical relevance landmark: prototypes are getting powerful enough to overpass classical computers. Consequently, quantum software development is becoming a crucial industrial short-term need;
- Quantum software deals with an entirely new programming paradigm. Upon its main particularities are the dual nature of information (either classical or quantum), the destructive measurement and irreducibly probabilistic computations;
- These specificities make programming particularly non-intuitive and prone to error. Furthermore, they make it very hard to directly import usual debugging methods from the classical practice (based on test and assertion checking). The technique presented in Section 7.8.4 might bring some hope, but this is still so far at a very preliminary stage;
- Formal methods appear as the privileged alternative for debugging strategies. Apart from providing solutions to the destructive measurement challenge, they have additional decisive advantages: mainly, they provide absolute guarantee of the correction of programs, and they hold for any instance of programs they verify;
- During the last ten years (the *genesis* of formal quantum programming), this new field has shown promising results in the different stages of software development: high-level program designs, circuit building languages, verification, compilation, optimization, *etc*.

7.9.2 MAIN CURRENT CHALLENGES

Although encouraging, these early successes draw the road map for rising the field from academic proof of concepts to practically usable programming solutions. We present the main coming challenges for this development in three categories: providing relevant integrated development solutions for the NISQ era, offering practical wide spreadable user experience and developing a full-fledged formally verified quantum compilation toolchain.

7.9.2.1 Provide Relevant Integrated Development Solutions for the NISQ Era

So far, quantum formal verification mainly proved its relevance by offering solutions for the unitary core of quantum computations. As a matter of fact, illustrations and concrete implementations using these techniques primarily treated historical emblematic algorithms such as Shor [174], QPE [112] or Grover search [84]. The classical treatment in these algorithms can be completely decoupled from their quantum core.

However, the first generation of quantum computing machine (the *NISQ era*, see Section 7.2.3.2) hints toward a radically distinct mode of operation. NISQ machines will have limited, noisy resources. A major consequence is that these quantum processors are too small to support the error correction mechanisms required for Shor's and Grover's algorithms. Such NISQ processors aim instead at different kinds of algorithms: hybrid algorithms such as variational algorithms [136]. Hybrid algorithms tightly mix quantum and classical data treatments: one cannot decouple the quantum part of the algorithm from its classical part.

Adapting the quantum formal methods to the NISQ setting to support hybrid quantum/classical computation is a challenge and an active current research avenue.

7.9.2.2 Offer Practical Wide Spreadable User Experience: Formalism, Language Design, Automation

In the present state of development, programming languages enabling formal verification usually sacrifice their expressivity to formal reasoning. In particular, all the solutions that have been probed against actual parametric quantum algorithms fail to satisfy essential elements of the requirements listed in Section 7.6.1.2 for scalable quantum programming languages: in addition to classical processing mentioned above, they cannot manipulate quantum registers and wire references. Verified programming should address these limitations.

Another issue, while concerning any method of quantum programming, is especially critical in the case of formally verified programming. It concerns the level of qualifications required from developers. The interpretation of quantum computations indeed requires unusual and non-intuitive mathematical formalism (including Kronecker products, complex phase amplitudes, probabilistic reasoning, *etc*). While the need for qualified programmers is prone to grow rapidly in the coming years, integrating formal verification should come with the development of user-friendly specification languages and highly automated mathematical reasoning engines.

7.9.2.3 Formally Verified Quantum Compilation Toolchain

In addition to the preceding considerations, in its early academic ages, quantum formal verification focused on idealized representations of quantum circuits, directly extracted from algorithm descriptions. As introduced in Section 7.2.3.2, these logical qubits are merely abstract models for actual computations to be run. A major addition that is left for future works in quantum formal verification is error correction, with formal verification that the state of the system preserves the functional correctness of computations (assuming a given error model, and possibly with probabilistic specifications). Presented in Section 7.5.4, VOQC is a first step toward this goal.

Another future direction concerns integration in a widespread classical development environment. Recall from Section 7.6.1.3 that many widespread quantum programming languages benefit from embeddings in usual programming languages–such as Python. Today most formal verification solutions are embedded in a more academic functional development environment (such as Haskell, Ocaml, proof assistants or deductive verification environments). Interfaces should be developed to integrate formally verified quantum computations into comprehensive projects.

At the other extremity of the development stack, formal verification should accompany the implementation of compiled programs on concrete machines. This induces verified solutions for the qubit mapping problem and gate simulation (see Section 7.2.3.3), depending on the particular target material and its proper constraints.

ACKNOWLEDGMENTS

This work was supported in part by the French National Research Agency (ANR) under the research project SoftQPRO ANR-17-CE25-0009-02, by the DGE of the French Ministry of Industry under the research project PIA-GDN/QuantEx P163746-484124, by the STIC-AmSud project Qapla' 21-SITC-10 and by the Carnot project Qstack. We thank the anonymous reviewers for helpful comments on earlier drafts of the manuscript.

REFERENCES

1. Scott Aaronson and Alex Arkhipov. The computational complexity of linear optics. In *Proceedings of the forty-third annual ACM symposium on Theory of computing*, pages 333–342, 2011.

2. Samson Abramsky and Bob Coecke. Categorical quantum mechanics. In *Handbook of Quantum Logic and Quantum Structures*, pages 261–323.

Elsevier, 2009. URL: https://doi.org/10.1016%2Fb978-0-444-52869-8.50010-4, `doi:10.1016/b978-0-444-52869-8.50010-4`.

3. Thorsten Altenkirch and Alexander S Green. The quantum IO monad. *Semantic Techniques in Quantum Computation*, pages 173–205, 2010.

4. Matthew Amy. *Formal Methods in Quantum Circuit Design*. PhD thesis, University of Waterloo, Ontario, Canada, 2019. URL: http://hdl.handle.net/10012/14480.

5. Matthew Amy. Towards large-scale functional verification of universal quantum circuits. In Peter Selinger and Giulio Chiribella, editors, *Proceedings 15th International Conference on Quantum Physics and Logic, QPL 2018*, volume 287 of *Electronic Proceedings in Theoretical Computer Science*, pages 1–21, Halifax, Canada, 2019. EPTCS. `doi:10.4204/EPTCS.287.1`.

6. Matthew Amy, Jianxin Chen, and Neil J. Ross. A finite presentation of CNOT-dihedral operators. In Bob Coecke and Aleks Kissinger, editors, *Proceedings 14th International Conference on Quantum Physics and Logic, Nijmegen, The Netherlands, 3-7 July 2017*, volume 266 of *Electronic Proceedings in Theoretical Computer Science*, pages 84–97, 2018. `doi:10.4204/EPTCS.266.5`.

7. Matthew Amy, Dmitri Maslov, and Michele Mosca. Polynomial-time T-depth optimization of Clifford+T circuits via matroid partitioning. *IEEE Transactions on Computer-Aided Design of Integrated Circuits and Systems*, 33(10):1476–1489, 2014.

8. Krzysztof R. Apt and Ernst-Rüdiger Olderog. Fifty years of Hoare's logic. *Formal Aspects Comput.*, 31(6):751–807, 2019. `doi:10.1007/s00165-019-00501-3`.

9. Frank Arute, Kunal Arya, Ryan Babbush, Dave Bacon, Joseph C. Bardin, Rami Barends, Rupak Biswas, Sergio Boixo, Fernando G. S. L. Brandao, David A. Buell, et al. Quantum supremacy using a programmable superconducting processor. *Nature*, 574(7779):505–510, 2019.

10. Miriam Backens. The ZX-calculus is complete for stabilizer quantum mechanics. In *New Journal of Physics*, volume 16, page 093021. IOP Publishing, Sep 2014. URL: https://doi.org/10.1088%2F1367-2630%2F16%2F9%2F093021, `doi:10.1088/1367-2630/16/9/093021`.

11. Miriam Backens, Hector Miller-Bakewell, Giovanni de Felice, Leo Lobski, and John van de Wetering. There and back again: A circuit extraction tale. *Quantum*, 5:421, 2021. `doi:10.22331/q-2021-03-25-421`.

12. Thomas Ball, Byron Cook, Vladimir Levin, and Sriram K. Rajamani. Slam and static driver verifier : Technology transfer of formal methods inside Microsoft. In *Integrated Formal Methods, 4th International Conference, IFM 2004*. Springer, 2004.

13. Alexandru Baltag, Jort Bergfeld, Kohei Kishida, Joshua Sack, Sonja Smets, and Shengyang Zhong. PLQP & Company: decidable logics for quantum algorithms. *International Journal of Theoretical Physics*, 53(10):3628–3647, 2014.

14. Alexandru Baltag and Sonja Smets. LQP: the dynamic logic of quantum information. *Mathematical Structures in Computer Science*, 16(3):491–525, 2006. doi:10.1017/S0960129506005299.

15. Alexandru Baltag and Sonja Smets. The logic of quantum programs. 2021. URL: https://arxiv.org/abs/2109.06792, doi:10.48550/ARXIV.2109.06792.

16. Henk P. Barendregt. *The Lambda-Calculus, its Syntax and Semantics*, volume 103 of *Studies in Logic and the Foundation of Mathematics*. North Holland, second edition, 1984.

17. Mike Barnett, Manuel Fähndrich, K. Rustan M. Leino, Peter Müller, Wolfram Schulte, and Herman Venter. Specification and verification: the SPEC# experience. *Commun. ACM*, 54(6):81–91, 2011.

18. Clark Barrett and Cesare Tinelli. Satisfiability modulo theories. In *Handbook of Model Checking*, pages 305–343. Springer, 2018.

19. Gilles Barthe, Justin Hsu, Mingsheng Ying, Nengkun Yu, and Li Zhou. Relational proofs for quantum programs. *Proc. ACM Program. Lang.*, 4(POPL):21:1–21:29, 2020. doi:10.1145/3371089.

20. Patrick Behm, Paul Benoit, Alain Faivre, and Jean-Marc Meynadier. Météor: A successful application of B in a large project. In *In Proceedings of the World Congress on Formal Methods in the Development of Computing Systems (FM'99)*. Springer, 1999.

21. John S Bell. On the Einstein Podolsky Rosen paradox. *Physics Physique Fizika*, 1(3):195–200, 1964.

22. C. H. Bennett and G. Brassard. Quantum cryptography: Public key distribution and coin tossing. In *Proceedings of IEEE International Conference on Computers, Systems and Signal Processing*, pages 175–179, Bengalore, India, 1984.

23. Nick Benton. A mixed linear and non-linear logic: Proofs, terms and models (extended abstract). In Leszek Pacholski and Jerzy Tiuryn, editors, *Computer Science Logic, Eighth International Workshop, CSL'94, Selected Papers*, volume 933 of *Lecture Notes in Computer Science*, pages 121–135, 1994. doi:10.1007/BFb0022251.

24. Jort Martinus Bergfeld and Joshua Sack. Deriving the correctness of quantum protocols in the probabilistic logic for quantum programs. *Soft Comput.*, 21(6):1421–1441, 2017. doi:10.1007/s00500-015-1802-6.

25. Yves Bertot and Pierre Castéran. *Interactive Theorem Proving and Program Development: Coq'Art: the Calculus of Inductive Constructions*. Springer Science & Business Media, 2013.

26. Jacob Biamonte, Peter Wittek, Nicola Pancotti, Patrick Rebentrost, Nathan Wiebe, and Seth Lloyd. Quantum machine learning. *Nature*, 549(7671):195, 2017. doi:10.1038/nature23474.

27. Benjamin Bichsel, Maximilian Baader, Timon Gehr, and Martin T. Vechev. Silq: a high-level quantum language with safe uncomputation and intuitive semantics. In Alastair F. Donaldson and Emina Torlak, editors, *Proceedings of the 41st ACM SIGPLAN International Conference on Programming Language Design and Implementation, PLDI 2020, London, UK, June 15-20, 2020*, pages 286–300. ACM, 2020. doi:10.1145/3385412.3386007.

28. Jaap Boender, Florian Kammüller, and Rajagopal Nagarajan. Formalization of quantum protocols using Coq. In Chris Heunen, Peter Selinger, and Jamie Vicary, editors, *Proceedings of the 12th International Workshop on Quantum Physics and Logic (QPL 2015)*, volume 195 of *Electronic Proceedings in Theoretical Computer Science*, pages 71–83, Oxford, UK, 2015. EPTCS. doi:10.4204/EPTCS.195.6.

29. Adam D Bookatz. QMA-complete problems. *Quantum Information & Computation*, 14(5&6):361–383, 2014.

30. Anthony Bordg, Hanna Lachnitt, and Yijun He. Certified quantum computation in Isabelle/HOL. *Journal of Automated Reasoning*, 65(5):691–709, 2021.

31. E. Bounimova, P. Godefroid, and D. Molnar. Billions and billions of constraints : Whitebox fuzz testing in production. In *35th International Conference on Software Engineering (ICSE)*, pages 122–131. IEEE/ACM, 2013.

32. Hans J. Briegel, Dan E. Browne, Wolfgang Dür, Robert Raußendorf, and Maarten Van den Nest. Measurement-based quantum computation. *Nature Physics*, 5(1):19–26, January 2009. arXiv:0910.1116, doi:10.1038/nphys1157.

33. Daniel E Browne, Elham Kashefi, Mehdi Mhalla, and Simon Perdrix. Generalized flow and determinism in measurement-based quantum computation. *New Journal of Physics*, 9(8):250–250, aug 2007. URL: https://doi.org/10.1088%2F1367-2630%2F9%2F8%2F250, doi:10.1088/1367-2630/9/8/250.

34. Olivier Brunet and Philippe Jorrand. Dynamic quantum logic for quantum programs. *International Journal of Quantum Information*, 2(01):45–54, 2004.

35. Cristian Cadar and Koushik Sen. Symbolic execution for software testing: three decades later. *Commun. ACM*, 56(2):82–90, 2013.

36. Titouan Carette, Emmanuel Jeandel, Simon Perdrix, and Renaud Vilmart. Completeness of Graphical Languages for Mixed States Quantum Mechanics. In Christel Baier, Ioannis Chatzigiannakis, Paola Flocchini, and Stefano Leonardi, editors, *46th International Colloquium on Automata, Languages, and Programming (ICALP 2019)*, volume 132 of *Leibniz International Proceedings in Informatics (LIPIcs)*, pages 108:1–108:15, Dagstuhl, Germany, 2019. Schloss Dagstuhl–Leibniz-Zentrum fuer Informatik. URL: http://drops.dagstuhl.de/opus/volltexte/2019/10684, doi:10.4230/LIPIcs.ICALP.2019.108.

37. Rohit Chadha, Paulo Mateus, and Amílcar Sernadas. Reasoning about imperative quantum programs. *Electronic Notes in Theoretical Computer Science*, 158:19–39, 2006.

38. Christophe Chareton, Sébastien Bardin, François Bobot, Valentin Perrelle, and Benoît Valiron. An automated deductive verification framework for circuit-building quantum programs. In Nobuko Yoshida, editor, *Programming Languages and Systems - 30th European Symposium on Programming, ESOP 2021, Luxembourg City, Luxembourg, March 27 - April 1, 2021, Proceedings*, volume 12648 of *Lecture Notes in Computer Science*, pages 148–177. Springer, 2021. doi:10.1007/978-3-030-72019-3_6.

39. Lily Chen, Lily Chen, Stephen Jordan, Yi-Kai Liu, Dustin Moody, Rene Peralta, Ray Perlner, and Daniel Smith-Tone. *Report on post-quantum cryptography*, volume 12. US Department of Commerce, National Institute of Standards and Technology, 2016.

40. John Chiaverini, Dietrich Leibfried, Tobias Schaetz, Murray D Barrett, RB Blakestad, J Britton, Wayne M Itano, Juergen D Jost, Emanuel Knill, Christopher Langer, et al. Realization of quantum error correction. *Nature*, 432(7017):602–605, 2004.

41. Cirq Developers. Cirq, July 2018. See full list of authors on Github: https://github.com/quantumlib/Cirq/graphs/contributors. URL: https://github.com/quantumlib/Cirq, doi:10.5281/zenodo.4062499.

42. Edmund M. Clarke and E. Allen Emerson. Design and synthesis of synchronization skeletons using branching-time temporal logic. In *Logics of Programs, Workshop, LNCS 131*, pages 52–71. Springer, 1981.

43. Edmund M. Clarke, Thomas A. Henzinger, Helmut Veith, and Roderick Bloem. *Handbook of Model Checking*. Springer, 2018.

44. Edmund M. Clarke and Jeannette M. Wing. Formal methods: State of the art and future directions. *ACM Computing Surveys (CSUR)*, 28(4):626–643, 1996. doi:10.1145/242223.242257.

45. John F Clauser, Michael A Horne, Abner Shimony, and Richard A Holt. Proposed experiment to test local hidden-variable theories. *Physical Review Letters*, 23(15):880, 1969.

46. Bob Coecke and Ross Duncan. Interacting quantum observables: Categorical algebra and diagrammatics. *New Journal of Physics*, 13(4):043016, Apr 2011. URL: https://doi.org/10.1088%2F1367-2630%2F13%2F4%2F043016, doi:10.1088/1367-2630/13/4/043016.

47. Bob Coecke and Aleks Kissinger. *Picturing quantum processes*. Cambridge University Press, Cambridge, United Kingdom, 2017.

48. Bob Coecke and Simon Perdrix. Environment and classical channels in categorical quantum mechanics. *Log. Methods Comput. Sci.*, 8(4), 2010. doi:10.2168/LMCS-8(4:14)2012.

49. Patrick Cousot and Radhia Cousot. Abstract interpretation : A unified lattice model for static analysis of programs by construction or approximation of fixpoints. In *Proceedings of the Fourth ACM Symposium on Principles of Programming Languages (POPL)*, pages 238–252. ACM, 1977.

50. Patrick Cousot, Radhia Cousot, Jerôme Feret, Laurent Mauborgne, Antoine Miné, David Monniaux, and Xavier Rival. The ASTRÉE analyzer. In *European Symposium on Programming Languages and Systems, ESOP 2005*. Springer, 2005.

51. Andrew W. Cross, Lev S. Bishop, John A. Smolin, and Jay M. Gambetta. Open quantum assembly language, 2017. arXiv:1707.03429.

52. Vincent Danos and Elham Kashefi. Determinism in the one-way model. *Phys. Rev. A*, 74:052310, Nov 2006. URL: https://link.aps.org/doi/10.1103/PhysRevA.74.052310, doi:10.1103/PhysRevA.74.052310.

53. Niel de Beaudrap, Xiaoning Bian, and Quanlong Wang. Fast and effective techniques for T-count reduction via spider nest identities. In Steven T. Flammia, editor, *15th Conference on the Theory of Quantum Computation, Communication and Cryptography, TQC 2020, June 9-12, 2020,*

Riga, Latvia, volume 158 of *LIPIcs*, pages 11:1–11:23. Schloss Dagstuhl - Leibniz-Zentrum für Informatik, 2020. doi:10.4230/LIPIcs.TQC.2020.11.

54. Niel de Beaudrap, Ross Duncan, Dominic Horsman, and Simon Perdrix. Pauli fusion: a computational model to realise quantum transformations from ZX terms. In *QPL'19 : International Conference on Quantum Physics and Logic*, Los Angeles, United States, June 2019. URL: https://hal.archives-ouvertes.fr/hal-02413388.

55. Niel de Beaudrap and Dominic Horsman. The ZX calculus is a language for surface code lattice surgery. *Quantum*, 4:218, January 2020. doi:10.22331/q-2020-01-09-218.

56. Ellie D'Hondt and Prakash Panangaden. Quantum weakest preconditions. *Mathematical Structures in Computer Science*, 16(3):429–451, 2006. doi:10.1017/S0960129506005251.

57. Alejandro Díaz-Caro. A lambda calculus for density matrices with classical and probabilistic controls. In Bor-Yuh Evan Chang, editor, *Proceedings of the 15th Asian Symposium on Programming Languages and Systems (APLAS'17)*, volume 10695 of *Lecture Notes in Computer Science*, pages 448–467, Suzhou, China, 2017. Springer. doi:10.1007/978-3-319-71237-6_22.

58. Alejandro Díaz-Caro and Octavio Malherbe. A concrete categorical semantics of lambda-S. In Beniamino Accattoli and Carlos Olarte, editors, *Proceedings of the 13th Workshop on Logical and Semantic Frameworks with Applications, LSFA 2018, Fortaleza, Brazil, September 26-28, 2018*, volume 344 of *Electronic Notes in Theoretical Computer Science*, pages 83–100. Elsevier, 2019. doi:10.1016/j.entcs.2019.07.006.

59. Edsger W. Dijkstra. *A Discipline of Programming*. Prentice-Hall, 1976.

60. Ross Duncan, Aleks Kissinger, Simon Perdrix, and John van de Wetering. Graph-theoretic simplification of quantum circuits with the ZX-calculus. *Quantum*, 4:279, 2020. doi:10.22331/q-2020-06-04-279.

61. Mnacho Echenim. Quantum projective measurements and the CHSH inequality. *Arch. Formal Proofs*, 2021, 2021.

62. Albert Einstein, Boris Podolsky, and Nathan Rosen. Can quantum-mechanical description of physical reality be considered complete? *Physical review*, 47(10):777, 1935.

63. Mohamed Elboukhari, Mostafa Azizi, and Abdelmalek Azizi. Verification of quantum cryptography protocols by model checking. *Int. J. Network Security & Appl*, 2(4):43–53, 2010.

64. Edward Farhi, Jeffrey Goldstone, and Sam Gutmann. A quantum approximate optimization algorithm. Technical Report MIT-CTP/4610, MIT, 2014.

65. Edward Farhi, Jeffrey Goldstone, Sam Gutmann, Joshua Lapan, Andrew Lundgren, and Daniel Preda. A quantum adiabatic evolution algorithm applied to random instances of an NP-complete problem. *Science*, 292(5516):472–475, 2001. doi:10.1126/science.1057726.

66. Yuan Feng, Runyao Duan, Zhengfeng Ji, and Mingsheng Ying. Proof rules for the correctness of quantum programs. *Theoretical Computer Science*, 386(1-2):151–166, 2007. doi:10.1016/j.tcs.2007.06.011.

67. Yuan Feng and Mingsheng Ying. Quantum hoare logic with classical variables. *ACM Transactions on Quantum Computing*, 2(4):16, December 2021. arXiv:2008.06812, doi:10.1145/3456877.

68. Verónica Fernández, María-José García-Martínez, Luis Hernández-Encinas, and Agustín Martín. Formal verification of the security of a free-space quantum key distribution system. In *Proc. World Congr. Comput. Sci. Comput. Eng. Appl. Comput.(WORLDCOMP) Int. Conf. Security Manag.(SAM)*, 2011.

69. Richard P. Feynman. Simulating physics with computers. *International Journal of Theoretical Physics*, 21(6–7):467–488, 1982. doi:10.1007/BF02650179.

70. Jean-Christophe Filliâtre and Andrei Paskevich. Why3 - where programs meet provers. In Matthias Felleisen and Philippa Gardner, editors, *Proceedings of the 22nd European Symposium on Programming Languages and Systems (ESOP 2013), Held as Part of the European Joint Conferences on Theory and Practice of Software (ETAPS 2013)*, volume 7792 of *Lecture Notes in Computer Science*, pages 125–128, Rome, Italy, 2013. Springer. doi:10.1007/978-3-642-37036-6_8.

71. Jean-Christophe Filliâtre. Deductive software verification. *STTT*, 13(5):397–403, 2011.

72. Michael J Fischer and Richard E Ladner. Propositional dynamic logic of regular programs. *Journal of Computer and System Sciences*, 18(2):194–211, 1979.

73. Melvin Fitting. *First-order logic and automated theorem proving*. Springer Science & Business Media, 2012.

74. R. W. Floyd. Assigning meanings to programs. In *Mathematical Aspects of Computer Science, Proceedings of Symposia in Applied Mathematics*, pages 19–32. American Mathematical Society, 1967.

75. Austin G. Fowler, Matteo Mariantoni, John M. Martinis, and Andrew N. Cleland. Surface codes: Towards practical large-scale quantum computation. *Phys. Rev. A*, 86:032324, Sep 2012. URL: https://link.aps.org/doi/10.1103/PhysRevA.86.032324, doi:10.1103/PhysRevA.86.032324.

76. Michael Freedman, Alexei Kitaev, Michael Larsen, and Zhenghan Wang. Topological quantum computation. *Bulletin of the American Mathematical Society*, 40(1):31–38, 2003.

77. Peng Fu, Kohei Kishida, and Peter Selinger. Linear dependent type theory for quantum programming languages. In *Proceedings of the 35th Annual ACM/IEEE Symposium on Logic in Computer Science*, pages 440–453, 2020.

78. Simon J. Gay, Rajagopal Nagarajan, and Nikolaos Papanikolaou. QMC: a model checker for quantum systems. In Aarti Gupta and Sharad Malik, editors, *Proceeding of the 20th International Conference on Computer Aided Verification (CAV 2008)*, volume 5123 of *Lecture Notes in Computer Science*, pages 543–547, Princeton, NJ, USA, 2008. Springer. doi:10.1007/978-3-540-70545-1_51.

79. Jean-Yves Girard. Linear logic. *Theoretical Computer Science*, 50(1):1–101, 1987.

80. Georges Gonthier. Formal proof – the four-color theorem. *Notices of the AMS*, 55(11):1382–1393, 2008.

81. Daniel Gottesman. *Stabilizer Codes and Quantum Error Correction*. PhD thesis, Caltech, 1997.

82. Alexander S. Green, Peter LeFanu Lumsdaine, Neil J. Ross, Peter Selinger, and Benoît Valiron. An introduction to quantum programming in Quipper. In Gerhard W. Dueck and D. Michael Miller, editors, *Proceedings of the 5th International Conference on Reversible Computation (RC'13)*, volume 7948 of *Lecture Notes in Computer Science*, pages 110–124, Victoria, BC, Canada, 2013. Springer. doi:10.1007/978-3-642-38986-3_10.

83. Alexander S. Green, Peter LeFanu Lumsdaine, Neil J. Ross, Peter Selinger, and Benoît Valiron. Quipper: A scalable quantum programming language. In Hans-Juergen Boehm and Cormac Flanagan, editors, *Proceedings of the ACM SIGPLAN Conference on Programming Language Design and Implementation, (PLDI'13)*, pages 333–342, Seattle, WA, USA, 2013. ACM. doi:10.1145/2491956.2462177.

84. Lov K. Grover. A fast quantum mechanical algorithm for database search. In Gary L. Miller, editor, *Proceedings of the Twenty-Eighth Annual ACM Symposium on the Theory of Computing (STOC)*, pages 212–219, Philadelphia, Pennsylvania, USA, 1996. ACM. doi:10.1145/237814.237866.

85. Amar Hadzihasanovic, Kang Feng Ng, and Quanlong Wang. Two complete axiomatisations of pure-state qubit quantum computing. In *Proceedings of the 33rd Annual ACM/IEEE Symposium on Logic in Computer Science*, LICS '18, pages 502–511, New York, NY, USA, 2018. ACM. URL: http://doi.acm.org/10.1145/3209108.3209128, doi:10.1145/3209108.3209128.

86. David Harel, Dexter Kozen, and Jerzy Tiuryn. Dynamic logic. In *Handbook of Philosophical Logic*, pages 99–217. Springer, 2001.

87. Aram W. Harrow, Avinatan Hassidim, and Seth Lloyd. Quantum algorithm for linear systems of equations. *Physical Review Letters*, 103:150502, Oct 2009. doi:10.1103/PhysRevLett.103.150502.

88. Ichiro Hasuo and Naohiko Hoshino. Semantics of higher-order quantum computation via geometry of interaction. *Annals of Pure and Applied Logic*, 168(2):404 – 469, 2017. URL: http://www.sciencedirect.com/science/article/pii/S0168007216301336, doi:https://doi.org/10.1016/j.apal.2016.10.010.

89. Thomas A. Henzinger, Ranjit Jhala, Rupak Majumdar, and Grégoire Sutre. Software verification with Blast. In *Proceedings of the 10th International Conference on Model Checking Software, SPIN'03*. Springer, 2003.

90. Kesha Hietala, Robert Rand, Shih-Han Hung, Liyi Li, and Michael Hicks. Proving quantum programs correct. In Liron Cohen and Cezary Kaliszyk, editors, *12th International Conference on Interactive Theorem Proving, ITP 2021, June 29 to July 1, 2021, Rome, Italy (Virtual Conference)*, volume 193 of *LIPIcs*, pages 21:1–21:19, 2021. doi:10.4230/LIPIcs.ITP.2021.21.

91. Kesha Hietala, Robert Rand, Shih-Han Hung, Xiaodi Wu, and Michael Hicks. A verified optimizer for quantum circuits. *Proc. ACM Program. Lang.*, 5(POPL):1–29, 2021. doi:10.1145/3434318.

92. Anne Hillebrand. Quantum protocols involving multiparticle entanglement and their representations. Master's thesis, University of Oxford, 2011. URL: https://www.cs.ox.ac.uk/people/bob.coecke/Anne.pdf.

93. R. Hindley. The principal type-scheme of an object in combinatory logic. *Transactions of the American Mathematical Society*, 146:29–60, 1969.

94. C. A. R. Hoare. An axiomatic basis for computer programming. *Commun. ACM*, 12(10):576–580, 1969.

95. Clare Horsman, Austin G. Fowler, Simon Devitt, and Rodney Van Meter. Surface code quantum computing by lattice surgery. *New Journal of Physics*, 14(12):123011, December 2012. arXiv:1111.4022, doi:10.1088/1367-2630/14/12/123011.

96. Yipeng Huang and Margaret Martonosi. Statistical assertions for validating patterns and finding bugs in quantum programs. In Srilatha Bobbie Manne, Hillery C. Hunter, and Erik R. Altman, editors, *Proceedings of the 46th International Symposium on Computer Architecture (ISCA 2019)*, pages 541–553, Phoenix, AZ, USA, 2019. ACM. doi:10.1145/3307650.3322213.

97. Shih-Han Hung, Kesha Hietala, Shaopeng Zhu, Mingsheng Ying, Michael Hicks, and Xiaodi Wu. Quantitative robustness analysis of quantum programs. *Proceedings of the ACM on Programming Languages*, 3(POPL):1–29, 2019.

98. Dominik Janzing, Pawel Wocjan, and Thomas Beth. "Non-identity-check" is QMA-complete. *International Journal of Quantum Information*, 03(03):463–473, 2005. arXiv:https://doi.org/10.1142/S0219749905001067, doi:10.1142/S0219749905001067.

99. Ali Javadi-Abhari, Arvin Faruque, Mohammad Javad Dousti, Lukas Svec, Oana Catu, Amlan Chakrabarti, Chen-Fu Chiang, Seth Vanderwilt, John Black, Frederic T. Chong, Margaret Martonosi, Martin Suchara, Ken Brown, Massoud Pedram, and Todd Brun. Scaffold: Quantum programming language. Technical Report TR-934-12, Princeton University, June 2012. URL: https://www.cs.princeton.edu/research/techreps/TR-934-12.

100. Ali Javadi-Abhari, Shruti Patil, Daniel Kudrow, Jeff Heckey, Alexey Lvov, Frederic T. Chong, and Margaret Martonosi. ScaffCC: Scalable compilation and analysis of quantum programs. *Parallel Computing*, 45:2–17, 2015. doi:10.1016/j.parco.2014.12.001.

101. Emmanuel Jeandel. The rational fragment of the ZX-calculus, 2018. arXiv:arXiv:1810.05377.

102. Emmanuel Jeandel, Simon Perdrix, and Renaud Vilmart. A complete axiomatisation of the ZX-calculus for Clifford+T quantum mechanics. In *Proceedings of the 33rd Annual ACM/IEEE Symposium on Logic in Computer Science*, LICS '18, pages 559–568, New York, NY, USA, 2018. ACM. URL: http://doi.acm.org/10.1145/3209108.3209131, doi:10.1145/3209108.3209131.

103. Emmanuel Jeandel, Simon Perdrix, and Renaud Vilmart. Diagrammatic reasoning beyond Clifford+T quantum mechanics. In *Proceedings of the 33rd Annual ACM/IEEE Symposium on Logic in Computer Science*, LICS '18, pages 569–578, New York, NY, USA, 2018. ACM. URL: http://doi.acm.org/10.1145/3209108.3209139, doi:10.1145/3209108.3209139.

104. Emmanuel Jeandel, Simon Perdrix, and Renaud Vilmart. A generic normal form for ZX-diagrams and application to the rational angle completeness. In *2019 34th Annual ACM/IEEE Symposium on Logic in Computer Science (LICS)*, pages 1–10, 2019. doi:10.1109/LICS.2019.8785754.

105. Emmanuel Jeandel, Simon Perdrix, and Renaud Vilmart. Completeness of the ZX-calculus. *Logical Methods in Computer Science*, 16(2), 2020. URL: https://lmcs.episciences.org/6532, doi:10.23638/LMCS-16(2:11)2020.

106. Yoshihiko Kakutani. A logic for formal verification of quantum programs. In *Annual Asian Computing Science Conference*, pages 79–93. Springer, 2009.

107. Nathan Killoran, Josh Izaac, Nicolás Quesada, Ville Bergholm, Matthew Amy, and Christian Weedbrook. Strawberry Fields: A software platform for photonic quantum computing. *Quantum*, 3:129, 2019. doi:10.22331/q-2019-03-11-129.

108. Florent Kirchner, Nikolai Kosmatov, Virgile Prevosto, Julien Signoles, and Boris Yakobowski. Frama-C : A software analysis perspective. *Formal Asp. Comput.*, 27(3):573–609, 2015.

109. Aleks Kissinger and John van de Wetering. PyZX: Large scale automated diagrammatic reasoning. *Electronic Proceedings in Theoretical Computer Science*, 318:229–241, May 2020. URL: http://dx.doi.org/10.4204/EPTCS.318.14, doi:10.4204/eptcs.318.14.

110. Aleks Kissinger and John van de Wetering. Reducing the number of non-clifford gates in quantum circuits. *Phys. Rev. A*, 102:022406, Aug 2020. URL: https://link.aps.org/doi/10.1103/PhysRevA.102.022406, doi:10.1103/PhysRevA.102.022406.

111. Aleks Kissinger and Vladimir Zamdzhiev. Quantomatic: A proof assistant for diagrammatic reasoning. In Amy P. Felty and Aart Middeldorp, editors, *Proceedings for the 25th International Conference on Automated Deduction (CADE-25)*, volume 9195 of *Lecture Notes in Computer Science*, pages 326–336, Berlin, Germany, 2015. Springer. doi:10.1007/978-3-319-21401-6_22.

112. A Y Kitaev. Quantum measurements and the abelian stabilizer problem. Available online as arXiv:quant-ph/9511026, 1995.

113. Erwin Klein, June Andronick, Kevin Elphinstone, Gernot Heiser, David Cock, Philip Derrin, Dhammika Elkaduwe, Kai Engelhardt, Rafal Kolanski, Michael Norrish, Thomas Sewell, Harvey Tuch, and Simon Winwood. seL4: formal verification of an operating-system kernel. *Commun. ACM*, 53(6):107–115, 2010.

114. Emmanuel Knill. Conventions for quantum pseudocode. Technical Report LA-UR-96-2724, Los Alamos National Laboratory, June 1996. doi:10.2172/366453.

115. Dexter Kozen. Results on the propositional μ-calculus. *Theoretical computer science*, 27(3):333–354, 1983.

116. Saul A. Kripke. Semantical considerations on modal logic. *Acta Philosophica Fennica*, 16:83–94, 1963.

117. Daniel Kroening and Michael Tautschnig. CBMC - C bounded model checker. In *Tools and Algorithms for the Construction and Analysis of Systems - 20th International Conference, TACAS 2014*, pages 389–391. Springer, 2014.

118. Takahiro Kubota, Yoshihiko Kakutani, Go Kato, Yasuhito Kawano, and Hideki Sakurada. Semi-automated verification of security proofs of quantum cryptographic protocols. *Journal of Symbolic Computation*, 73:192–220, 2016.

119. Hidenori Kuwakado and Masakatu Morii. Quantum distinguisher between the 3-round Feistel cipher and the random permutation. In *2010 IEEE International Symposium on Information Theory*, pages 2682–2685. IEEE, 2010.

120. Hidenori Kuwakado and Masakatu Morii. Security on the quantum-type Even-Mansour cipher. In *2012 International Symposium on Information Theory and its Applications*, pages 312–316. IEEE, 2012.

121. Ugo Dal Lago, Claudia Faggian, Benoît Valiron, and Akira Yoshimizu. The geometry of parallelism: classical, probabilistic, and quantum effects. In Giuseppe Castagna and Andrew D. Gordon, editors, *Proceedings of the 44th ACM SIGPLAN Symposium on Principles of Programming Languages (POPL'17)*, pages 833–845, Paris, France, 2017. ACM. doi:10.1145/3009837.3009859.

122. Joachim Lambek. The mathematics of sentence structure. *The American Mathematical Monthly*, 65(3):154–170, 1958.

123. Martin Lange. Model checking propositional dynamic logic with all extras. *Journal of Applied Logic*, 4(1):39–49, 2006.

124. Louis Lemonnier, John van de Wetering, and Aleks Kissinger. Hypergraph Simplification: Linking the Path-sum Approach to the ZH-calculus. In Benoît Valiron, Shane Mansfield, Pablo Arrighi, and Prakash Panangaden, editors, *Proceedings 17th International Conference on Quantum Physics and Logic, Paris, France, June 2 - 6, 2020*, volume 340 of *Electronic Proceedings in Theoretical Computer Science*, pages 188–212. Open Publishing Association, 2021. doi:10.4204/EPTCS.340.10.

125. Xavier Leroy. Formal verification of a realistic compiler. *Commun. ACM*, 52(7):107–115, 2009.

126. Jerzy Lewandowski. Volume and quantizations. *Classical and Quantum Gravity*, 14(1):71, 1997.

127. Gushu Li, Li Zhou, Nengkun Yu, Yufei Ding, Mingsheng Ying, and Yuan Xie. Projection-based runtime assertions for testing and debugging quantum programs. *Proceedings of the ACM on Programming Languages*, 4(OOPSLA):1–29, 2020.

128. Yangjia Li and Mingsheng Ying. Algorithmic analysis of termination problems for quantum programs. *Proceedings of the ACM on Programming Languages*, 2(POPL):1–29, 2017.

129. Sheng-Kai Liao, Wen-Qi Cai, Wei-Yue Liu, Liang Zhang, Yang Li, Ji-Gang Ren, Juan Yin, Qi Shen, Yuan Cao, Zheng-Ping Li, et al. Satellite-to-ground quantum key distribution. *Nature*, 549(7670):43–47, 2017.

130. Daniel A Lidar and Todd A Brun. *Quantum Error Correction*. Cambridge University Press, 2013.

131. Bert Lindenhovius, Michael Mislove, and Vladimir Zamdzhiev. Enriching a linear/non-linear lambda calculus: A programming language for string diagrams. In *Proceedings of the 33rd Annual ACM/IEEE Symposium on Logic in Computer Science*, pages 659–668, 2018.

132. Junyi Liu, Bohua Zhan, Shuling Wang, Shenggang Ying, Tao Liu, Yangjia Li, Mingsheng Ying, and Naijun Zhan. Formal verification of quantum algorithms using quantum Hoare logic. In Isil Dillig and Serdar Tasiran, editors, *Computer Aided Verification*, pages 187–207, Cham, 2019. Springer International Publishing.

133. Tao Liu, Yangjia Li, Shuling Wang, Mingsheng Ying, and Naijun Zhan. A theorem prover for quantum Hoare logic and its applications. Available as `arXiv:1601.03835`, 2016.

134. Seth Lloyd, Masoud Mohseni, and Patrick Rebentrost. Quantum algorithms for supervised and unsupervised machine learning. *arXiv preprint arXiv:1307.0411*, 2013.

135. Hoi-Kwong Lo, Xiongfeng Ma, and Kai Chen. Decoy state quantum key distribution. *Physical Review Letters*, 94(23):230504, 2005.

136. Jarrod R McClean, Jonathan Romero, Ryan Babbush, and Alán Aspuru-Guzik. The theory of variational hybrid quantum-classical algorithms. *New Journal of Physics*, 18(2):023023, 2016.

137. Robin Milner. A theory of type polymorphism in programming. *Journal of Computer and System Sciences*, 17(3):348–375, 1978. `doi:10.1016/0022-0000(78)90014-4`.

138. Rajagopal Nagarajan and Simon Gay. Formal verification of quantum protocols. Available online as `arXiv:quant-ph/0203086`, 2002.

139. Yunseong Nam, Neil J Ross, Yuan Su, Andrew M Childs, and Dmitri Maslov. Automated optimization of large quantum circuits with continuous parameters. *npj Quantum Information*, 4(1):1–12, 2018.

140. Michael A. Nielsen and Isaac L. Chuang. *Quantum Computation and Quantum Information (10th Anniversary edition)*. Cambridge University Press, 2016. URL: https://www.cambridge.org/de/academic/subjects/physics/quantum-physics-quantum-information-and-quantum-computation/quantum-computation-and-quantum-information-10th-anniversary-edition?format=HB.

141. Tobias Nipkow, Lawrence C. Paulson, and Markus Wenzel. *Isabelle/HOL: a Proof Assistant for Higher-Order Logic*. Springer, 2002.

142. Berhnard Ömer. *Structured Quantum Programming*. PhD thesis, TU Wien, 2003.

143. Michele Pagani, Peter Selinger, and Benoît Valiron. Applying quantitative semantics to higher-order quantum computing. In *Proceedings of the 41st ACM SIGPLAN-SIGACT Symposium on Principles of Programming Languages*, POPL '14, page 647–658, New York, NY, USA, 2014. Association for Computing Machinery. doi:10.1145/2535838.2535879.

144. Luca Paolini and Margherita Zorzi. qPCF: A language for quantum circuit computations. In T. V. Gopal, Gerhard Jäger, and Silvia Steila, editors, *Proceedings of the 14th Annual Conference on Theory and Applications of Models of Computation (TAMC'17)*, volume 10185 of *Lecture Notes in Computer Science*, pages 455–469, Bern, Switzerland, 2017. doi:10.1007/978-3-319-55911-7_33.

145. Christine Paulin-Mohring. Introduction to the calculus of inductive constructions. In Bruno Woltzenlogel Paleo and David Delahaye, editors, *All about Proofs, Proofs for All*, volume 55 of *Studies in Logic (Mathematical logic and foundations)*. College Publications, 2015.

146. Jennifer Paykin. *Linear/non-Linear Types For Embedded Domain-Specific Languages*. PhD thesis, University of Pennsylvania, 2018.

147. Jennifer Paykin, Robert Rand, and Steve Zdancewic. QWIRE: a core language for quantum circuits. In Giuseppe Castagna and Andrew D. Gordon, editors, *Proceedings of the 44th ACM SIGPLAN Symposium on Principles of Programming Languages (POPL'17)*, pages 846–858, Paris, France, 2017. ACM. doi:10.1145/3009837.3009894.

148. Simon Perdrix. Quantum entanglement analysis based on abstract interpretation. In *International Static Analysis Symposium*, pages 270–282. Springer, 2008.

149. Benjamin C. Pierce. *Types and Programming Languages*. MIT Press, 2002.

150. Amir Pnueli. The temporal logic of programs. In *18th Annual Symposium on Foundations of Computer Science (sfcs 1977)*, pages 46–57. ieee, 1977.

151. John Preskill. Quantum computing in the NISQ era and beyond. *Quantum*, 2:79, 2018. doi:10.22331/q-2018-08-06-79.

152. Qiskit Community. Qiskit: An open-source framework for quantum computing, March 2017. URL: https://github.com/Qiskit/qiskit, doi:10.5281/zenodo.2562110.

153. Quantum Computing Report. List of tools. Available online[17], 2019.

154. Quingo Development Team. Quingo: A programming framework for heterogeneous quantum-classical computing with nisq features. *ACM Transactions on Quantum Computing*, 2(4):19, December 2021. URL: https://github.com/Quingo, doi:10.1145/3483528.

155. Robert Rand, Kesha Hietala, and Michael Hicks. Formal verification vs. quantum uncertainty. In *3rd Summit on Advances in Programming Languages (SNAPL 2019)*. Schloss Dagstuhl-Leibniz-Zentrum fuer Informatik, 2019.

156. Robert Rand, Jennifer Paykin, and Steve Zdancewic. QWIRE practice: Formal verification of quantum circuits in Coq. In Bob Coecke and Aleks Kissinger, editors, *Proceedings 14th International Conference on Quantum Physics and Logic (QPL 2017)*, volume 266 of *Electronic Proceedings in Theoretical Computer Science*, pages 119–132, Nijmegen, The Netherlands, 2017. EPTCS. doi:10.4204/EPTCS.266.8.

157. Robert Raußendorf, Daniel E Browne, and Hans J Briegel. Measurement-based quantum computation on cluster states. *Physical Review A*, 68(2):022312, 2003.

158. John C Reynolds. Separation logic: A logic for shared mutable data structures. In *Proceedings 17th Annual IEEE Symposium on Logic in Computer Science*, pages 55–74. IEEE, 2002.

159. Francisco Rios. *On a Categorically Sound Quantum Programming Language for Circuit Description*. PhD thesis, Dalhousie University, Halifax, Nova Scotia, Canada, August 2021. URL: https://dalspace.library.dal.ca/handle/10222/80771.

160. Francisco Rios and Peter Selinger. A categorical model for a quantum circuit description language. In Bob Coecke and Aleks Kissinger, editors, *Proceedings 14th International Conference on Quantum Physics and*

[17] https://quantumcomputingreport.com/resources/tools/

Logic (QPL 2017), volume 266 of *Electronic Proceedings in Theoretical Computer Science*, pages 164–178, Nijmegen, The Netherlands, 2018. doi:10.4204/EPTCS.266.11.

161. Xavier Rival and Kwangkeun Yi. *Introduction to Static Analysis: An Abstract Interpretation Perspective*. The MIT Press, 2020.

162. Neil J. Ross. *Algebraic and Logical Methods in Quantum Computation*. PhD thesis, Dalhousie University, 2015. Available as arxiv:1510.02198.

163. Thomas Santoli and Christian Schaffner. Using Simon's algorithm to attack symmetric-key cryptographic primitives. *Quantum Inf. Comput.*, 17(1&2):65–78, 2017. doi:10.26421/QIC17.1-2-4.

164. Valerio Scarani, Helle Bechmann-Pasquinucci, Nicolas J Cerf, Miloslav Dušek, Norbert Lütkenhaus, and Momtchil Peev. The security of practical quantum key distribution. *Reviews of Modern Physics*, 81(3):1301, 2009.

165. Artur Scherer, Benoît Valiron, Siun-Chuon Mau, Scott Alexander, Eric Van den Berg, and Thomas E Chapuran. Concrete resource analysis of the quantum linear-system algorithm used to compute the electromagnetic scattering cross section of a 2d target. *Quantum Information Processing*, 16(3):60, 2017.

166. Maria Schuld. *Supervised Learning with Quantum Computers*. Springer, 2018.

167. Dana Scott and Jacobus Willem de Bakker. A theory of programs. *Unpublished manuscript, IBM, Vienna*, 1969.

168. Peter Selinger. Towards a quantum programming language. *Mathematical Structures in Computer Science*, 14(4):527–586, 2004. doi:10.1017/S0960129504004256.

169. Peter Selinger. Generators and relations for n-qubit Clifford operators. *Logical Methods in Computer Science*, 11(2), Jun 2015. URL: https://lmcs.episciences.org/1570, doi:10.2168/LMCS-11(2:10)2015.

170. Peter Selinger and Benoît Valiron. A lambda calculus for quantum computation with classical control. In Paweł Urzyczyn, editor, *Typed Lambda Calculi and Applications*, pages 354–368, Berlin, Heidelberg, 2005. Springer Berlin Heidelberg.

171. Peter Selinger and Benoît Valiron. A linear-non-linear model for a computational call-by-value lambda calculus (extended abstract). In Roberto M. Amadio, editor, *Proceedings of the 11th International Conference on Foundations of Software Science and Computational Structures (FOSSACS'08)*, volume 4962 of *Lecture Notes in Computer Science*, pages 81–96, Budapest, Hungary, 2008. Springer. doi:10.1007/978-3-540-78499-9_7.

172. Peter Selinger and Benoît Valiron. Quantum Lambda Calculus. In Simon J. Gay and Ian Mackie, editors, *Semantic Techniques in Quantum Computation*, pages 135–172. Cambridge University Press, Cambridge, November 2009. URL: https://www.mscs.dal.ca/~selinger/papers/qlambdabook.pdf, doi:10.1017/CBO9781139193313.005.

173. Yunong Shi, Runzhou Tao, Xupeng Li, Ali Javadi-Abhari, Andrew W Cross, Frederic T Chong, and Ronghui Gu. Certiq: A mostly-automated verification of a realistic quantum compiler. *arXiv preprint arXiv:1908.08963*, 2019.

174. Peter W. Shor. Algorithms for quantum computation: Discrete log and factoring. In *Proceedings of the 35th Annual Symposium on Foundations of Computer Science (FOCS'94)*, pages 124–134, Santa Fe, New Mexico, US., 1994. IEEE, IEEE Computer Society Press. doi:10.1109/SFCS.1994.365700.

175. Peter W Shor. Scheme for reducing decoherence in quantum computer memory. *Physical Review A*, 52(4):R2493, 1995.

176. Daniel R Simon. On the power of quantum computation. *SIAM Journal on Computing*, 26(5):1474–1483, 1997.

177. Kartik Singhal and John Reppy. Quantum Hoare Type Theory: Extended abstract. In Benoît Valiron, Shane Mansfield, Pablo Arrighi, and Prakash Panangaden, editors, *Proceedings 17th International Conference on Quantum Physics and Logic, QPL 2020, Paris, France, June 2 - 6, 2020*, volume 340 of *EPTCS*, pages 291–302, 2021. doi:10.4204/EPTCS.340.15.

178. Seyon Sivarajah, Silas Dilkes, Alexander Cowtan, Will Simmons, Alec Edgington, and Ross Duncan. t|ket⟩: a retargetable compiler for NISQ devices. *Quantum Science and Technology*, 6(1):014003, 2020.

179. Sam Staton. Algebraic effects, linearity, and quantum programming languages. In *Proceedings of the 42nd Annual ACM SIGPLAN-SIGACT Symposium on Principles of Programming Languages*, POPL '15, pages 395–406, New York, NY, USA, January 2015. Association for Computing Machinery. URL: http://www.cs.ox.ac.uk/people/samuel.staton/papers/popl2015.pdf, doi:10.1145/2676726.2676999.

180. Damian S Steiger, Thomas Häner, and Matthias Troyer. ProjectQ: an open source software framework for quantum computing. *Quantum*, 2(49):10–22331, 2018.

181. Krysta M. Svore, Alan Geller, Matthias Troyer, John Azariah, Christopher E. Granade, Bettina Heim, Vadym Kliuchnikov, Mariia Mykhailova, Andres Paz, and Martin Roetteler. Q#: Enabling Scalable Quantum Computing and Development with a High-level DSL. In *Proceedings of the Real World Domain Specific Languages Workshop 2018*, RWDSL '18, page 7, New

York, NY, USA, February 2018. Association for Computing Machinery. arXiv:1803.00652, doi:10.1145/3183895.3183901.

182. Nikhil Swamy, Cătălin Hriţcu, Chantal Keller, Aseem Rastogi, Antoine Delignat-Lavaud, Simon Forest, Karthikeyan Bhargavan, Cédric Fournet, Pierre-Yves Strub, Markulf Kohlweiss, et al. Dependent types and multi-monadic effects in F*. In *Proceedings of the 43rd annual ACM SIGPLAN-SIGACT Symposium on Principles of Programming Languages*, pages 256–270, 2016.

183. Runzhou Tao, Yunong Shi, Jianan Yao, Xupeng Li, Ali Javadi-Abhari, Andrew W. Cross, Frederic T. Chong, and Ronghui Gu. Giallar: Push-button verification for the qiskit quantum compiler. In *Proceedings of the 43rd ACM SIGPLAN International Conference on Programming Language Design and Implementation*, PLDI '22, New York, NY, USA, 2022. Association for Computing Machinery. To appear at PLDI 2022. arXiv:1908.08963.

184. Dominique Unruh. Quantum Hoare logic with ghost variables. In *2019 34th Annual ACM/IEEE Symposium on Logic in Computer Science (LICS)*, pages 1–13. IEEE, 2019.

185. Dominique Unruh. Quantum relational Hoare logic. *Proc. ACM Program. Lang.*, 3(POPL):33:1–33:31, 2019. doi:10.1145/3290346.

186. Benoît Valiron. Generating reversible circuits from higher-order functional programs. In Simon J. Devitt and Ivan Lanese, editors, *Reversible Computation - 8th International Conference, RC 2016, Bologna, Italy, July 7-8, 2016, Proceedings*, volume 9720 of *Lecture Notes in Computer Science*, pages 289–306. Springer, 2016. doi:10.1007/978-3-319-40578-0_21.

187. Wim Van Dam, Sean Hallgren, and Lawrence Ip. Quantum algorithms for some hidden shift problems. *SIAM Journal on Computing*, 36(3):763–778, 2006.

188. André van Tonder. A lambda calculus for quantum computation. *SIAM Journal on Computing*, 33(5):1109–1135, 2004. doi:10.1137/S0097539703432165.

189. Renaud Vilmart. A near-minimal axiomatisation of ZX-calculus for pure qubit quantum mechanics. In *2019 34th Annual ACM/IEEE Symposium on Logic in Computer Science (LICS)*, pages 1–10, June 2019. doi:10.1109/LICS.2019.8785765.

190. Renaud Vilmart. The structure of sum-over-paths, its consequences, and completeness for clifford. In Stefan Kiefer and Christine Tasson, editors, *Foundations of Software Science and Computation Structures*, pages 531–550, Cham, 2021. Springer International Publishing.

191. Willem Visser, Corina S. Pasareanu, and Sarfraz Khurshid. Test input generation with Java PathFinder. In *2004 ACM SIGSOFT International Symposium on Software Testing and Analysis, ISSTA '04*. ACM, 2004.

192. Chuan Wang, Fu-Guo Deng, Yan-Song Li, Xiao-Shu Liu, and Gui Lu Long. Quantum secure direct communication with high-dimension quantum superdense coding. *Phys. Rev. A*, 71:044305, Apr 2005. URL: https://link.aps.org/doi/10.1103/PhysRevA.71.044305, `doi:10.1103/PhysRevA.71.044305`.

193. Dave Wecker and Krysta M Svore. LIQUi$|\rangle$: A software design architecture and domain-specific language for quantum computing. Available online as `arXiv:1402.4467`, 2014.

194. Mingsheng Ying. Floyd-Hoare logic for quantum programs. *ACM Transactions on Programming Languages and Systems (TOPLAS)*, 33(6):19:1–19:49, 2011. `doi:10.1145/2049706.2049708`.

195. Mingsheng Ying. *Foundations of Quantum Programming*. Morgan Kaufmann, 2016.

196. Mingsheng Ying and Yuan Feng. *Model Checking Quantum Systems: Principles and Algorithms*. Cambridge University Press, 2021.

197. Mingsheng Ying, Yangjia Li, Nengkun Yu, and Yuan Feng. Model-checking linear-time properties of quantum systems. *ACM Transactions on Computational Logic*, 15(3):22:1–22:31, 2014. `doi:10.1145/2629680`.

198. Mingsheng Ying, Shenggang Ying, and Xiaodi Wu. Invariants of quantum programs: characterisations and generation. In Giuseppe Castagna and Andrew D. Gordon, editors, *Proceedings of the 44th ACM SIGPLAN Symposium on Principles of Programming Languages, POPL 2017, Paris, France, January 18-20, 2017*, pages 818–832. ACM, 2017. `doi:10.1145/3009837.3009840`.

199. Nengkun Yu and Jens Palsberg. Quantum abstract interpretation. In *Proceedings of the 42nd ACM SIGPLAN International Conference on Programming Language Design and Implementation*, pages 542–558, 2021.

200. Shengyu Zhang. BQP-complete problems. In Grzegorz Rozenberg, Thomas Bäck, and Joost N. Kok, editors, *Handbook of Natural Computing*, pages 1545–1571. Springer, 2012.

201. Li Zhou, Gilles Barthe, Justin Hsu, Mingsheng Ying, and Nengkun Yu. A quantum interpretation of bunched logic for quantum separation logic. In *2021 36th Annual ACM/IEEE Symposium on Logic in Computer Science (LICS)*, pages 1–14. IEEE, 2021.

Index

Printed in the United States
by Baker & Taylor Publisher Services